Atlas of the Lunar Terminator
John E. Westfall

This Atlas is based on high-resolution electronic images of the terminator area of the Moon under 47 different illuminations. Each image is displayed across two or three pages. Features as small as 1–2 kilometers can be seen. For each illumination, accompanying text describes the major features shown in each view, along with larger-scale images of selected areas, which are indexed with their positions and dimensions.

Two text chapters describe and illustrate the techniques of lunar observing and the types of landforms. Observing data are presented through to 2010. Experienced observers will find the Atlas an invaluable planning tool, while beginners will gain an understanding of lunar geography and geology. There is an index of 1000 named lunar formations, cross-referenced to the images.

No other atlas shows the Moon under such a variety of observing conditions.

PROFESSOR JOHN WESTFALL is with the Department of Geography and Human Environmental Studies, San Francisco State University. From 1985–95 he served as Director of the Association of Lunar and Planetary Observers, whose Journal he continues to edit.

Frontispiece. Waxing crescent Moon, age 3.4 days. CCD image mosaic, 1995 MAY 03, 03h32m UT. 28-cm Schmidt-Cassegrain, f/10, 0.15 s, W58 (green) Filter. Colongitude 308°.9, Solar latitude 0°.2N, Librations 2°.4W/5°.2N. North at top.

Atlas of the
Lunar Terminator

John E. Westfall
San Francisco State University

PUBLISHED BY THE PRESS SYNDICATE OF THE UNIVERSITY OF CAMBRIDGE
The Pitt Building, Trumpington Street, Cambridge, United Kingdom

CAMBRIDGE UNIVERSITY PRESS
The Edinburgh Building, Cambridge CB2 2RU, UK   www.cup.cam.ac.uk
40 West 20th Street, New York, NY 10011–4211, USA   www.cup.org
10 Stamford Road, Oakleigh, Melbourne 3166, Australia
Ruiz de Alarcón 13, 28014 Madrid, Spain

© John Westfall 2000

This book is in copyright. Subject to statutory exception
and to the provisions of relevant collective licensing agreements,
no reproduction of any part may take place without
the written permission of Cambridge University Press.

First published 2000

Printed in the United Kingdom at the University Press, Cambridge

Typeset by the Author in QuarkXPress®

A catalogue record for this book is available from the British Library

Library of Congress Cataloguing in Publication data

Self-organized biological dynamics and nonlinear control: exploring biological complexity, stochasticity, chaos, and electromagnetic interactions in cell signaling systems/edited by Jan Walleczek.
    p.   cm.
  Includes index.
  ISBN 0 521 624336 3 (hb)
  1. Cellular signal transduction. 2. Self-organizing systems. 3. Nonlinear systems.
  I. Walleczek, Jan, 1964–

QP517.C45 S45 2000
571.6–dc21   99-044949

ISBN 0 521 59002 7 hardback

Contents

Preface and Acknowledgements vii

Section I. Terminator Observation and Interpretation

    Chapter 1. Observing the Moon and Its Terminator 1
    Chapter 2. The Moon's Surface Features 26

Section II. Atlas

    Arrangement of the Atlas 57
    Making the Mosaics 57
    Colongitude 291° 61
    Colongitude 303° 64
    Colongitude 309° 68
    Colongitude 315° 72
    Colongitude 323° 76
    Colongitude 328° 80
    Colongitude 334° 84
    Colongitude 343° 88
    Colongitude 351° 92
    Colongitude 357° 96
    Colongitude 003° 100
    Colongitude 009° 104
    Colongitude 018° 108
    Colongitude 024° 112
    Colongitude 029° 116
    Colongitude 036° 120
    Colongitude 040° 124
    Colongitude 048° 128
    Colongitude 055° 132
    Colongitude 061° 136
    Colongitude 067° 140
    Colongitude 075° 144
    Colongitude 080° 148
    Near Full 152
    Colongitude 099° 162
        Colongitude 100°–N 165
    Colongitude 105° 166
        Colongitude 106°–N 169
    Colongitude 112° 170
    Colongitude 119° 174
    Colongitude 124° 178
    Colongitude 131° 182

Colongitude 135° 186
Colongitude 143° 190
Colongitude 148° 194
Colongitude 154° 198
Colongitude 161° 202

Colongitude 167° 206
Colongitude 174° 210
Colongitude 179° 214
Colongitude 183° 218
Colongitude 191° 222

Colongitude 201° 226
Colongitude 205° 230
Colongitude 213° 234
Colongitude 225° 238
Colongitude 233° 242

Colongitude 237° 246
Colongitude 247° 250

Appendix: Lunar Physical Ephemeris, 2000-2010 255

Bibliography 265

Mosaic Feature Index 269

Subject Index 289

Individual Features Index 290

PREFACE AND ACKNOWLEDGEMENTS

Although there are not a great many atlases of the Moon currently in print, the reader may still wonder why another one has been produced; after all, the Moon hardly changes very rapidly. This atlas differs from others in that it concentrates on the Moon's terminator zone; the area of low solar elevation where the glancing illumination highlights low relief features that are either subdued or invisible under higher lighting. Curiously, it is difficult to find published photographs or other images that show low-lighting areas appropriately exposed. The NASA Lunar Orbiter 4 images that covered almost the entire Moon in 1966-67 showed most areas when the Sun was 16° above their horizon. The Clementine Mission in 1994 obtained many thousand multiband images, but taken at local noontime for each area, so they showed low sun angles only near the Moon's poles! Thus this publication should fill a gap and be of especial use in investigating such low-lying lunar features as ancient basins, ghost craters, domes, and ridges.

Another feature of this atlas is that it shows the earthside face of the Moon at frequent phase intervals, with 47 different illuminations (from about 1.7 days after New Moon to 1.9 days before). Although not perfectly evenly spaced, on the average the views are at intervals of 6°-7° in solar lighting. Thus whenever you observe the Moon, you should be able to find a view in this atlas where the Sun angle differs by no more than a few degrees from what you see in the telescope. This helps in planning observations as well as identifying what one sees in the telescope.

The heart of this book consists of mosaics of CCD images of the Moon. These views have the virtue of being visually realistic, but this means that they are not on any standard map projection (such as the Orthographic, widely used for lunar maps), so that measurements of positions on them will not be accurate. However, the graphic scale with each mosaic will allow the approximate measurement of the unforeshortened dimensions of features. In order to cover the entire terminator at 47 different phases in a book of reasonable size, a compromise has had to be made in terms of resolution. The pixel size of the original images ranges between about 1-2 km. This means that craters smaller than about 4 km will not usually be visible, nor rilles narrower than that value. Higher-resolution views accompanying the descriptive text for each mosaic provide greater detail for selected features.

Following the section of terminator views, an Appendix gives an ephemeris of lunar lighting and libration for 2000-2010, for planning future observations. After a short bibliography is an index of the features that are shown by name on the CCD mosaics. These names reflect International Astronomical Union nomenclature as of the IAU 1997 Congress; but of course new names may be added, and old names changed or deleted, in future Congresses. The feature index also gives categories, and approximate positions and dimensions, for the features listed.

The writer is solely to be blamed for any defects in this atlas. Nonetheless, he is grateful to many lunar aficionados who have talked or corresponded with him, either recently or sometimes decades in the past, giving information and inspiration as a result of their own enthusiasm for the study of a celestial body that is so accessible that sometimes it hardly seems "celestial" at all. Sadly, some who kindled the writer's initial interest in the Moon are no longer with us: Dinsmore Alter, Joseph Ashbrook, David Barcroft, James Barlett, Chesley Bonestell, Peter Hédervári, Alika Herring, Gerard Kuiper, Eugene Shoemaker, Clyde Tombaugh, and H.P. Wilkins. Other such persons, still active in lunar studies, are: Winifred Cameron, Nancy Cox, Bill Davis, Jean Dragesco, Francis Graham, Bob Garfinkle, Harry Jamieson, Patrick Moore, Georges Viscardy, and Don Wilhelms, many of them members of the Association of Lunar and Planetary Observers. I hope that new and future students of the Moon will in some small degree be encouraged by this atlas.

I here express my appreciation to San Francisco State University, which granted me Sabbatical Leave in Fall Semester, 1997, giving me the time needed to winnow through several thousand CCD images and from them create the mosaics that appear in this book. Finally, the patience and assistance of my wife Elizabeth has been essential at all stages in this production.

John E. Westfall
San Francisco, California

Dedication

To Walter H. Haas,
Founder of the Association of Lunar and Planetary Observers.
For most of my lifetime, a continuing inspiration for my lunar and planetary studies.

Section I. Terminator Observation and Interpretation

Chapter 1. Observing the Moon and Its Terminator

Lunar Visibility Factors

The Moon's Changing Lighting

A line encircles the Moon and rotates about it every 29.5 Earth days, moving westward along the equator at a leisurely 15 kilometers per hour. It has circled the Moon tens of thousands of millions of times in the past and should continue to do so at least as many times in the future. This line is called the *terminator*, and is defined as the great circle where the Sun is on the local horizon at any particular time. One half of the terminator marks lunar sunrise, moving so as to expose more and more of the Moon's surface to sunlight. The other half of the line defines where the Sun is setting, where areas become gradually engulfed by the lunar night. The area of the terminator is of prime interest because its glancing illumination highlights subtle, low-lying relief features.

The movement of the terminator from one night to the next is obvious even when viewing the Moon with the naked eye. In the telescope, the changing lighting conditions are apparent in a matter of hours, and near the terminator itself in minutes, as is shown in the time sequence of CCD frames in Figure 1.1 (page 2).

There are several ways that the lighting conditions of the Moon can be described for any particular time. The first three are approximate; phase, age, and proportion illuminated. *Phase* simply describes lighting with the approximate phrase: New Moon, Waxing Crescent, First Quarter, Waxing Gibbous, Full Moon, Waning Gibbous, Last (or Third) Quarter, and Waning Crescent. To be more precise, one can cite the Moon's *age*, the number of days and fractional days since the last New Moon; this may sound quite accurate, but is affected by small variations in the length of the month and by libration (variations in the Moon's orientation toward the Earth). A third approach, quantitative but still approximate, is to give the proportion of the Moon that appears to be sunlit (*phase coefficient*), varying from 0.00 to 1.00.

To describe the Moon's global lighting, as with measurements of the integrated brightness of the entire satellite, *phase angle* is a useful measure, defined as the angle between the Sun and the Earth as seen from the Moon, ranging from 0° to 180°. Unfortunately, due to libration, it is not an accurate description of the lighting conditions for specific lunar locations.

What has proved the most useful way of describing the surface lighting of the Moon is the solar *colongitude*. This quantity is defined as the lunar longitude of the sunrise terminator, measured westward along the equator, 0°-360°, from the apparent mean center of the Moon's disk. To explain this concept more clearly, it is necessary to describe the Moon's latitude/longitude coordinate system, which is similar to that of the Earth. First, the Moon has an axis of rotation, which defines its north and south poles. Midway between the poles is the great circle that represents the lunar equator. Latitudes on the Moon are measured in terms of the angle subtended between the equator and a location, measured northward or southward as the case may be. Northern latitudes are considered positive, and southern negative.

Defining a point's lunar longitude requires a definition of where zero longitude falls. That is, the Moon's prime meridian, defined as the great circle that passes through the poles and the appar-

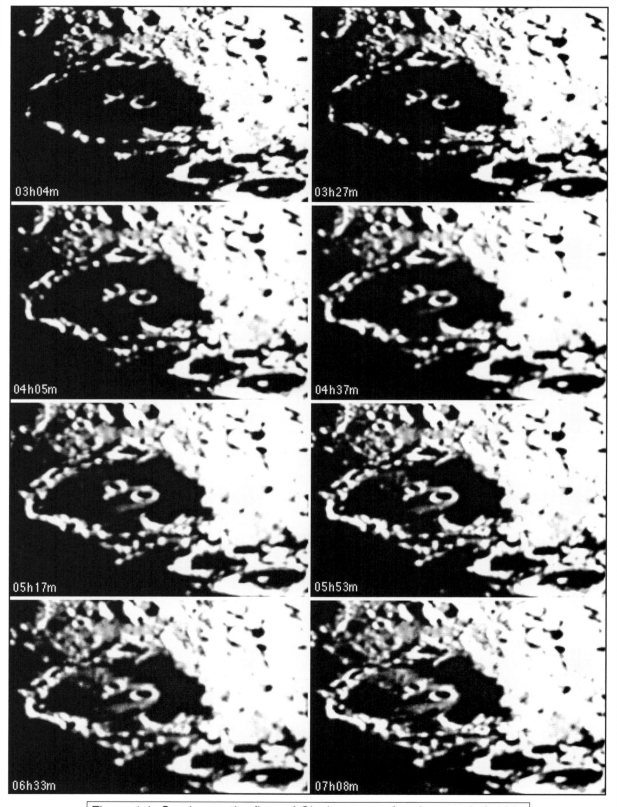

Figure 1.1. Sunrise on the floor of Clavius over a four-hour period. 1995 FEB 09, UT times as indicated. 28-cm Sch.-Cass., f/10, 0.20 s. Colongitudes for each frame are: 03h04m, 017°.46; 03h27m, 017°.66; 04h05m, 017°.98; 04h37m, 018°.25; 05h17m, 018°.58; 05h53m, 018°.89; 06h33m, 019°.23; 07h08m, 019°.52. Selenocentric solar latitude +1°.56.

1. Observing the Moon

Table 1.1. Relationship Among Lunar Lighting Parameters.

Phase Description	Age	Phase Angle	Phase Coefficient	Solar Colongitude	Longitude Sun	Terminator Sunrise	Sunset
New Moon	0.0d	180°	0.000	270°	180°	90°E	90°W
Waxing Crescent	3.7d	135°	0.146	315°	135°E	45°E	135°W
First Quarter	7.4d	90°	0.500	000°	90°E	0°	180°
Waxing Gibbous	11.1d	45°	0.854	045°	45°E	45°W	135°E
Full Moon	14.8d	0°	1.000	090°	0°	90°W	90°E
Waning Gibbous	18.5d	45°	0.854	135°	45°W	135°W	45°E
Last Quarter	22.1d	90°	0.500	180°	90°W	180°	0°
Waning Crescent	25.8d	135°	0.146	225°	135°W	135°E	45°W

ent disk center when the libration in longitude is at its mean value (0°.00), in the Moon's appropriately named *Sinus Medii*. Longitude is measured eastward from the prime meridian, and often longitudes are given in the range 0°-180° east or west of the prime meridian. Longitudes in the Moon's eastern hemisphere are taken as positive; those in the western hemisphere, negative. Neglecting librations, the Earthside hemisphere is that zone lying within 90° of the prime meridian, with the farside lying within 90° of the 180° meridian.

Colongitude is the single most important description of the global lighting conditions for the Moon. Colongitude advances approximately 360° over the course of a lunation, with a mean change of 12°.19 per day or 0°.508 per hour. Ignoring librations for the sake of simplicity, and assuming for this example that Crescent Phase is midway between Quarter and New and Gibbous is midway between Quarter and Full, colongitude is related to the previous means of describing lunar lighting as is shown in Table 1.1 (above).

Although colongitude is paramount in describing the global lighting conditions of the Moon, the description is not complete without specifying the *selenocentric solar latitude*, which varies by an average of ±1°.54 due to the tilt of the Moon's equator to the ecliptic, which causes the Moon's rather minor "seasons." The effect of solar latitude on the lighting angle at a specific lunar location is minor near the equator, but becomes more noticeable in the higher latitudes.

To describe the solar lighting at a specific place on the Moon, the Sun's local *altitude* (H) and *azimuth* (A) are calculated from the point's longitude (λ) and latitude (β) and the colongitude (c_s) and solar latitude (β_s) by formulae (1), (2), and (3), where $L = \lambda - c_s + 90°$:

(1) $\quad \sin H = \sin \beta_s \sin \beta + \cos \beta_s \cos \beta \sin (c_s + \lambda)$,

(2) $\quad \sin A = \dfrac{\sin L \cos \beta_s}{\cos H}$,

(3) $\quad \cos A = \dfrac{\sin \beta_s - \sin \beta \sin H}{\cos \beta \cos H}$,

where both formulae (2) and (3) are needed to specify in which quadrant the azimuth falls.

Librations

The often-quoted statement that "the Moon keeps the same side toward the Earth" is not quite true; it actually wobbles both east-west and north-south to our line of sight. The east-west oscillation, called *libration in longitude*, is due chiefly to the eccentricity of the lunar orbit. Although the Moon's mean period of revolution about the Earth equals that of its rotation (27.32 days), the lunar orbit's eccentricity causes differences between rotation and revolution that result in the apparent center of the disk wobbling east-west by up to about ±7°.95.

The Moon also has a north-south libration in latitude, caused by the inclination of its equator to its orbital plane, in a range averaging ±6°.85.

The Moon's orbital motion is quite complicated (Newton said it gave him headaches), and the Moon's axis also wobbles in space in a complex manner by a few hundredths of a degree, a phenomenon called *physical libration*. The result is that the timing, amount, and pattern of our satellite's librations vary considerably over time.

Were the above not enough, all the librations described above are *geocentric*, as seen from the center of the Earth. An observer's *selenocentric* position may be displaced from the geocentric by up to about one degree, affecting his or her *topocentric libration* by an equal amount.

When all the forms of libration are taken into account, there remains an extensive portion of the Moon that is always visible from the Earth. This zone is centered on the mean center of the disk and amounts to approximately 41 percent of the Moon's area. There is a corresponding zone of equal size, never visible from the Earth, centered on the center of the lunar farside. Between these two zones there is a significant area, called the *libratory zone* or the *marginal zone,* about 18 percent of the Moon's surface, whose occasional visibility depends on libration.

Perspective Effects

Because the Moon's shape closely resembles a sphere, rather than a flat disk perpendicular to our line of sight, it follows that we are looking vertically only at the apparent center of its disk. The farther we look toward the Moon's limb, or apparent edge, the more obliquely we see its features. The result is a perspective effect called *foreshortening*, which compresses apparent distances that are radial to the lunar limb, but leaves distances undistorted parallel to the limb. This naturally modifies the apparent shape of features, for example making circular craters appear as ellipses whose major axes are parallel to the limb.

However, the Moon is not a smooth, perfect sphere. The elevation of a point above, or depression below, the mean lunar surface creates an effect called *relief displacement*, which again is most pronounced near the limb. Elevated features are displaced limbward, while depressions are displaced toward the apparent disk center. Craters near the limb appear asymmetric, their inner walls narrower on their earthward side than their limbward side, while peaks appear to lean toward the limb. Indeed, this effect makes the lunar limb appear irregular, with protrusions and indentations, rather than a smooth circle.

Were foreshortening the only distortion that affects the appearance of the Moon's features, it would be simple to correct a lunar image to a "true" overhead view. However, *relief displacement* is caused by the local relief of the Moon's surface and is far more difficult to remove, requiring stereoscopic orbital photography. Table 1.2, to the left, indicates the magnitudes of the effects of foreshortening and relief displacement.

Besides distorting the shape of objects, these perspective effects result in the loss of detail near the limb. First, features near the limb appear smaller, measured either in square arc-seconds (e.g., for visual observers) or in pixels (e.g, for CCD cameras); this causes a loss in

Table 1.2. Effects of Foreshortening and Relief Displacement.

Feature's Distance from Apparent Center of Disk		Apparent Dimensions of a 100-km Circular Crater	Limbward Displacement of a 1-km mountain	
Radial*	Angular		Apparent	In Projection
0.000	0°.0	100 × 100 km	0.00 km	0.00 km
0.200	11°.5	100 × 98 km	0.20 km	0.20 km
0.400	23°.6	100 × 92 km	0.40 km	0.44 km
0.600	36°.9	100 × 80 km	0.60 km	0.75 km
0.800	53°.1	100 × 60 km	0.80 km	1.33 km
0.900	64°.2	100 × 44 km	0.90 km	2.06 km
0.950	71°.8	100 × 31 km	0.95 km	3.04 km
0.990	81°.9	100 × 14 km	0.99 km	7.02 km
1.000	90°.0	100 × 0 km	1.00 km	---

* Relative to an apparent lunar radius of 1.000 units.

information. Second, relief displacement causes elevations to block our view of features beyond them, and depressions to be blocked from view. Thus, particularly near the limb, it is desirable to observe a feature when favorable libration displaces the area toward the apparent center of the disk.

Lunar Altitude

We earthbound observers have to view the Moon through our own atmosphere, which both absorbs some light (the amount depending on wavelength, giving the Moon false coloration), and reducing resolution due to atmospheric turbulence; a phenomenon called *seeing*. Lunar observers obviously prefer highly transparent skies, and sometimes observe from considerable elevations above sea level to achieve this. They also prefer steady air, again often selecting sites where this situation is common, or otherwise waiting patiently for conditions to steady. Everything else being equal, though, it is desirable to observe the Moon when it is as high above one's horizon as possible, as this reduces the optical path length through the atmosphere. How high the Moon appears in the sky depends on several factors, including lunar phase, time of day or night, the observer's latitude, and the Moon's declination. Ideally one should observe the Moon when it is near the local meridian, but for about a week either side of New Moon this necessitates observation during daylight or twilight.

Because the Moon's path approximately follows the ecliptic, its declination depends in part on its right ascension (or, alternatively, on its celestial longitude), as shown in Figure 1.2, below, where the ecliptic is shown as a solid line. Note, though, that the Moon's orbit is actually inclined by 5°.15 to the ecliptic. In addition, the nodes of the Moon's orbit move around the ecliptic in an 18.6-year cycle, affecting how much the Moon moves north and south of the celestial equator. For example, in 1997, the Moon's declination ranged between ±18° (see dashed line in Figure 1.2), but in 2006 will vary by almost ±29° (dotted line in Figure 1.2).

Whatever the positions of the lunar nodes, the much larger effect of right ascension upon the Moon's declination means that some terrestrial seasons are more favorable for observing particular phases of the Moon than are others, as is apparent from Figure 1.3 (page 6), which shows the pattern of declination versus phase for the sample year 2001 (note also how the extremes of declination increase during the year due to the gradual motion of the Moon's nodes). Table 1.3 (page 7) indicates approximately when the best seasons occur for particular phases, correspondences which are the same for the terrestrial north and south hemispheres.

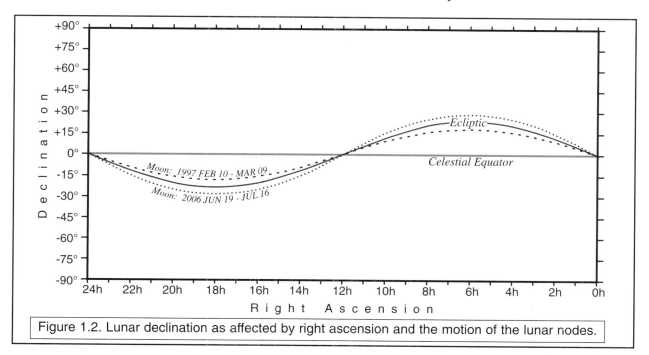

Figure 1.2. Lunar declination as affected by right ascension and the motion of the lunar nodes.

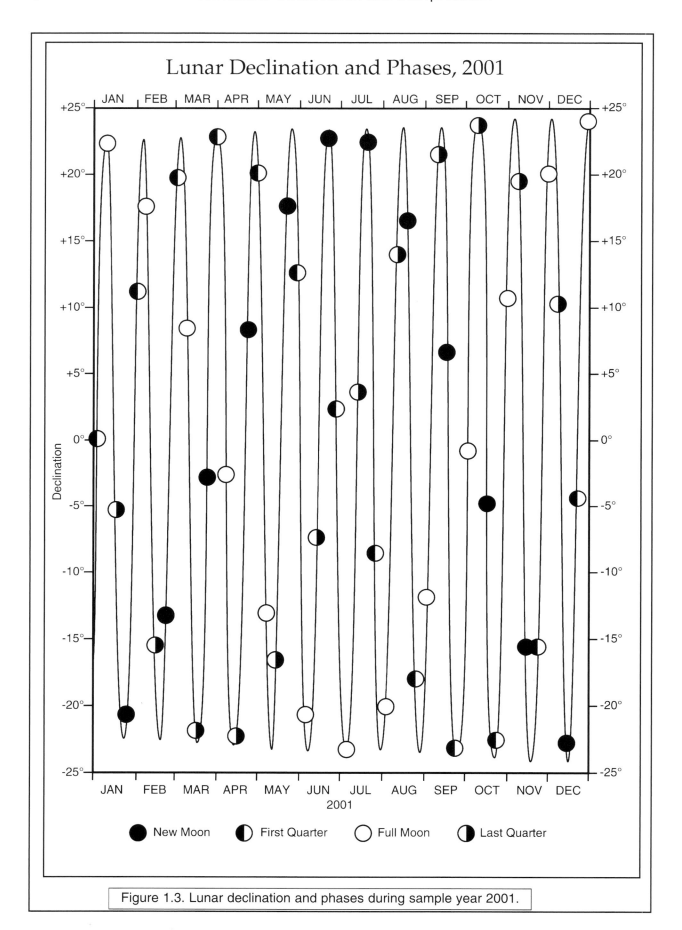

Figure 1.3. Lunar declination and phases during sample year 2001.

1. Observing the Moon

Table 1.3. Preferred Terrestrial Season for Viewing Specific Lunar Phases.

Phase	Season
Waxing Crescent	Late Spring
First Quarter	Early Spring
Waxing Gibbous	Late Winter
Full Moon	Early Winter
Waning Gibbous	Late Fall
Third Quarter	Early Fall
Waning Crescent	Late Summer

Observing Programs and Visibility Cycles

To make a casual lunar observation, the requirements are simple; the Moon should be up and the Sun down. A systematic program of lunar study, however, has more stringent conditions. Of several additional factors, probably colongitude is the most important. In studying features at all near the lunar limb, libration is normally the next consideration; but there are two librations, libration in longitude and libration in latitude, and these may only rarely combine satisfactorily with each other and with colongitude to view a specific limb area under low lighting. Then, if other conditions are satisfied, it is preferable to have the Moon high in the sky; near maximum northerly declination for observers in the northern hemisphere and near maximum southerly declination for their colleagues in the southern hemisphere. In addition to all this are two lesser factors: the distance from us to the Moon varies by about 11 percent between perigee and apogee, and the selenocentric solar latitude by about $\pm 1°.5$.

Table 1.4. Mean Periods of Selected Lunar Visibility Cycles.*

Visibility Factor	Name	Mean Period (Days)
Libration in Latitude	Draconic Month	27.212221
Lunar Latitude	"	"
Lunar Declination	Tropical Month	27.321582
Libration in Longitude	Anomalistic Month	27.554550
Lunar Distance	"	"
Colongitude (phase; lunation)	Synodic Month	29.530589
Selenocentric Solar Latitude	Eclipse Year	346.620075
Colong./Libration Cycle†	Saros Cycle	~6585.
Lunar Declination Range	Revolution of Nodes	6798.410101

* Periods are for 1999.0; values from U.S. Naval Observatory, 1998.

† Based on the approximate correspondence between 223 lunations (6585.321347 d) and 19 eclipse years (6585.781245 d).

All these visibility factors vary in cycles, some with many components. Also, the cycles of the visibility factors are not of equal length, nor even commensurate. The mean periods of the several significant lunar visibility cycles are given in Table 1.4, to the left.

This means that a particular aspect of the Moon—the combination of its colongitude, solar latitude, librations, distance, and declination—will never exactly repeat, although if one waits long enough several of these factors may be close to their previous values. The most basic cycle is the synodic month. Because its mean period is close to 29.5 days, the colongitude at the same Universal Time (UT; "Greenwich Time") will differ by about 6° for successive lunations. Fortuitously, two synodic months total very nearly 59 days, so that lunar lighting closely repeats every two lunations. Indeed, given clear skies and good seeing every night, this atlas could have been made in just two lunations, covering the Earthside from waxing crescent to waning crescent at 6° lighting intervals!

Combinations of colongitude, libration in longitude, and libration in latitude recur closely at an interval of one Saros Cycle (18.03 years), with approximate recurrences at one-third of that period (*ca.* 6.01 years). For example, the view of the lunar southwest limb near full phase in this atlas was taken on 1997 MAY 22, 05h39m UT. The colongitude at the time was 093°.4 and the geocentric librations were 5°.3 W/5°.8 S, favorable for viewing that region. Table 1.5 (page 8) gives some observing dates in the past and the future when the lighting and libration conditions on 1997 MAY 22 will approximately recur (the dates and times in the table have been selected so that the exact colongitude will recur). At about six-year intervals there is an approximate recurrence of conditions, but a much closer correspondence every 18 years (e.g., 2009 and 1991, 2003 and 1986, and 1997 and 1979). Because 19 solar latitude periods (6585.78 d) is very close to 223 lunations (6585.32 d), selenocentric solar latitude also tends to repeat at six-year intervals. Note, however, that it may not be possible to observe the Moon from the same terrestrial location on all the dates given in Table 1.5, in which case it would be necessary to accept a weaker correspon-

Table 1.5. Past and Future Recurrences of Colongitude/Libration Combinations of 1997 May 22.				
		Geocentric Libration		Solar
UT Date and Time	Colongitude	Longitude	Latitude	Latitude
2009 Oct 04, 03.8h	093°.4	4°.6 W	6°.2 S	1°.5 S
2003 Jan 18, 07.5h	093°.4	5°.2 W	4°.9 S	1°.2 S
1997 May 22, 05.6h	*093°.4*	*5°.3 W*	*5°.8 S*	*1°.4 S*
1991 Sep 23, 20.1h	093°.4	4°.7 W	6°.2 S	1°.5 S
1986 Feb 24, 13.9h	093°.4	4°.8 W	5°.5 S	1°.3 S
1979 May 11, 22.4h	093°.4	5°.4 W	5°.9 S	1°.4 S

dence between librations and observe one lunation earlier or later than the dates given in the table.

Because the periods of libration in longitude and latitude differ, their relationship differs for every lunation. Their pattern during the first complete lunation in 2001, along with the dates of lunar phases, is shown in Figure 1.4, below.

Figure 1.4 demonstrates a frequent irony that occurs when planning an observing program. The lunar northwest limb will be particularly well presented on 2001 Jan 30 - Feb 02, being tilted toward Earth by as much as 10°. However, that limb will be in darkness for those dates! The same situation pertains for the favorable presentation of the southeast limb on 2001 Feb 12 - 15. At the time of Full Moon, when limb-area observations are best conducted, there is only a moderate southerly libration favoring the south limb.

As a matter of fact, if one wishes to have a combination of favorable longitude and latitude libration for observing a specific limb area near Full Moon, one may have to wait for a long time.

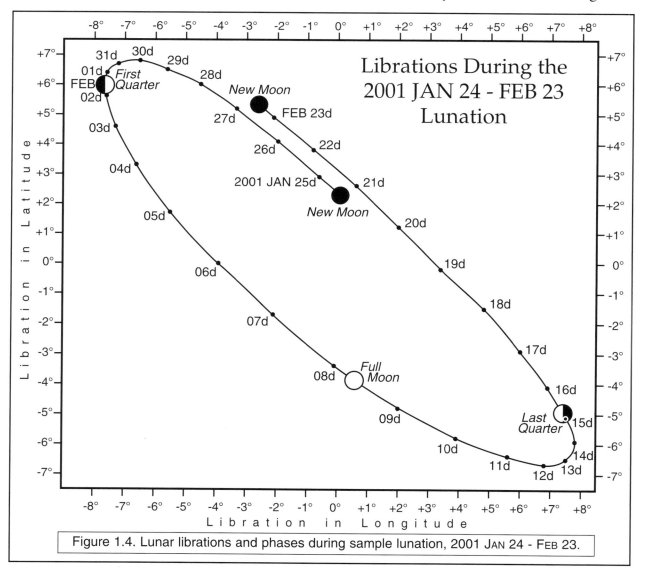

Figure 1.4. Lunar librations and phases during sample lunation, 2001 Jan 24 - Feb 23.

1. Observing the Moon

Figure 1.5, below, looks at the librations at Full Moon through a 6-year period (1999 through 2004). Note that the northwest limb can be favorably observed only in Fall, 1999-2001; the northeast limb in Fall, 2002-2004; the southeast limb in Spring, 2000-2001; and the southwest limb in Winter, 2003-2004. For each of those years, the particular limb is well-presented in only two or three lunations. If one wishes to observe the east or the west limb, one is concerned only with the libration in longitude; for the north or south limb, only the libration in latitude; and favorable dates are more frequent.

When planning the timing of one's observations for the best conditions, the rather stringent requirements described above apply to studies of the limb areas only. For areas nearer the center

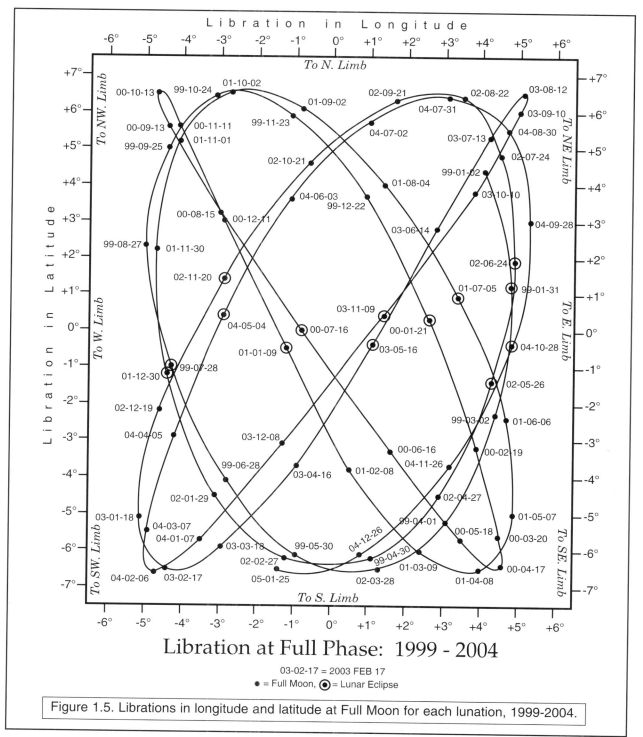

Figure 1.5. Librations in longitude and latitude at Full Moon for each lunation, 1999-2004.

of the disk, an observer need "only" be concerned with selecting the desired lighting conditions and with whether the Moon is then reasonably high above the horizon in a dark sky.

Illumination Near the Lunar Terminator

The area of the Moon's terminator, the focus of this atlas, is a region where the glancing illumination on a flat surface results in a very dimly lit scene. In addition, this is a region of great contrast, both in terms of brightness differences between areas very near the terminator and areas farther away and also between areas sloping away from the Sun and those inclined toward the direction of light.

Even in terms of relatively flat areas, the *photometric function* of the Moon is complex. Certainly, relative brightness is not constant throughout the sunlit hemisphere. Brightness does not even follow the well-known "cosine law" for smooth surfaces, where an area's brightness would be expected to be proportional to the cosine of solar zenith distance (or the sine of the Sun's elevation). Besides the *angle of incidence* (i.e., solar zenith distance), perceived brightness depends also on the *angle of emission*, the angle between the perpendicular to a surface and the observer. Some overall "laws" have been formulated to describe changes in the apparent brightness of most level areas on the Moon: (1) Brightness depends on the phase angle (see page 1) and the *brightness longitude* (the angle at a site between the local surface normal and the observer, measured in the plane that contains the Sun, the site, and the observer). (2) Neighboring areas of the same albedo (i.e., reflecting power on a scale from 0.00 representing pure black to 1.00 for pure white; the lunar mean is 0.12 at full phase) tend to have the same ratio of brightness throughout a lunation. (3) The brightness of a feature increases rapidly near Full Moon; most features are brightest at full phase (although some light rays peak shortly after Full Moon). (4) At Full Moon, features of similar albedos have similar brightness regardless of their position on the Moon; also, the Moon shows neither limb darkening nor limb brightening.

Figure 1.6 (page 11) graphs the theoretical relative brightness of a level area near the Moon's terminator as a function of brightness longitude and phase angle, based on the photometric model of Hapke [1966]. Near the terminator at least, the function is approximately linear until within about 45° of full phase. Note how rapidly brightness increases as Full Moon is approached; conversely how faint the illumination is in the crescent phases. Finally, brightness longitude can be thought of as longitude measured on an "illumination equator" that passes through the subsolar point and the apparent center of the disk; this means that the brightness longitude changes very rapidly near the north and south limbs and brightness thus changes very rapidly in those areas as well. It may go against one's intuition, but is an observed fact, that when two areas north and south of each other are compared at the same phase, their brightness ratio depends on their respective brightness longitudes, rather than on their solar elevations.

Figure 1.6 assumes a theoretical flat surface and thus exhibits smooth curves. In reality, the actual surface relief in an area, along with differences in albedo, makes the actual brightness function much less regular. Figure 1.7 (page 12) graphs the observed brightness function, as a profile from two unmodified CCD images, taken only six minutes apart with the same instrument, filter, and exposure time (for further information see the caption on the terminator mosaic for colongitude 174°). In each case the height of the lowest (darkest) portions of the profile represent the "dark count" inherent in the CCD camera, a quantity which would be subtracted from the brightness of all pixels in an image intended for pictorial purposes. Small-scale fluctuations in the profile, particularly in shadow areas, represent camera "noise." The left-hand image is of a relatively-smooth maria area, Mare Vaporum. Its brightness increases almost linearly with distance from the terminator until a range of hills is reached near the left margin of the frame. As one moves away from the terminator in the right-hand frame, the line of the brightness profile passes through the craters Airy (on the terminator), Parrot P, and Parrot C, and finally through the central peak of Arzachel. The rough terrain in this highlands area results in an extremely irregular brightness profile in which the effect of increasing solar elevation is obscured by the much greater effect caused by local terrain containing slopes differing in steepness and orientation with respect to the

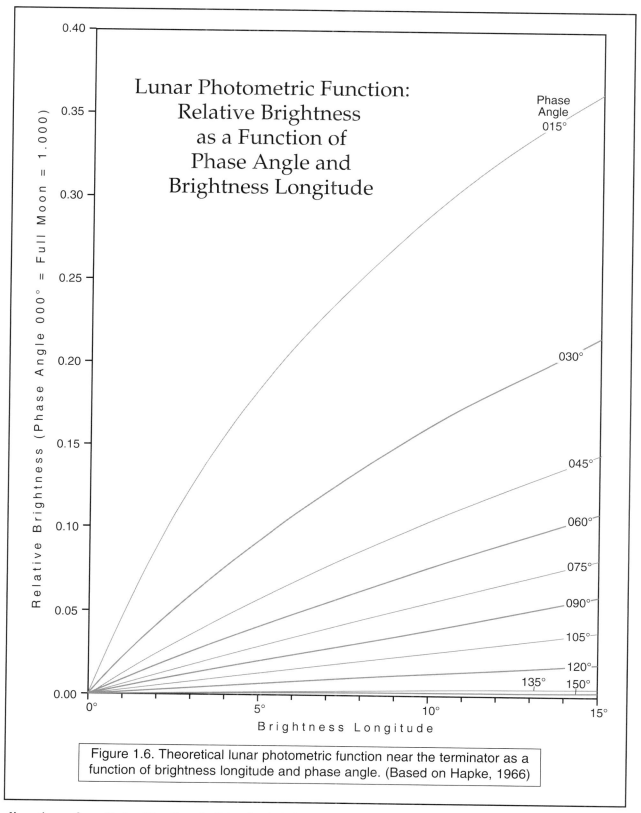

Figure 1.6. Theoretical lunar photometric function near the terminator as a function of brightness longitude and phase angle. (Based on Hapke, 1966)

direction of sunlight. The "peaks" in the brightness profile represent sunward-facing crater walls (i.e., east inner walls or west outer walls), while the "valleys" are areas in shadow (the west inner walls and floors or east outer walls of the craters).

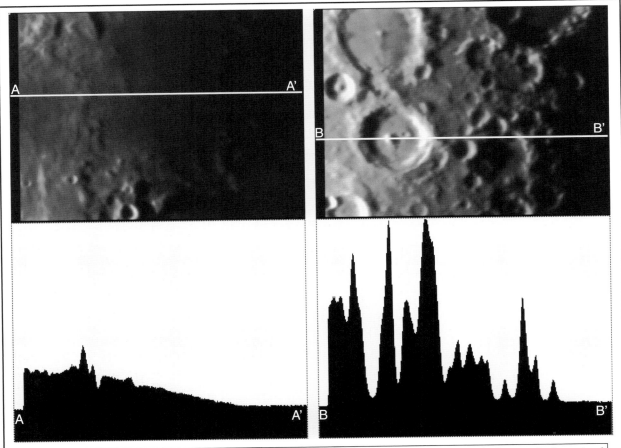

Figure 1.7. Brightness profiles of CCD images; Mare Vaporum (left) and through the central peak of Arzachel (right); taken on 1995 Sep 16, 11h42m and 11h48m respectively. The white lines (upper portion) represent the locations of the profiles (lower portion). Sunlight is from the left; north at top.

Equipment and Techniques

Telescopes and Mountings

Recommending a telescope for observing the Moon is to walk a mine field. Most observers agree on some aspects of the question: (1) Larger apertures are preferable to smaller, although atmospheric "seeing" means that one eventually reaches a point of diminishing returns; (2) whatever the optical design, the optics need be of high quality; (3) the optics need to be clean (and coated if a refractor, or with a fresh layer of aluminum for a reflector); and (4) the instrument should be well mounted.

Beyond these generalities, lunar observers begin to disagree. Their common goals are the highest possible resolution and the highest possible contrast in the image. Where they disagree is as to how these goals are achieved. Clearly, for a given aperture, an unobstructed optical design is to be preferred, restricting one to a refractor or an off-axis reflector such as a Schiefspiegler (tilted-mirror) design. However, for a given cash outlay, one can gain more aperture with an obstructed design. Naturally, the smaller the obstruction, the better; long-focus Newtonians or Cassegrains are also popular, with central obstructions (their secondary mirrors) diameters no more than 10-15 percent of the aperture. Well down many observers' preference lists are catadioptric systems (i.e., those using both primary and secondary mirrors, as well as a correcting lens), such as Schmidt-Cassegrains, because they have central obstructions with diameters of the order of 25-35 percent of the aperture. Maksutovs, with somewhat smaller central obstructions, and arguably of better optical design, are often preferred to Schmidt-Cassegrains.

The above comments reflect the views of the majority of lunar observers. However, the writer acquired all the terminator mosaics in this book using Schmidt-Cassegrain systems! Basically, he chose this design because of the fairly high aperture/cost ratio, its compactness, and its convenience of use for visual, photographic, video, and CCD observing. In an overwhelming majority of observing sessions, his resolution has proven seeing-limited rather than instrument-limited. Also, he has found (from both ray-tracing simulation and actual observation) that the sharpness of his images has improved when he installed an "apodizing ring"; an opaque annulus mounted and centered on the correcting plate of his 28-cm Schmidt-Cassegrain, which has a 32-percent central obstruction. The inner and outer radii of the ring are 72 percent and 86 percent of the aperture, respectively, optimum values determined from a ray-tracing program (for an unobstructed system, an annulus of radii 73 percent and 88 percent also improves performance). The resolution of optical systems may be compared in terms of the radii of circles around point (e.g., stellar) images that would contain a given percentage of the original light; clearly the smaller such radii for a given aperture, the better. Table 1.6, to the left, makes such a comparison for 28-cm aperture optical systems, unobstructed and with a 32-percent central obstruction, and without and with apodizing rings.

Table 1.6 Resolution Comparison of Unobstructed and Obstructed Optical Systems, Without and With Apodizing Rings.

(A perfect optical system of 28-cm aperture is assumed, at 5600Å.)

Amount of Central Obstruction	Apodizing Ring?	Radius of Circle (in arc-seconds) Containing: 50% of Light	80% of Light	Relative Transmission
0 %	No	0".15	0".56	100 %
0 %	Yes	0".12	0".48	76 %
32 %	No	0".39	0".59	90 %
32 %	Yes	0".17	0".55	68 %

Table 1.6 makes it clear why observers prefer unobstructed systems; using the 50-percent-of-total-light criterion an instrument with a 32-percent central obstruction has the resolution of an unobstructed telescope of only 38 percent of its aperture (although this increases to 95 percent with the 80-percent-of-total-light criterion). On the other hand, using an apodizing ring brings the performance of the obstructed instrument up to 88 percent of the unobstructed system (50 percent criterion; actually 105 percent using the 80-percent criterion). In summary, a central obstruction removes light from the central Airy disk and places it in the diffraction rings, while an apodizing ring has the opposite effect. An apodizing ring can also improve the performance of an unobstructed system. Using an apodizing ring creates some light loss, but with the Moon this is scarcely a problem.

If cost is unimportant, the writer recommends refractors for observers who plan to make visual observations only. Refractors are unobstructed systems, are convenient to use, and tend to remain in collimation. They have just two drawbacks. First, they are expensive per inch of aperture and an observer can buy much more aperture for his or her money by selecting a reflecting system. Second, most two-element refractors are achromatized for the portion of the spectrum centered on the human eye's maximum sensitivity (about 5500Å). With a "minus-violet" filter, achromatic refractors often give good photographic images as well. Unfortunately, the sensitivity of video and CCD cameras extends well into the infrared and it is impossible to achieve a sharp focus for their images when using a conventional refractor unless fairly narrow-band filters are used. It is possible that this chromatic aberration problem can be reduced with (yet more expensive) apochromatic systems, but the writer has heard no reports yet from video- or CCD-camera users on the results of lunar video or CCD imaging with such systems.

With a limited budget, or for video or CCD imaging, a reflecting system is recommended. Many observers favor a Newtonian reflector with focal ratio f/6 or greater. Again, the writer has found catadioptric systems quite satisfactory, but many respected colleagues disagree.

Whatever the optical design chosen, the tube assembly must be well-mounted because of the high magnifications used in lunar studies. Some observers favor the German form of mounting to the fork (English), largely because the former is perceived as being easier to balance. Again disagreeing, the writer prefers the fork type because of its compactness and because it does not have to be reversed when an object crosses the meridian.

Visual observing, including making drawings, is possible without a clock drive or even an equatorial mounting, but each is a great convenience. For photography, videography, or CCD imaging a clock-drive equatorial mounting is essential. It is also helpful to have a lunar drive rate option because the Moon moves an average of about 0.5 arc-seconds eastward per second of time. Unfortunately this rate itself is variable and the Moon moves in declination as well. It is fortunate that modern films and CCD cameras permit short lunar exposures, and that a video frame is exposed for only 1/30 second, so that using a mean lunar rate in right ascension, and no correction in declination, should not cause blurring.

Visual Lunar Observation

The human eye is a very suitable instrument for observing the Moon, particularly its terminator region, because of the great brightness range it can accommodate simultaneously. The one essential accessory for visual observation is of course an eyepiece, or really a set of eyepieces to allow a choice of magnification. Given that a telescope's magnification is found by dividing its focal length by the focal length of the eyepiece used, and that the maximum useful magnification will be of the order of 50× per inch of aperture (2× per millimeter), one can easily calculate the shortest focal length for a useful eyepiece. For example, this would be about 3 mm for an f/6 system, 5 mm for an f/10, and 7.5 mm for an f/15 telescope. For all but the best seeing conditions, longer-focus (i.e., lower magnification) eyepieces should also be available.

Lunar observing does not require a wide field of view, and relatively long focal ratios are preferred. This means that the wide-field multielement eyepieces favored by deep-sky observers are not needed; and indeed are even undesirable because of contrast loss caused by the multiple reflections among their multiple elements. This writer favors the relatively simple Orthoscopic eyepiece design, although some advocate the even simpler design of the triplet. Whatever the design, coated eyepiece optics are highly recommended.

There are two accessories for visual lunar observing that the writer has found desirable, although not absolutely essential. One is the binocular viewer; a device that uses a beam splitter to permit viewing with both eyes. Although this unit doubles one's eyepiece needs, it allows more familiar (and thus more comfortable) viewing with both eyes, and also reduces the problem of "floaters" (particles floating in one's eye) which is quite annoying when using high magnifications, and thus small exit pupils.

A second useful accessory is a Barlow lens, which is a negative lens that increases a telescope's effective focal length, typically by a factor of 2-3×. This permits the use of more comfortable longer focal length eyepieces to achieve high magnifications. If purchasing a Barlow, make sure that it is achromatic. Also check for internal reflections; every Barlow the writer has used has had to be baffled to prevent contrast loss and even false images with photography or CCD and video imaging; some makes even require internal baffles for visual work.

Although the writer is not among them, it is only fair to say that some lunar observers find the Moon near full phase too bright for comfortable viewing and thus they then use filters. Given the weak colors of the Moon, there is no general agreement as to what color these filters should be, if not actually neutral in hue, such as with an adjustable polarizing filter. At any rate, if any filter makes one's lunar viewing more comfortable, he or she should use it. The writer prefers to use suitably high magnification to reduce surface brightness; remember that he habitually uses a binocular adapter, which automatically reduces image brightness by one-half. Also, excess image brightness is definitely not a problem when viewing near the lunar terminator.

Visual lunar observations are of no use, even to oneself, unless they are recorded (and of no use to others unless they are communicated). One can simply make written notes of what one sees through the telescope, but long experience as an editor has taught the writer that even the best notes are sometimes ambiguous and hard to interpret. Thus the preferred means of recording a visual observation is by making a drawing. One advantage of making a drawing, instead of a photographic or electronic image, is that in drawing one has to carefully study and interpret the lunar terrain; a process done only well after the fact with photographs, video, or CCD images. Several experienced observers have published guidelines for making lunar drawings. [Hill, 1991; Will, 1998; Price, 1988]

There are several means of making lunar drawings; they all require some care to be useful, but none require actual artistic talent. Four approaches to lunar drawing are shown in Figure 1.8 (page 16). Each of the four renditions is a simulated drawing based on the same CCD view of the crater Petavius, taken on 1994 Nov 20, which accompanies the Colongitude 112° terminator mosaic. South is at the top to conform with the usual view from the Earth's northern hemisphere in an inverting telescope.

A problem common to all the drawing techniques is placing features in their correct relative positions, and showing each feature in its correct size, shape, and orientation. It is a great help to draw upon a previously prepared outline chart of the area, traced from a photograph or map. The drawback to this method is that, in areas away from the center of the disk, librations greatly change features' shapes and positions. In limb areas it is necessary to prepare the outline chart from a photograph taken with approximately the same librations as pertain when one plans to observe.

Recording any form of lunar observation is without value unless the appropriate supporting data are added: date (in Universal Time), Universal Time to one-minute precision (for drawings, UTs of beginning and ending the drawing), telescope type and aperture, magnification (if a drawing), and atmospheric seeing and transparency. Seeing can be expressed either in the Pickering or the Antoniadi Scales, while transparency is the limiting naked-eye stellar magnitude in the portion of the sky being observed. Note which lunar direction is toward the top of the drawing, and be sure to indicate if a diagonal or other image-reversing device is used.

The schematic type of drawing (Figure 1.8.a) requires the least artistic ability, but still must be done carefully. It is purely a black-and-white rendition, with shadows black and everything else on a white background. This technique concentrates on relief and usually ignores albedo variations, although bright spots or rays are sometimes enclosed by dotted lines. Relief is shown by using dashed lines for positive slope breaks (i.e., the bases of slopes) and solid lines for negative slope breaks (i.e., ridges or summits). Deciding where slopes change, and in what direction, is the aspect of this drawing method that forces the observer to study the terrain carefully.

The stipple method (Figure 1.8.b) requires care and takes time. In fact, one may wish to prepare a visual-photometry rendition (Figure 1.8.c) at the telescope first, and then later convert it to a stipple drawing. In stippling, the density of black dots on a white background reflects the darkness of an area; in dark areas one may prefer to use white dots on a black background. The effect is somewhat similar to a halftone photograph, except that the dots should be randomly placed, rather than in rows and columns. This concept sounds simple but in practice can be very tedious and is not for everyone; however, in the right hands it is very effective in showing the subtle shadings near the lunar terminator.

The visual photometry approach is shown in Figure 1.8.c. Here, the observer outlines areas that are homogenous in tone internally, but which differ in tone from their surroundings. The result is a drawing that is covered by patches, each identified by a particular tone designation. Many observers use a 0-to-10 tonal scale, based on that originally devised by Schröter and modified by Elger; 0 stands for black shadow and the scale ranges up to 10 for the brightest lunar feature (usually taken to be the central peak of Aristarchus). However, the writer has difficulty with this scale because it was intended to describe albedo features at Full Moon, not the range of brightness that one sees in a limited area near the terminator. In the example in Figure 1.8.c, the

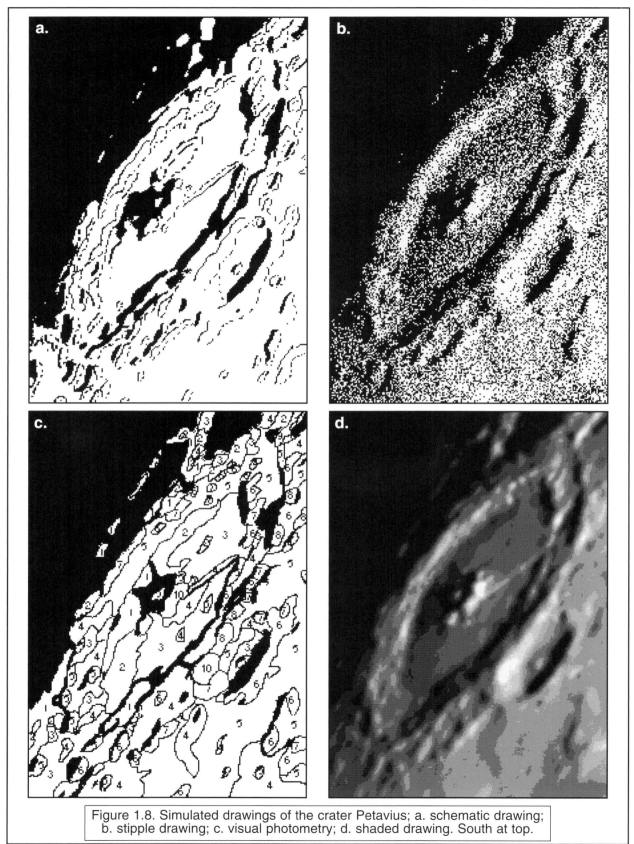

Figure 1.8. Simulated drawings of the crater Petavius; a. schematic drawing; b. stipple drawing; c. visual photometry; d. shaded drawing. South at top.

writer has made two modifications to the scale. First, rather than designating areas as "0", he has simply drawn them in black. More significantly, he has designated the brightest features in the area as "10", and numbered the remaining values 1-9 accordingly.

Visual-photometry renditions are difficult to make in areas of complex topography because of their seemingly innumerable distinct tonal areas. The writer considers the sample area in Figure 1.8 as really too large and too complex for this method to show all the details visible. Either the size of the area should be reduced, or it be observed under higher lighting because albedo variations are usually larger in scale than relief variations. Nonetheless, a well-done visual photometry diagram can be converted later to either a stipple or a shaded drawing.

Figure 1.8.d shows such a shaded drawing. Each portion of the drawing is shaded appropriately for its apparent brightness. Ten different shadings were used in this example, to conform approximately to the ten steps used for the visual-photometry diagram. Those with sufficient confidence can, of course, prepare a shaded drawing directly at the telescope without preparing a visual-photometry diagram at all. Whichever approach is employed, the most frequently used shading medium is a soft-lead artist's pencil, although charcoal is also sometimes chosen.

Photography

With the results now being achieved with lunar video and CCD imaging, film photography of the Moon may no longer appear a very useful means of recording detail. Nonetheless, film has some advantages. First, it is a "mature" technology, so that a choice of suitable cameras and films is readily available. Second, photographic equipment is self-contained and portable; neither a computer, line current, or even bulky batteries are required to take photographs. Photographic film remains perhaps the densest information-recording medium available; in terms of information content per square inch, cubic inch, or possibly even per dollar. Color photographs are not much more difficult than black-and-white, and do not suffer the resolution loss one finds when comparing black-and-white with color video or CCD images.

There are several useful recent books that describe the techniques of lunar astrophotography [Dragesco, 1995; Dobbins *et al.*, 1988; Wallis and Provin, 1988], so this section is only a short summary. It is assumed that the most popular film format, 35 mm, will be used.

First, a suitable camera is needed, and the single-lens reflex type is the most popular. Most models have removable lenses, making it possible to attach the camera directly to the telescope's eyepiece drawtube using a T-adapter. This type of camera also has through-the-lens focussing, so one can use the camera's viewscreen to focus at the prime or extended focus of the telescope. Finally, the range of shutter speeds available includes the range needed for lunar photography. Avoid over-automated cameras as the Moon is a tricky subject and you should be able to manually set exposure times. Deep-sky astrophotographers avoid cameras that drain an internal battery during time exposures, but this is not really a problem for lunar photography.

There are three drawbacks to single-lens reflex cameras, though. First, they are focussed by means of a behind-the-lens mirror that flips upward, out of the light path, when the exposure is made. This can cause serious vibration, and thus image blurring, on all but the heavier, more steadily mounted telescopes. The solution is to use a camera model where the mirror can be locked out of the way before the exposure is made. Even so, these cameras also have fairly heavy roller-blind focal-plane shutters that themselves can cause vibration. If this is the case, the only solution is to have an additional shutter somewhere ahead of the camera in the optical path; close this "front shutter" prior to exposure, then set the camera's shutter to "bulb" or "time" and then expose through the front shutter. The additional shutter may be a leaf-type shutter, for example in the eyepiece drawtube, or even a cap over the front of the telescope, although the latter is not suitable for short exposures. The third problem is that the conventional ground-glass viewing screen provides faint, hard-to-focus images at long focal ratios. If possible, the ground glass screen should be replaced by a clear glass screen with etched crosswires for focussing.

For high-resolution lunar photography, the prime focus of the telescope will not provide sufficient magnification to fully exploit the resolution of the instrument. Experienced lunar photographers often work with focal ratios in the f/40-f/100 range. One way to achieve this is with eyepiece projection, where a commercially available adapter is attached to the telescope, holding an eyepiece which projects an enlarged image to the camera. The degree of enlargement can be controlled by the choice of eyepiece focal length, adjusting the eyepiece-film distance, or both. Another enlargement method, preferred by the writer, uses modest f/20-f/50 focal ratios, with an achromatic Barlow lens, baffled in order to prevent internal reflections.

The choice of a suitable black-and-white film is simple; Kodak Technical Pan 2415 is almost universally preferred. This film has three advantages: it has extremely fine grain and high definition; it is very flexible in that light sensitivity and contrast can be varied by choice of developer and development time and temperature; it also has "extended red sensitivity" reaching to the boundary of the near-infrared.

The choice of a color film for lunar photography is less simple, chiefly because new films are entering, and old films leaving, the market all the time. The photographer may prefer prints to slides, or *vice versa*. Another factor is that there is a tradeoff between film speed and image quality. Slow, high-resolution films are good choices for the bright Gibbous or Full Moon, or at relatively short focal ratios, say f/15 or less. Some popular print films for these circumstances include Agfacolor Ultra 50, Kodak Royal Gold 25, and Konica Impresa 50; for dimmer conditions Kodak Royal Gold 100 or even Fuji Super G+ 400 can be used (the numbers refer to their ISO light-sensitivity rating). For slides, when there is sufficient light, Kodachrome 25 and Kodachrome 64 give excellent results, but in the UK can be processed only at the Kodak laboratories in Wimbledon; thus the writer prefers Fuji Velvia (ISO 50) or Kodak Ektachrome Elite II 50, going to Ektachrome Elite II 100 if there is less light, or even to Elite II 200 or 400. (For more information, see the article by Dyer, 1998; as films change frequently it is a good idea to look at the most recent results published in the major astronomy magazines.)

Discussions of developing and enlarging techniques can be found in the sources listed at the beginning of this section. Certainly, film must still be processed. However, film scanning now offers a more flexible alternative to enlarging and, indeed, has given new life to old photographs. Fairly inexpensive scanners can scan either 35-mm negatives or mounted slides at resolutions over 100 lines/mm, exceeding all but the highest-resolution films, in black-and-white or color, positive or negative, with a brightness resolution of 12 bits (i.e., 4096 shades of grey or brightness levels for each color in an RGB image). Once it is converted to digital form, a photograph may be enhanced using all the techniques applicable to CCD images; contrast stretching, sharpening, band arithmatic, and so forth. Some of these techniques are demonstrated with scanned photographs in Figures 1.9 and 1.10 (page 19).

In Figure 1.9 an original photograph of Mare Serenitatis has been scanned and is shown on the left (Figure 1.9.a). On the right (Figure 1.9.b) the same photograph has undergone a non-linear contrast stretch and unsharp masking, both used to enhance the ridges of the mare. Below it, on Figure 1.10, a grey-scale version of a color photograph is shown on the left (Figure 1.10.a), while the right-hand version (Figure 1.10.b) shows the color difference found by subtracting the blue channel from the red channel, so that the bluer areas are dark and the redder areas are bright.

Video

Video recording of the Moon has some distinct advantages. Unlike the case with film, results can be seen immediately. Thirty full frames per second are exposed so that those frames exposed during moments of excellent seeing can be selected. Although individual frames may have objectionable noise, particularly in dark areas, video frames may be read into a computer and any number averaged to improve image quality. A video digitizer is needed to input video images into a computer. However, once this is done, a video image can be enhanced using the same techniques that are used with scanned photographs or CCD images.

Conventional camcorders can be used to image the Moon, but as the lenses usually cannot be removed, they must somehow be held up to the eyepiece of the telescope. The usual camcorder

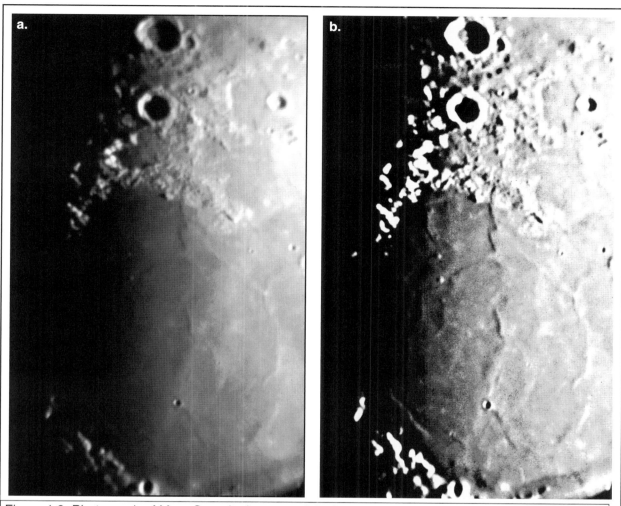

Figure 1.9. Photograph of Mare Serenitatis area; original and enhanced. a. Original; 1985 Nov 19, 02h 55m UT, 25-cm Cassegrain, f/45, 2 sec on Ektachrome 200. Col. 349°.2, north at top. b. Photograph after nonlinear contrast stretching and unsharp masking to enhance the ridges of the mare.

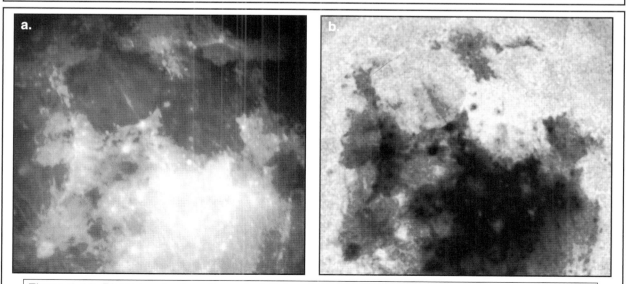

Figure 1.10. Photograph of north-central disk; original and color-difference. a. Grey-scale version of color photograph; 1972 Apr 28, 07h 02m UT. 25-cm Cassegrain, f/16.5, 1/4 sec on Kodak SO-456 photomicrography film. Col. 087°.3., north at top. b. Red-minus-blue color-difference version of same photograph; dark areas are bluer and bright areas are redder.

also has limited, if any, operator control over gain and iris settings, so the automatic exposure that results may not be what the user wants (e.g., an overexposed planet in a noisy grey sky). Color cameras tend to have poorer resolution than black-and-white ones because the former require three adjacent red, green, and blue photosites to form one color pixel. Finally, if the camcorder is connected to the telescope, vibrations from the tape drive's motor may blur the image.

Because one usually doesn't need color imaging for the Moon, but wishes to attach the camera directly to the telescope and to adjust the exposure manually, the lunar videographer actually has more convenience and better results with a relatively inexpensive black-and-white surveillance camera than a feature-loaded camcorder. The camcorder still has a role in recording the camera's video signal. The black-and-white video camera need not even have a lens; if it comes with one, detach it. A T-adapter then connects the camera to the drawtube, eyepiece projection adapter, or Barlow lens, just as with a film camera. An optional unit is a "video enhancer", which performs the analog equivalent of unsharp masking, producing a sharper image.

However the videotape is obtained, individual frames will appear grainy due to electronic noise from the uncooled CCD chip; this effect is more apparent at lower light levels. Most off-the-shelf cameras cannot make exposures longer than 1/30 second. When viewed at the normal 30 frames per second, the eye integrates successive images and the grainy effect is less obvious, and may not even be noticeable.

In order to read the analog video signal into a computer, a video digitizer ("frame grabber") is needed. Such devices are available for the major types of computer for $200 or less. Once the video image is in the computer's memory the entire range of image-enhancement techniques can be applied to it. Since one usually has a large number of video frames, 30 for every second of taping, several (or many) frames can be selected, brought into register, and averaged. This procedure both reduces noise (residual noise is inversely proportional to the square root of the number of frames averaged) and averages out image shifts due to seeing.

An example of a "manipulated" video frame is shown in Figure 1.11 (page 21). The original frame is at the top (Figure 1.11.a). A 16-frame average composite is in the center (Figure 1.11.b); and an enhanced version of the composite, using contrast stretching and unsharp masking, is at the bottom (Figure 1.11.c).

Despite the limitations of its 1/30-second exposure time, a conventional video camera or camcorders can record highly detailed lunar images [Dobbins, 1996]. The writer recommends the Super-VHS video format, which produces more detailed images than VHS; some users favor the Beta format over Super-VHS [Dantowitz, 1998]. Consumer video is about to undergo a major change, with digital-format video cameras now becoming affordable and HDTV (high-definition TV) on the horizon, along with DTV disk recording. Whatever the video format used, now and in future, one's images can easily be enhanced by computer processing.

CCD Imaging

CCD cameras are growing in popularity, and several references give considerable information about their workings and usage. [Berry, 1992; Buil, 1991; Howell, 1992; Jacoby, 1990] These notes will concentrate on CCD use for lunar imaging.

CCD cameras often have the same CCD chips that are employed in consumer video cameras, but the former have three advantages over video: (1) their chips are refrigerated, reducing electronic noise; (2) exposure times of seconds or even minutes can be used; (3) CCD cameras usually are linked to an analog-to-digital converter so that images are automatically digitized and can be stored on the computer's hard disk. The chief disadvantages follow from the advantages. First, it takes a minimum of several seconds to acquire, digitize, and store each image, reducing one's time resolution. Second, most CCD cameras are not self-contained, but must be used with a computer.

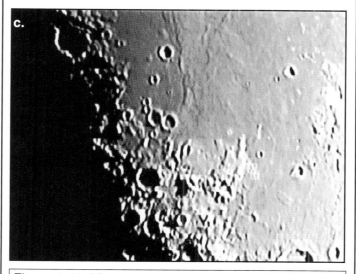

Figure 1.11. Video image of Mare Tranquillitatis; original, composite, and enhanced. a. Original image, 1990 MAY 01, 03h49m UT, 25-cm Cassegrain, f/16.5, Sanyo black-and-white video camera, recorded in S-VHS format. North at top. b. Composite of 16 video images. c. Composite enhanced using Abode Photoshop.

The heart of the CCD camera is the CCD chip, a tiny array of rows and columns of light-sensitive photosites; often called "pixels" (picture elements), although technically the pixels are the components of the resulting image. When photons strike a photosite, they cause it to release electrons, often with about a 60-percent efficiency (i.e., 100 photons result in 60 electrons).

CCD cameras differ from each other in several significant aspects. One is the number of pixels, the product of the number of rows and the number of columns. Another is the physical size of each pixel, and whether it is square or rectangular. The third is a parameter called pixel depth; how many binary bits are used to code the brightness level ("DN") of the pixel. Finally, there is the question of the wavelength sensitivity of the chip. For example, the writer uses a SpectraSource Lynxx 2000 camera with a TC255 chip that measures 336 columns×243 rows of pixels, resulting in 81,648 pixels total. Each pixel measures 10 micrometers (0.01 mm) on a side, so the entire chip measures only 3.3×2.4 mm. Twelve binary bits are used to digitize the brightness of each pixel, resulting in 4096 brightness levels (2^{12}); as it takes two bytes to code 12 bits, each image takes about 160 kilobytes of computer memory. The unit's spectral sensitivity is high in the red and near-infrared out to about 1.1 micrometers, peaking at about 0.7 micrometers, but drops toward the shorter wavelengths, and is quite low in the near ultraviolet.

The above values are typical for amateur CCD cameras. Deep-sky imagers prefer larger pixel dimensions because large pixels collect more photons; but lunar and planetary imagers want higher resolution and do not need so much light sensitivity. Particularly for lunar work, a larger pixel array size is desirable, covering more of the Moon in each frame, and there is a tendency toward using CCD cameras with larger array sizes. Sixteen-bit depth (65,536 brightness levels) is preferable to 12 bits

(4096 levels), and the former are becoming more common. Larger arrays and more bits per pixel create larger image files, taking more time to download and to process, and which occupy more disk space. Most astronomical CCD cameras produce black-and-white images, although some color units are appearing. The usual practice for obtaining a color image, however, is to take successive black-and-white images through colored filters and then to combine the separate images into an RGB (red-green-blue) color composite with a computer.

Each photosite in an ideal CCD camera would give a DN of 0 when there was no light, and a value increasing perfectly linearly with increasing light until it became saturated; in addition, the response of every photosite would be identical. Actual CCD cameras differ from this ideal. First, the electron count for a photosite is a function of the number of photons that strike it *plus* the number of electrons due to thermal noise. *Thermal noise* lives up to its name and typically is reduced by half for every 5°C decrease in temperature; this is why the chips are cooled. In order to insure that the camera temperature is brought down to its normal operating level, one should wait about one half-hour after turning on the cooler before the first exposure is made. A second degradation is due to *readout noise*, caused by statistical errors in reading out the number of electrons per photosite. This relative sampling error decreases inversely with the square root of a pixel's DN. The final problem is that the pixels are not equal in their light sensitivity, with typical variations of 1-2 percent among the photosites in an array.

For photometric applications, it is essential to correct the above problems as much as possible. To do so, one needs to expose at least two calibration frames in addition to the actual image frame. One such is the *dark frame*, an exposure of the same duration as the image frame, and as close in time to the image frame as possible. The dark frame records both *bias noise*, caused by systematic noise in the camera electronics, and thermal noise, described above. The second calibration is the *flat-field frame*, which is an exposure of a uniform light source using the same optical system (telescope, camera orientation, focus, filter, and eyepiece or Barlow lens) as was used for the image frame. Achieving a uniform light background is not trivial; the twilight sky, a projection screen uniformly illuminated in front of the telescope, or a uniformly lit diffusing screen over the telescope aperture have all been used. The purpose of the flat-field frame is to record the differences in light sensitivity (*quantum efficiency*) of the photosites; it also records the effect of dust or other optical imperfections, which is why the optical setup must be identical to that used for the image frame.

The software provided with one's CCD camera should allow creating a calibrated image from the raw image with the dark and flat-field frames. This is necessary for photometry, but can affect the pictorial quality of an image by creating more noise. This is because the calibration frames contain random noise of their own. For the highest-quality calibrated frames, the dark and flat-field frames used for calibration should each be the means of several dark and flat-field frames.

It must now be said that the writer did not use this calibration process with the CCD frames used in this atlas. One reason for this is that a large number of frames need to be exposed to cover the entire lunar terminator, and because the terminator moves, these need to be taken in as short a time period as possible. Another is that, the Moon being a bright short-exposure object, the pictorial quality of the frames was satisfactory without this calibration, which itself would introduce more noise. The image calibration that the writer did employ used the fact that each frame of the terminator inevitably includes areas in darkness beyond the terminator. The darkest DN in this zone was measured for each frame, along with that of the brightest portion of the sunlit area. Then, the contrast was stretched so that the darkest frame was set to zero and the brightest frame unchanged. This procedure alone helped to correct for scattered light in the optics and atmosphere, something not accomplished by the standard calibration technique.

When used with a telescope, the field of view of CCD cameras is tiny. Even at the f/10 focus of the writer's 28-cm telescope, his camera covers a field only 240×180 arc-seconds. With a 2.8× Barlow lens at f/29 the field drops to about 87×63 arc-seconds. When imaging planets, this means it is difficult to locate them in the first place, using the camera's focussing mode.

Fortunately the Moon is a big target, although locating and centering a particular feature may be an instructive exercise in lunar geography!

Once the chosen area is centered, the next step is critical focussing, which may take some time when seeing is mediocre. After the best focus position has been achieved, the exposure time is adjusted. If one is interested in obtaining useful detail throughout the frame, a good rule of thumb is that the brightest areas in the frame should be about 80 percent of maximum brightness (e.g., about 3300 on a 0-4095 scale). In terminator work, though, it may be necessary to saturate the brighter areas in order to bring out faint terminator detail. If the intention is to take multiple frames in order to construct a mosaic, the exposure for all frames must be identical.

Unless seeing is really excellent, one now begins repetitively to expose frames, saving to disk only those that correspond to moments of superior seeing. It is not unusual for the writer to expose 10-20 frames for every one saved. A major advantage of CCD imaging is that one has immediate feedback on whether a frame is acceptable or not.

As stated in the Preface, the purpose of the CCD mosaics used in this atlas was to cover the entire Earthside hemisphere of the Moon under many different lightings. Most frames were exposed at f/10, giving a pixel size of 0.74 arc-seconds with a 28-cm reflector. If one wishes to exploit the resolution potential of one's telescope, much higher f-ratios are needed. A rule called the *Nyquist sampling theorem* states that the photosite size should be one-half the size of the Airy diffraction disk. Given this, the formula to find the focal ratio needed is: focal ratio = 2 × photosite size/wavelength [Berry, 1992, p. 38]. With a 10-micrometer photosite size and a sensitivity peak at 0.7 micrometers, this indicates a focal ratio of about f/29 for the writer's camera to obtain maximum resolution. There is little point in using yet longer focal ratios as, the light intensity being inversely proportional to the square of the focal ratio, undesirably long exposures would be required. After all, another advantage of the CCD camera is that relatively short exposures may be used, less subject to seeing degradation than is the case with film photography.

After the observing session is ended, the next step is computer processing of the CCD images. An example of the successive modifications to a CD image is given in Figure 1.12 (page 24). The writer's procedure is to first use the camera-specific software to measure each frame's maximum and minimum brightness and then perform a linear contrast stretch (Figure 1.12.b; the original image is Figure 1.12.a). He does the remaining processing with Adobe Photoshop (currently version 5.0), although a variety of other software can be used.

For views of the terminator, it is usually best to begin further image enhancement with a non-linear contrast stretch ("Image - Adjust - Curves" in Photoshop), increasing the brightness of the darker areas more than for the brighter areas (Figure 1.12.c and d). The next, and possibly final, stage is to sharpen detail using unsharp masking, adjusting the command's three parameters to obtain the best result (Figure 1.12.e).

Two or more adjoining images may be mosaiced with most image-processing software, by selecting one frame and moving it to correspond with the adjoining frame. Variations in atmospheric transparency or camera shutter speed may cause small brightness differences between adjoining frames, which must be rectified to make them correspond. A "joint" may still be apparent due to atmospheric seeing shifting the relative positions of features between exposures, and little can be done about this except to acquire images under the best possible seeing conditions.

A further refinement, not used for this atlas, is to geometrically correct images or mosaics for the effects of foreshortening (see page 4). For example, if a circular selection marque is accurately fitted to the Moon's limb, the Photoshop "Spherize" command, using a parameter of -100 percent, converts our usual Orthographic Projection view to an truer-perspective Azimuthal Equidistant Projection, as shown in Figure 1.13 (page 25; note that relief displacement causes distortions near the limb). With a collection of lunar CCD images available in computer memory, the opportunity for image enhancement is almost limitless.

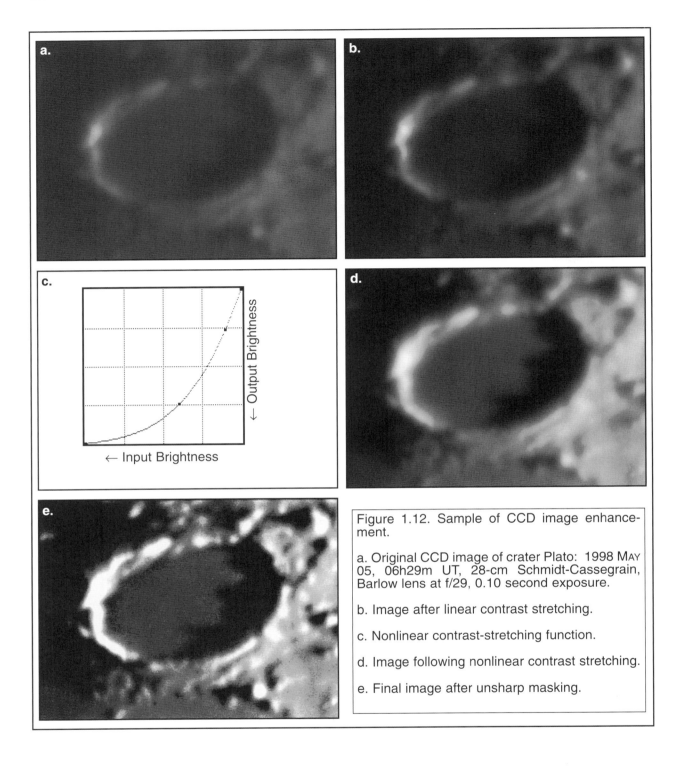

Figure 1.12. Sample of CCD image enhancement.

a. Original CCD image of crater Plato: 1998 MAY 05, 06h29m UT, 28-cm Schmidt-Cassegrain, Barlow lens at f/29, 0.10 second exposure.

b. Image after linear contrast stretching.

c. Nonlinear contrast-stretching function.

d. Image following nonlinear contrast stretching.

e. Final image after unsharp masking.

1. Observing the Moon

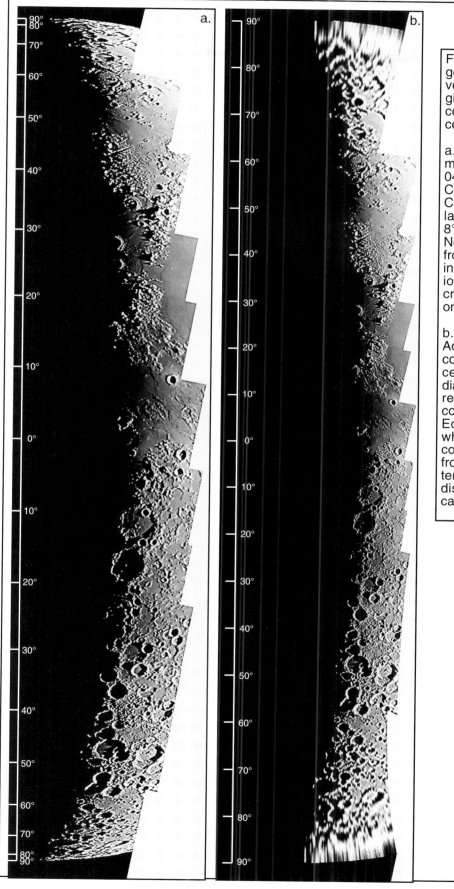

Figure 1.13. The effect of geometric rectification. The vertical scales on each view give the degrees of selenocentric arc from the apparent center of the disk.

a. Original CCD-image mosaic; 1994 A_PR 18, 04h26m UT, 28-cm Schmidt-Cassegrain, f/10, 0.07 s. Colongitude 356°.8, solar latitude 0°.7 N, librations 8°.4 W/6°.0 N, 1.96 km/pixel. North at top. Being taken from the Earth, this image is in the Orthographic Projection and thus displays increasing foreshortening as one approaches the limb.

b. Mosaic rectified with Adobe Photoshop "Spherize" command, setting -100 percent, with a lunar image diameter of 1774 pixels. The rectified image has been converted into an Azimuthal Equidistant Projection, in which the radial scale is constant in terms of distance from the apparent disk center. At the extreme limb, relief displacement continues to cause distortion.

Chapter 2. The Moon's Surface Features

Processes and Periods

The Visibility of Features

The Moon bridges the gap between the terrestrial landscapes we observe at first hand and the celestial objects we look at from afar. Over one hundred times as close as the nearest planet, we can see features on the Moon with the naked eye that would require a telescope to see on any other body. With a telescope, our resolution of detail on the Moon is comparable to that of spacecraft images of the planets and their satellites.

It is thus not surprising that the Moon has been our training ground for planetary science. The phases of the Moon demonstrated that the Moon was an approximately spherical world like the Earth itself. Beyond this, the first features noticed upon the Moon were its dark tonal or *albedo features*, generally called *maria* or "seas." Shadows or terminator extensions giving proof of the existence of lunar relief features are at the limits of naked-eye visibility, such as terminator "dents" caused by the crater Clavius or the Montes Apenninus. Theories about processes significant for the other solid bodies of the solar system, such as impact craters and basin formation, and crustal differentiation, were first developed for the Moon before they were applied to farther worlds.

Lunar landforms are obvious in any telescope and are the most prominent type of feature near the terminator, where tonal differences are obscured by the play of light and shadow. The tonal and color differences of the Moon's surface, along with their changes with lighting, are second only to surface samples returned to Earth as a source of information about the mineralogical composition of the Moon. Also, specific types of tonal features are of interest in themselves; rays, swirls, and dark-haloed and banded craters, for example. The tones and colors of the Moon deserve separate treatment, but cannot be well observed near the terminator and so will not be discussed here, where the focus is on the landforms of the Moon's surface. Particular sources that are valuable for the latter include: Basaltic Volcanism Study Project, 1981; Carr, 1984; Heiken, Vaniman and French, 1991; Mutch, 1970; Schultz, 1976a; Short, 1975; Spudis, 1993, 1996; Taylor, 1982; Wilhelms, 1987; and Wilhelms and McCauley, 1971.

Processes and Consequences

The varied features found on the Moon were created by a combination of four general types of surface process. A particular feature may have been created by one of the processes and subsequently modified by another, resulting in a varied landscape indeed.

The first such process is *impact cratering*, which creates approximately circular depressions ranging in size from pits less than a millimeter across to vast impact basins over a thousand kilometers in diameter. The idea that impacts of bodies ranging in size from micrometeorites to asteroids are responsible for the great majority of the Moon's craters has become the dominant, although not universal, theory among planetary scientists over roughly the last 50 years, due particularly to the efforts of Baldwin (1949, 1963) and Shoemaker (1962). Melosh (1989) and Spudis (1993) provide useful descriptions of this process.

Briefly put, the impact process initially excavates a depression, which is approximately circular unless caused by a very oblique impact. This process is not solely erosive, however, because all or almost all of the *ejecta* (excavated material) rains down and is deposited outside the crater itself; most of it relatively close, but some *ejecta blankets* may extend for hundreds of kilometers from a major impact. Ejected material may be in solid or liquid form; the latter is called *impact melt*. The ejecta may consist of mountain-size clumps, creating the jumbled mountain ranges found around many basins, such as the Montes Apenninus. In other cases, the ejecta may consist of fine solid material, or even be liquid, and form the light-toned *plains-forming units* found, for example, around the Orientale Basin. In addition, large bodies of ejecta can themselves carve radial valleys by grazing the lunar surface far from the impact site.

Although the impact process is clearly still going on, the ancient rate of cratering was much more intense than at present (the *heavy bombardment period*) so that many past craters undoubtedly have been obliterated by subsequent cratering or by other processes, while others have been greatly modified. Also, it is important to distinguish between *primary cratering*, caused by bodies striking the Moon from space, and *secondary cratering*, resulting from the impact of debris blasted from the surface by primary impacts.

The spatial pattern of primary impacts should be random in terms of location, although subsequent events often modify the crater density in an area. Also, it is possible that some of the more linear crater chains have been caused by the impact of strings of tidally fragmented bodies, as was the case with the impact of the Comet Shoemaker-Levy 9 fragments with the planet Jupiter. Secondary impact craters show greater densities nearer to their parent primary crater, and also often form clusters or approximate lines and arcs.

The second lunar surface process is volcanism. This term refers to the wide range of features ultimately caused by the Moon's internal heat. Little or no volcanism occurs at present; most or all of the Moon's volcanic activity occurred in the past, although significant volcanism continued well after the heavy bombardment period.

The most obvious results of lunar volcanism are *extrusive* in nature; where magma was extruded onto the surface. One could argue that the entire lunar crust is ultimately volcanic in origin. However, the most evident expressions of volcanism on the present Moon are the basaltic low-albedo lava flows that form the dark *maria* patches that are readily apparent under high lighting. Indeed, maria areas are also evident under a low Sun because they have a much lower crater density than highland areas. The lunar magma often had a very low viscosity, enabling lava flows to cover large areas before the magma solidified. The generic term *maria* is perhaps unfortunate, however, because maria material is not restricted to the large "oceans" or maria that we see with the naked eye; this material also constitutes the surface of the smaller units called *sinus*, *palus*, or *lacus*, as well as yet smaller unnamed units and the flooded floors of many craters. Another type of extrusive volcanic landform is the classic *lunar dome*, a low swelling that resembles terrestrial calderas like Mauna Loa on the island of Hawaii. In addition, large units like the Aristarchus Plateau and the Marius Hills may also be extrusive. Furthermore, many dark, relatively steep-sided hills may be *cinder cones* formed by solid material ejected from volcanic vents. However, magma *intrusion* can create surface swellings, called *laccoliths*, which may be the case for some domes. Likewise, some mare ridges ("wrinkle ridges" or *dorsa*) may be magma extrusions along fissures, although a tectonic origin due to compression is a more widely-held theory for ridge formation.

Volcanism has certainly caused some lunar craters. There are the *summit craters* found on many domes, for example, and rimless individual craters and crater chains, as well as less-regular depressions, that apparently are due to surface collapse as subsurface magma reservoirs have drained. Some medium-size craters, like Wolf, may be volcanic as well. Finally, the Moon's *sinuous rilles* are usually interpreted as collapsed lava tubes, making them yet another form of lunar volcanic feature.

The third general form of lunar surface process is *tectonism*, where the Moon's crust has been deformed vertically, horizontally, or in both directions. Although the Moon does not have continental plates, portions of its surface have shifted due to displacements caused by large impacts and earth tides, the latter much more important in the geologic past when the Earth and Moon were closer. Past tectonism may be the explanation for the polygonal outlines of many large ancient craters. Vertical tectonic uplift created crater central peaks. Other forms of crustal deformation include magma loading of the surface, magma intrusion or extrusion, or crustal contraction due to magma cooling and solidification, so that the clearest evidence of tectonic movement lies within or near the maria. Tectonic *compression* probably created the wrinkle ridges, for example. On the other hand, crustal stretching or *extension* produces faults, creating *scarps* or *graben*, the latter taking the form of the Moon's straight rilles. Although the causes of tectonism vary, in all cases they were far more significant in the Moon's distant past than at present, as is shown by the minimal activity recorded by the Apollo seismic experiments.

The final lunar surface process is *erosion*. The most important erosion process is micrometeorite bombardment, with fracturing due to thermal stress a distant second. These present-day processes operate slowly and on a small scale, so do not create results evident through the telescope. However, larger-scale erosive processes in the past produced the valleys radial to the impact basins as well as the sinuous rilles.

Chronology

The processes described above were all much more significant in the Moon's distant past than at present, so its rate of surface change has diminished greatly over time. A consequence of this is that the Moon's surface and its surface features are ancient by terrestrial standards; an intelligent dinosaur (particularly if not possessing a telescope) would have seen a Moon almost identical to the one we see.

However, the Moon's surface did not take its present form all at once, and earth-originated geologic concepts and techniques such as superimposition and stratigraphy allowed a relative lunar chronology to be established even before surface samples were available. The underlying principle for relative dating is simple. For example, a lava flow will bury or surround older features, including possible previous flows, but will have newer features superimposed on its surface. More useful yet, because it represents almost a moment in time, a major impact superimposes a crater, ejecta, and possible radial valleys upon an older surface. Thus, at least in the vicinity of the impact, we can identify features that were in place either before or after the impact. Noteworthy impacts that affected significant portions of the Moon's surface and which have been used to define the commencement of geologic periods were, in chronological order, the formations of the Nectarian Basin, the Imbrian Basin, and the craters Eratosthenes and Copernicus. Thus we speak of the *pre-Nectarian*, *Nectarian*, *Imbrian*, *Eratosthenian*, and *Copernican Periods*. When we refer to the groups of surface features associated with these time spans, the term *system* is used instead of period and sometimes the phrase *time-stratigraphic unit* is used to designate both a period and a system.

As an example of relative dating, consider the Imbrian Basin. Clearly an event like its formation erased any features that were inside its rim, so this event establishes a base time for the basin interior. It is also clear that the flooding of the basin with lava, creating Mare Imbrium itself, took place some time after the basin's formation because there exist post-impact features within the basin that were flooded or otherwise affected by the lava flows. These include isolated mountains such as Pico, the crater Archimedes with its flooded floor, the partly-obliterated partial crater Wallace, and the almost-erased "ghost crater" Lambert R.

Crater condition can sometimes be used to estimate age. Copernican craters usually have sharp rims, are bright, and often possess bright rays. Eratosthenian craters are slightly more subdued and have fainter rays, if any. Another relative-dating tool is crater-counting, where the numbers of primary craters in particular size ranges in an area are counted, and their densities calculated. Since even at present craters are created far more rapidly than they are erased by erosion or subsequent impacts, older surfaces will have higher crater densities than younger surfaces. Indeed, it is tempting to think that this quantitative approach could give us absolute dates. Unfortunately, we know that the Moon's *cratering rate* has varied greatly over time, so even here only relative dating is reliable.

These techniques allowed lunar geologic mapping to commence before surface samples were available, relying primarily on telescopic observation. Most of the Earthside was mapped at 1:1,000,000 scale, and the entire Moon has been mapped at 1:5,000,000 [Wilhelms and McCauley, 1971 ("Near Side"); Wilhelms and El-Baz, 1977 ("East Side"); Scott, McCauley and West, 1977 ("West Side"); Stuart-Alexander, 1978 ("Central Far Side"); Lucchitta, 1978 ("North Side"); Wilhelms, Howard and Wilshire, 1979 ("South Side")].

When the six Apollo expeditions that landed on the Moon returned rocks and soil samples in 1969-1972 the samples were subjected to radioisotope dating in the laboratory and the absolute dates of the samples found. This allowed absolute dates to be assigned to the lunar time-stratigraphic units, although with some uncertainty. The uncertainties in the dates apply not to the surface samples, but to the time-stratigraphic unit to which the samples are assigned, so there is still some debate about precise ages. Two sources of sample bias are that the Apollo expeditions landed at only six points and explored very limited areas, and also that these areas were all at low-latitude locations on the Moon's Earthside. Nonetheless, dates are firm enough that an approximate chronological table can be constructed, as in done in Table 2.1 below.

Table 2.1. Lunar Geologic Time Scale.*

Significant Events (Basins in chronological order; maria and craters in approximate chronological order)

Period	Approximate Dividing Age (million years)	Basin Formation	Volcanism	Sample Craters
Copernican		(none)	Lichtenberg flow	Aristarchus Tycho Copernicus
	1100			
Eratosthenian		(none)	Younger Imbrium flows Marius Hills	Hausen Eratosthenes
	3200			
Imbrian (*Late Imbrian* and *Early Imbrian Epochs* are often differentiated, with a dividing age about 3800 million years)		Orientale Schrödinger Imbrium	Aristarchus Plateau Older Imbrium flows Mare Crisium Mare Serenitatis Mare Humorum Mare Nubium Younger Tranquillitatis Older Tranquillitatis Mare Fecunditatis Mare Nectaris Highlands volcanism?	Sharp Atlas Humboldt Archimedes Plato Sinus Iridum Piccolomini Arzachel Cassini Petavius Maupertuis
	3850			
Nectarian		Serenitatis Crisium Humorum Humboldtianum Nectaris	Highlands volcanism?	Bailly Clavius Gauss Longomontanus
	3920			
pre-Nectarian		Grimaldi Schiller-Zucchius Smythii Nubium Fecunditatis Tranquillitatis Australe Pingré-Hausen Marginis Insularum South Pole-Aitken	Highlands volcanism?	Ptolemaeus Hipparchus Maginus Janssen Hommel Deslandres
	4550 (*Formation of Moon*)			

* After Wilhelms and McCauley (1971), Wilhelms (1987) and Heiken, Vaniman and French (1991).

It is clear from Table 2.1 that the rate of change of the lunar surface has varied greatly; rapid originally and then slowing down until today there is little activity. Most of the basins and 70 percent of its large Earthside craters (40 km in diameter and greater) were in place by the end of the Nectarian period, only about 700 million years after the Moon's formation. In the subsequent approximately equal length of time most of the maria had formed and so when the Moon was only 30 percent of its present age it closely resembled its present aspect.

Lunar Landforms

This section describes the types of features visible on the Moon from Earth, particularly relief features that stand out under low lighting. In contrast to the medium-scale earthbased views in the Atlas Section, the illustrations in this section use images from the Lunar Orbiter, Apollo, and Clementine space missions to show features visible from the Earth, but in greater detail than permitted in telescopic views. North will be at the top unless otherwise stated. A notation in brackets indicates a mosaic in the Atlas Section that shows a feature (e.g., "[C003°-c]" indicates the central portion of the Colongitude 003° mosaic).

Craters

The classical lunar literature uses special terms to distinguish between several types of more-or-less circular depressions. The more modern usage is to call all such features "craters." This section will largely conform to the modern usage, although occasionally the classical terminology will be used when a particular form of crater is discussed. When they are named at all, lunar craters are named after a person (e.g., "Copernicus"); a person's name is followed by one or two capital Roman letters for the smaller craters (e.g., "Copernicus H" or "Copernicus HA").

Figure 2.1. Oblique view over the crater Bullialdus and its neighbors. Mosaic of Apollo-16 Hasselblad frames 19095/19096 looking to the south-southwest over Mare Nubium. This view shows the characteristic features of several types of lunar craters. [C029°-Fig. 3]

This section begins with the typical morphologies of craters in particular size ranges; several unusual types of crater will be described later. The smaller lunar primary impact craters visible from the Earth are among the simplest of the Moon's landforms. Called "craterlets" and "craters" in the older literature, these features are basically paraboloidal in cross section, circular in plan, with a sharp rim elevated above the outside terrain. An example of one such feature is Lubiniezky F, in the foreground of Figure 2.1. This small crater measures 7.5 km in diameter and is 1350 meters deep; the height of its sharply-defined circular rim has not been measured but probably is no more than 100 meters above the mare surface. It has been dated to the Eratosthenian Period. These features are justly called *simple craters*.

As craters grow in diameter they gradually become more complex. As this happens, their walls tend to slump and fallen material starts to form either a level crater floor, or sometimes one or more floor mounds, destroying the simple profile of the smaller craters. Several authors have noted the approximate crater diameters at which simple craters begin to take on the characteristics of *complex craters* [Howard, 1974; Pike, 1980, pp. C30-C31; Spudis, 1996, pp. 24-25]. At diameters of about 20 km, floors start to become flat rather than bowl-shaped and the wall outline becomes scalloped instead of a simple arc. Above about 30 km we start to see central peaks and terraced walls. In the classical terminology these complex craters would be called "ring plains." Several examples of craters showing this transition appear in Figure 2.1, all of which have been dated to the Eratosthenian Period. Bullialdus B (20.9 km), Bullialdus A (26.2 km), and König (22.2 km) all show wall slumping, while the first two have a single terrace for at least part of the circuit of their inner walls. Their floors all contain mounds or ridges of debris derived from their walls; however, none show central peaks.

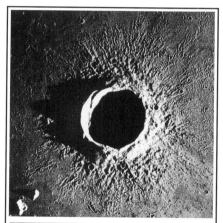

Figure 2.2. Ejecta blanket surrounding the crater Euler, seen shortly after sunrise. Euler is Eratosthenian and 27 km across; its rim is 740 meters above Mare Imbrium; comparison of its shadow length with those of the ejecta ridges shows them to be at most 50 meters high. Apollo-17 Mapping Camera frame 2292 [C029°-c]

Bullialdus (Figure 2.1) is an excellent example of a complex crater or "ring plain"; Eratosthenian in age, 59 km in diameter, 3510 meters deep, with a 590-meter high central elevation and a scalloped rim that rises 580 meters above the outside plain. Its inner walls show several complex terraces and its outer wall (*glacis*) blends into an encircling *ejecta blanket*. Ejecta blankets often extend one or more crater diameters from their parent crater, with a hummocky pattern near the crater, often irregular but frequently with some concentric structure, and showing radial ridges in their outer portions. Their low-lying relief is best seen under a low sun angle, such as with the view of the crater Euler in Figure 2.2 (left).

Indeed, when observing near the terminator, changes in sun angle can transform the appearance of both simple and complex craters. Figure 2.3 (page 32) shows, from sunrise to local noon, an area in Mare Imbrium with one complex crater and two simple craters. The complex crater near the left margin is Timocharis, Eratosthenian in age, 34 km in diameter, 3000 meters deep, with a 850-meter rim height and a 700-meter central elevation partly obliterated by a central crater. The simple craters form a pair in the figure's upper-right corner. The upper-left member is Feuillée (9.6 km, 1890 meters deep, 180-meter rim height) and its companion is Beer (9.4 km, 1670 meters deep, 280-meter rim height). Feuillée is also dated as Eratosthenian, but Beer is Copernican in age; the former crater is somewhat the darker of the two, has more debris on its floor, and its wall rim is deformed by a small crater.

Except for its central crater and smaller size, Timocharis closely resembles Bullialdus; Beer and Feuillée are just larger versions of Lubiniezky F. It is clear that a low sun angle is not conducive to investigating the interiors of fresh craters, although their outer walls and ejecta blankets are best seen with a solar elevation of 10 degrees or less. When the rising Sun is a few degrees high we begin to see the terraces of the west inner wall of Timocharis, and its floor is gradually revealed in the solar elevation range 10-20 degrees. A yet higher Sun is needed to highlight the terraces of Timocharis' east inner wall. The inner walls of the simple craters Beer and Feuillée are steeper than those of Timocharis, so still higher sun angles are needed to see much of their interiors. Under a sufficiently high Sun it is clear that even these small fresh craters have debris

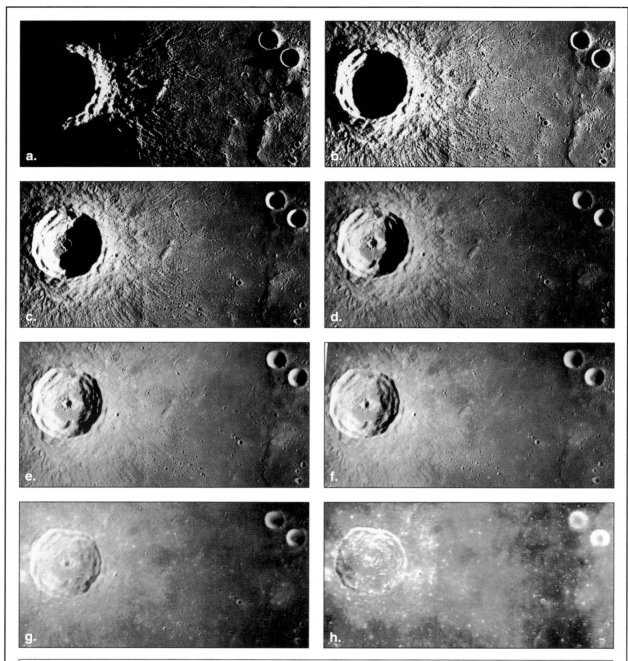

Figure 2.3. Effect of increasing solar elevation on visibility of relief and tonal features. In this area in Mare Imbrium, the large crater on the left is Timocharis; the crater pair in the upper right is Beer (lower) and Feuillé. The sources for these views are Apollo-15 Mapping Camera photography for a.- g. and the Clementine USGS Digital Mosaic for h. Frame numbers, colongitudes, and solar elevations are: a. AS15-0425, 013°, 0°; b. AS15-0597, 019°, +5°; c. AS15-1004, 027°, +12°; d. AS15-1146. 033°, +18°; e. AS15-1687, 038°, +22°; f. AS15-1829, 044°, +27°; g. AS15-2315. 063°, +43°; h. 103°, +63°. [C018°-n, C024°-n]

accumulations on their floors, marring their original geometrical profile. The difference in tone within all of these three craters shows that freshly-exposed material on steep slopes is much brighter than is older accumulated material; apparently an external process, such as micometeorite "tilling" of the lunar soil, the solar wind, or both, causes bright material to gradually darken. Although relief becomes gradually less apparent with increasing solar elevation, tonal features become more pronounced, and under noon lighting each of these three craters has a light *nimbus* surrounding it, extending out about two crater diameters in the case of Timocharis.

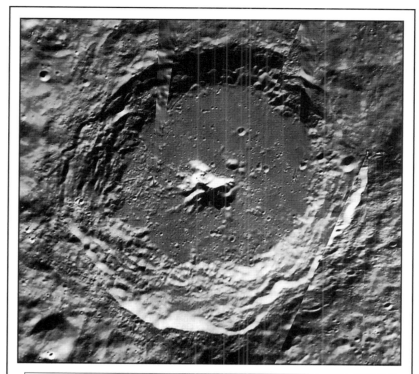

Figure 2.4. The complex crater Moretus; from USGS Digital Mosaic of Clementine images. [C018°-Fig. 3]

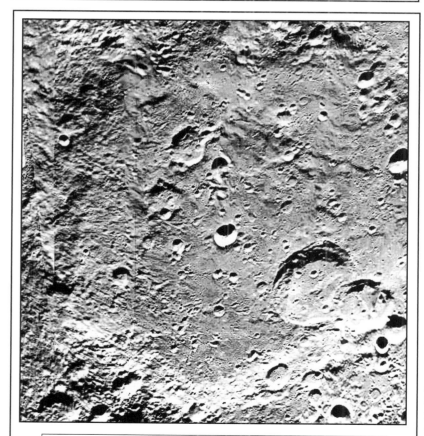

Figure 2.5. The peak ring basin Bailly; from USGS Digital Mosaic of Clementine Images. [C075°-Fig. 2]

The ring plains, or complex craters, do not change their qualitative characteristics as their diameters increase up to about 180-200 km, as shown by the large-scale examples of Hausen (167 km; Quadrant 3-sw), Gauss (177 km; Quadrant 1-ne), and Humboldt (201 km; Quadrant 4-se). They continue to have wall terraces and flat floors with central peaks. Quantitatively, their depths and rim heights tend to increase more slowly than their diameters. However, their floor diameter tends to increase gradually in relation to the rim diameter. Moretus, shown in Figure 2.4 to the left, is Eratosthenian in age, and a good example of a large complex crater with a multiple central peak, a flat floor with a smooth portion indicating impact melt, multiple terraces on its inner wall, and a scalloped rim. The diameter of Moretus is 114 km, its depth 5240 meters, and the central peak rises 2650 meters above the floor (the maximum rim height is 2010 meters, but Moretus lies in a rugged highlands area so this parameter here has little meaning).

Clavius, 225 km wide, is the largest Earthside crater that possesses a feature at all like a central peak [C024°-s, C191°-s]. Also, craters its size or larger have flat floors that occupy such large proportions of their rim diameters that their walls are relatively narrow and do not show well-defined terraces. Instead, these very large craters now start to show signs of a roughly circular inner ring of peaks on their floors. For the complex craters that exhibit central peaks, the term *central peak basin* is sometimes used; for those that contain inner rings, the term *peak ring basin* is preferred. Bailly, shown in Figure 2.5 to the left, is 284 km in diameter and 4130 meters deep, making it the largest true crater on the Earthside hemisphere, and is the most readily visible example of a peak ring basin. Bailly is Nectarian in age and is considerably battered by ejecta from the Oriental Basin, far to its north, and from more recent nearby craters such as Hausen.

Multiring Basins

There are larger circular depressions on the Moon than Bailly, but at a diameter above about 300 km they begin to show a pattern of one or more concentric rings outside the main rim. These features are called *multiring basins* (also "multi-ring basins", "multi-ring impact basins", "multi-ringed basins", or simply "basins") and range in size from Bailly as a marginal smallest example up to the South-Pole Aitken Basin, the latter falling mainly on the lunar Farside, whose main ring has a diameter of 2500 km. A multiring basin is so large that, to see all of it under low lighting, one must observe for several nights in succession

Although formed by asteroid impact, and really just very large craters, the formation of multiring basins was the dominant agency in shaping the face of the Moon. First, the depressions they excavated occupy much of the Moon's surface; perhaps most of the Moon's surface if we postulate very early impacts whose traces are now obliterated. Second, the ejecta from the impacts extend far beyond the basin boundaries, sometimes for thousands of kilometers, blanketing and sometimes burying earlier formations. Besides their size, their most obvious property is their concentric rings. The means of origin of multiple rings is not agreed upon. One theory is that the rings are large terraces ("megaterraces") created by slumping around a relatively small crater excavated by the impact. Another theory suggests that the energy of the impact "fluidized" the Moon's crust in the area, allowing waves to be formed. In a third theory, the rings reflect the excavation of layers in the crust that have different structural strengths.

Figure 2.6, on p. 35, shows the more agreed-upon impact basins on the Earthside and Marginal Zone; it also shows the maria, with their names given if not associated with a basin, together with this writer's highland-region names as used in the descriptions in Section II. There are, of course, other possible basins in the area visible from Earth that have been suggested, but are not recognized by all basin investigators. Some of these proposed basins are the Procellarum ("Gargantuan") Basin (15° W/26°N, 3200 km), Mutus-Vlacq Basin (21°E/51°S, 690 km), Insularum Basin (18°W/9°N, 600 km), Marginis Basin (84°E/20°N, 580 km), Flamsteed-Billy Basin (45°W/7°S, 570 km), Balmer Basin (70°E/15°S, 500 km), Werner-Airy Basin (12°E/24°S, 500 km), Amundsen-Ganswindt Basin (120°E/81°S, 335 km), Lorentz Basin (97°W/34°N, 365 km), Sikorsky-Rittenhouse Basin (111°E/68°S, 310 km), and Pingré-Hausen Basin (82°W/56°S, 300 km) (in parentheses are the coordinates of the basin center and the main rim diameter). For those impact basins shown on the map, Table 2.2, to the left, lists them in approximate ascen-

Table 2.2. Selected Multiring Impact Basins Visible from Earth.

Basin Name	Geological Period	Ring Diameters (km; most prominent underlined)
Orientale	Imbrian	320, 480, 620, <u>930</u>, 1300, 1900
Imbrium	Imbrian	670, 900, <u>1200</u>, 1700, 2250, 3200
Bailly	Nectarian	150, <u>300</u>
Serenitatis	Nectarian	420, <u>740</u>, 1300, 1800
Crisium	Nectarian	380, 500, <u>740</u>, 1060, 1600
Humorum	Nectarian	210, 325, <u>440</u>, 560, 820, 1195
Humboldtianum	Nectarian	275, 340, 460, <u>600</u>, 1050, 1350
Mendel-Rydberg	Nectarian	200, 300, 460, <u>630</u>
Nectaris	Nectarian	240, 350, 450, 600, <u>860</u>, 1320
Grimaldi	pre-Nectarian	<u>230</u>, 300, 430
Schiller-Zucchius	pre-Nectarian	165, <u>325</u>
Smythii	pre-Nectarian	260, 360, 480, 600, <u>840</u>, 1130
Nubium	pre-Nectarian	<u>690</u>
Fecunditatis	pre-Nectarian	<u>690</u>, 990
Tranquillitatis	pre-Nectarian	<u>775</u>, 950
Australe	pre-Nectarian	95, 275, 550, <u>880</u>
South Pole-Aitken	pre-Nectarian	900, 1900, <u>2500</u>

Sources: Heiken *et al.*, 1991, pp. 118-119; Melosh, 1989. p.169; Spudis, 1993, pp. 38 and 40; Wilhelms, 1987, pp. 64-65.

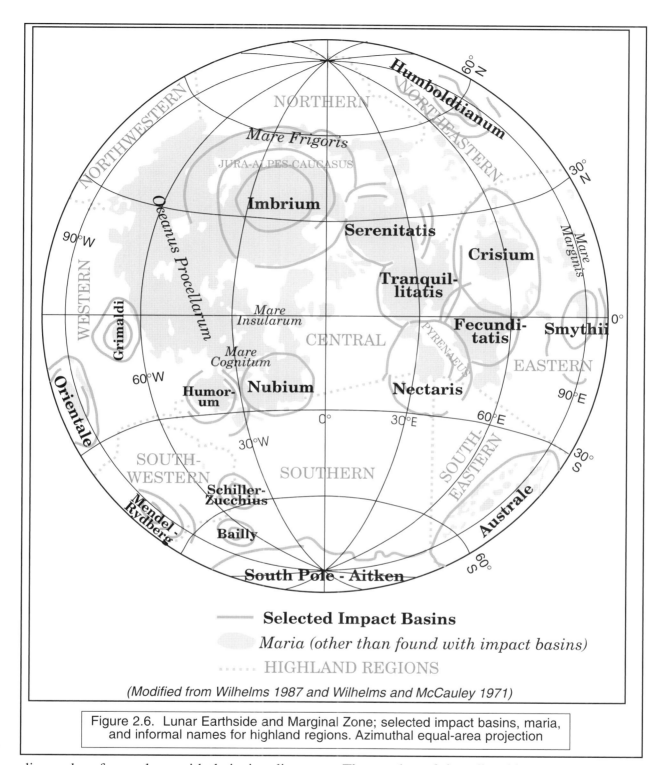

Figure 2.6. Lunar Earthside and Marginal Zone; selected impact basins, maria, and informal names for highland regions. Azimuthal equal-area projection

ding order of age, along with their ring diameters. The number of rings listed is generous because not all rings are agreed upon by all authors. Basin names are not sanctioned by the International Astronomical Union, but those in the table and map are widely used and reflect either associated craters or maria.

The Humorum Basin is shown in Figure 2.7 on page 36. This is a fairly representative basin in age and size and is readily visible from the Earth. The upper portion of the figure shows the mare and basin as imaged by Lunar Orbiter 4, while the inset in the lower left of the figure depicts the rings identified with that basin.

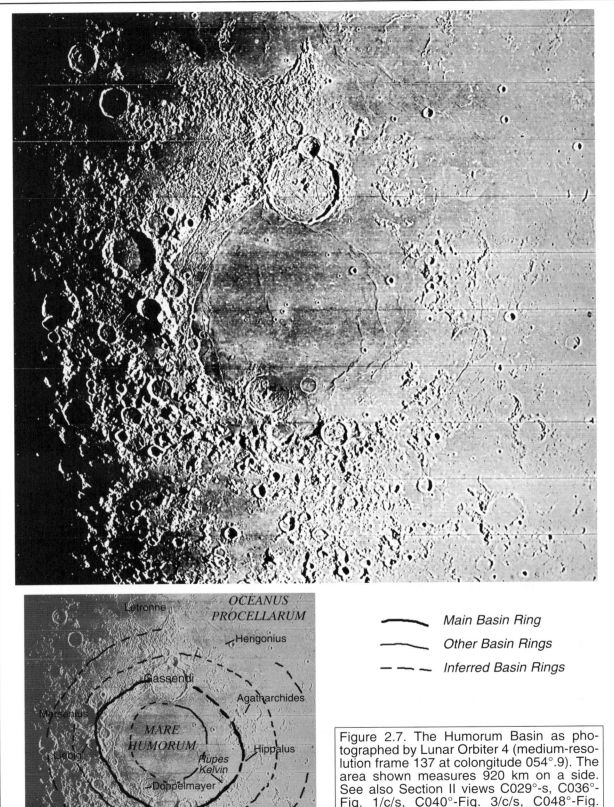

Figure 2.7. The Humorum Basin as photographed by Lunar Orbiter 4 (medium-resolution frame 137 at colongitude 054°.9). The area shown measures 920 km on a side. See also Section II views C029°-s, C036°-Fig. 1/c/s, C040°-Fig. 3/c/s, C048°-Fig. 2/c/s, C055°-Fig. 2/c/s, C061°-c/s, C205°-s, C213°-Fig. 1/c/s, and C225°-Fig. 2/c.

The Humorum Basin was formed in the Nectarian Period, and subsequently, in the Early Imbrian Epoch, was blanketed by ejecta from the Imbrian impact about 2000 km to the north. Then, in the Late Imbrian, most of the basin interior was flooded by mare lava. Nonetheless, the main ring of the basin is clearly defined by a 440-km diameter circle running through the craters Gassendi and Vitello. Portions of its rim are demarcated by named fault scarps, the Rupes Kelvin on the southeast and the Rupes Liebig on the west.

The innermost ring of the Humorum Basin is defined by mare ridges rather than the highlands crustal material that composes the main ring. Inner rings defined by mare ridges can also be found in the Imbrian, Serenitatis, and Crisium Basins, and may be caused by tectonic movement after the impact event itself. Another form of feature, also roughly concentric with the basin rim and due to tectonic movement, is rille systems. The Humorum Basin has several of these that have been named for nearby craters: the Rimae Herigonius, Agatharchides, Hippalus, Doppelmayer, and Mersenius. Indeed, several craters near or on the main Humorum Basin rim show signs of tectonic movement on their interiors: Gassendi, Hippalus, Vitello, Doppelmayer, Palmieri, de Gasparis, and Mersenius.

There are probably two rings outside the main Humorum ring, but they can be traced only intermittently because they pass through irregular terrain modified by Imbrian and possibly by Orientale ejecta and by the lava flooding of adjacent portions of Oceanus Procellarum and Mare Nubium in the Imbrian and Eratosthenian Periods.

Although the Humorum Basin is typical of lunar basins, at least those of similar age, there are significant differences among these features. The fresher basins, Orientale and Imbrium, have obvious ejecta blankets that overlay older features on their periphery. Radial valleys, gouged from the lunar crust by large chunks of debris cast out by the impact, are particularly associated with the Oriental and Imbrian Basins but also occur with the Crisium and Nectaris Basins.

The Imbrian Basin stands out among those basins whose interiors have been occupied by mare lava, in that isolated peaks of highlands material protrude from the mare and appear to trace out inner rings that were probably formed by the impact event itself. Other mare basins that differ from Humorum are those whose interiors have been only partially flooded so that they contain isolated patches of lava. With the Oriental basin, Mare Orientale occupies part of the inner basin, while Lacus Veris and Lacus Autumni are elongated and lie between successive rings. Indeed, the western two-thirds of Mare Frigoris appears to lie between two successive Imbrian Basin rings. The Humboldtianum and Marginis Basins are only partly flooded. Finally, within the Australe Basin, the ill-named Mare Australe is actually a group of many discontinuous lava patches, including several flooded craters.

The last category of impact basins is the ancient ones found in the highlands. Their presence is not signalled by mare flooding, so the chief evidence for their existence is their relief, which is often obscured in these crater-saturated areas. The Bailly, Schiller-Zucchius, Mendel-Rydberg and South Pole-Aitken Basins fall in this category, although they are fairly well-defined. More obscure, and thus debatable, are the highland Mutus-Vlacq, Balmer, Werner-Airy, Amundsen-Ganswindt, Lorentz, Sikorsky-Rittenhouse, and Pingré-Hausen Basins. Because their rims, if any, are so subdued, near-terminator earthbased observation is the best means of confirming their existence and possibly discovering other examples.

Crater Variants

Up to this point, the discussion has centered on what might be called the "normal" sequence of craters: simple craters—complex craters—central peak basins or peak ring basins—multiring basins. However, there are many craters, or at least "depressions," that do not fit this neat sequence. The latter often appear to be the result of more than one process rather than simple impact; indeed, some types of crater may not be due to impact at all.

Mare-Flooded Craters. With this class of feature, part or all of the floor appears almost absolutely flat. There are two main types, those where the level portion is dark and consists of mare material, and those with light floors formed of "plains material." The floors of the larger flooded craters may be studied with advantage when near the terminator. They are so shallow that the shadows of their walls never reach their center so that we can view their floors under grazing illumination.

Figure 2.8. The mare-flooded crater Endymion; 125 km in diameter and 4070 meters deep. USGS Digital Mosaic of Clementine-mission images. [C309°-Fig. 1/n, C315°-n, C100°-n, C105°-n, C106°-n, C112°-n, and C119°-n]

The mare-flooded craters can be placed in a sequence starting with those like Humboldt [C099°-c], and Petavius [C112°-Fig. 1] where only a small part of the floor is flooded, through half-flooded examples such as Riccioli [C080°-Fig. 1] and Schickard [C055°-Fig. 3], to completely flooded specimens such as Plato [C183°-Fig. 1], Archimedes [C009°-Fig. 1, C179°-Fig. 1], and Endymion, the last of which appears in Figure 2.8, to the left. Indeed, we can even extend the sequence to unusual craters like Mersenius [C055°-Fig. 2], with its convex floor, and on to Wargentin [C061°-Fig. 3], whose floor has been filled almost to its crater rim.

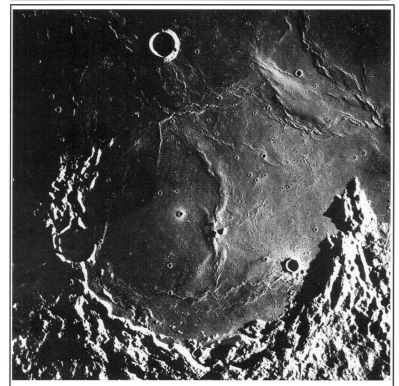

Pre-mare craters within or adjoining maria often have had their walls broken or even almost entirely submerged by lava flows originating from within the mare. Such breached craters lying on the "shores" of maria include Le Monnier [Mare Serenitatis; C334°-n]; Sinus Iridum [Mare Imbrium; C205°-n]; Hippalus, Doppelmayer, and Lee [Mare Humorum; C205°-s]; Pitatus [Mare Nubium; C191°-Fig. 3]; Fracastorius [Mare Nectaris; C328°-Fig. 3]; and Lick [Mare Crisium; C309°-c]. An excellent example of one of these craters is Letronne on the southern edge of Oceanus Procellarum, shown in Figure 2.9 to the left. Although it contains the remnants of a central peak, its northern walls are defined only vaguely by a system of mare ridges. Subsidence of the crater interior may be responsible for the north-south medial ridge and the irregular system of ridges on its southern floor.

Figure 2.9. The mare-flooded crater Letronne; diameter 157 km, 0-1190 meters deep. Apollo-16 Mapping Camera frame 2995. [C048°-Fig.2, C225°-Fig. 2]

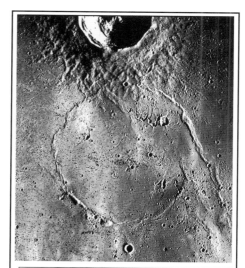

Figure 2.10. The "ghost crater" Lambert R (center); 30.3 km in diameter with walls 0-300 meters high. Apollo-17 Mapping Camera frame 2286. [C024°-Fig. 1]

When a pre-mare crater lies entirely within a mare, it may be almost obliterated by the mare lava. Undoubtedly, many such craters were entirely erased and thus are unknown to us. The surviving examples are often called *ghost craters*. The Flamsteed P ring in Oceanus Procellarum is an example, and was the landing site of the Surveyor 1 spacecraft in 1966 [C048°-c]. Its walls are a mixture of highland-material peaks and mare ridges. On the other hand, the walls of some of these formations are composed entirely of ridges, like the old "Newton" ring between Plato and Pico [C183°-Fig. 1]. Another such example is Lamont [C334°-Fig. 2]; in its case, one wonders if its impact happened to occur when the mare lava was still liquid. A third example of a ghost crater is Lambert R in Mare Imbrium, photographed in Figure 2.10 to the left.

Craters With Floors of Light-Plains Material. Another class of craters has interiors just as flat as the mare-filled variety, but the floors are light-toned, and often hard to differentiate from the walls under a high Sun. Such crater floors are formed of *light-plains* material (also called "highland plains", "Cayley plains", "terra plains", or just "plains"). Some geologists believe these units to be volcanic in origin, others consider them to be ejecta deposits. This material, whatever its origin, has been dated to the Imbrian Period and fills or mostly fills the interiors of many earlier craters, such as: Ptolemaeus, Alphonsus, and Flammarion [C003°-Fig. 2, C009°-Fig. 2]; Hipparchus and Albategnius [C357°-Fig. 3]; Fra Mauro, Guericke, Parry, and Bonpland [C024°-Fig. 2]; Maginus [C183°-Fig. 1], Clavius [C024°-s], Schiller [C048°-s], and Schickard. The last is an interesting case where part of its floor is flooded by mare material and the remainder by plains material.

These "light-plains floor" craters are particularly prevalent in the region north of Mare Frigoris, which is shown in Figure 2.11 on page 40. Some craters in this region, such as Anaximenes, Barrow, Meton [C148°-Fig. 1, C161°-Fig. 1], Kane, Arnold, Neison, Baillaud, and Byrd, have particularly smooth floors. Their covering is interpreted as Imbrian, or even Orientale, ejecta. The floors of others ancient craters, like South, J. Herschel, Birmingham, and W. Bond are hummocky [C174°-Fig. 1] and assigned to the "Fra Mauro Formation," a thick deposit of material thrown from the Imbrian Basin. The craters themselves mostly date back to the pre-Nectarian Period. Several are elongated or polygonal, suggesting tectonic movement since they were formed. The Meton complex is particularly interesting, as its scalloped outline results from the fusion of perhaps five separate craters. It is unclear how a thin layer of Imbrian ejecta could erase the adjoining walls of the crater group, so perhaps volcanism combined with tectonic movement and primary and secondary impact to create this landform.

Fractured-Floor Craters. The floors of many older craters have been distorted by tectonic movement, volcanism, or both. Their floors often contain ridges, rilles, or scarps that are roughly concentric with the crater rim. Sometimes the floor is broken up by complex systems of rilles. Some lunar geologists consider these features to be "ring dikes," which are concentric volcanic extrusions [Cameron and Padgett, 1974]. Others see signs of doming and crustal expansion due to igneous intrusions beneath an impact crater [Schultz, 1976b; Wichman and Schultz 1996]. Prominent examples of this type of crater are Humboldt [C099°-c], Gassendi [C048°-Fig. 2], and Posidonius [C343°-Fig. 1]. Figure 2.12, on page 40, shows the crater Vitello as an example.

These features are concentrated in particular regions of the Moon, especially around the shores of maria. They also occur in lesser numbers in the interiors of maria and even in highland areas. A map showing their distribution on the lunar Earthside, along with several other types of features, is given in Figure 2.13 (page 41).

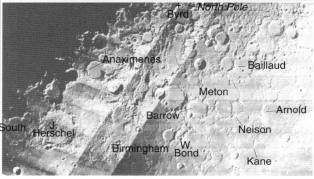

Figure 2.11. Ancient craters with floors buried in light-plains material north of Mare Frigoris. Mosaic of Lunar Orbiter 4 medium-resolution frames 116, 128, and 140 (right to left). The dimensions (diameter/depth) of the larger craters are: W. Bond, 157 km/1890 meters; J. Herschel, 156 km/900 meters; Meton, 122 km/1830 meters; South, 98 km/970 meters; Birmingham, 97 km/650 meters; Arnold, 95 km/950 meters; Barrow, 93 km/2380 meters; Baillaud, 90 km/1760 meters; Byrd, 83 km/1380 meters; Anaximenes, 80 km/880 meters; and Kane, 55 km/570 meters.

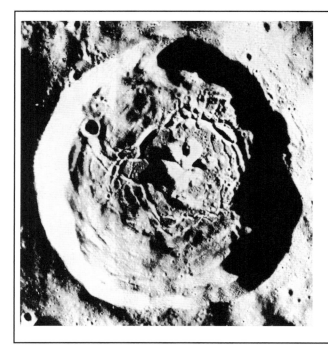

Figure 2.12 (left). Fractured-floor crater Vitello, photographed by Lunar Orbiter 5 (medium-resolution frame 168). Colongitude 049°.6. Vitello is 45 km in diameter, 1700 meters deep, with a 610-meter rim height above the exterior, and a multiple central peak rising 1100 meters above its floor. The floor contains ridges and rilles concentric with its wall rim. [C048°-s, C213°-Fig. 1] (Processing defects have been removed from the version of the photograph shown here.)

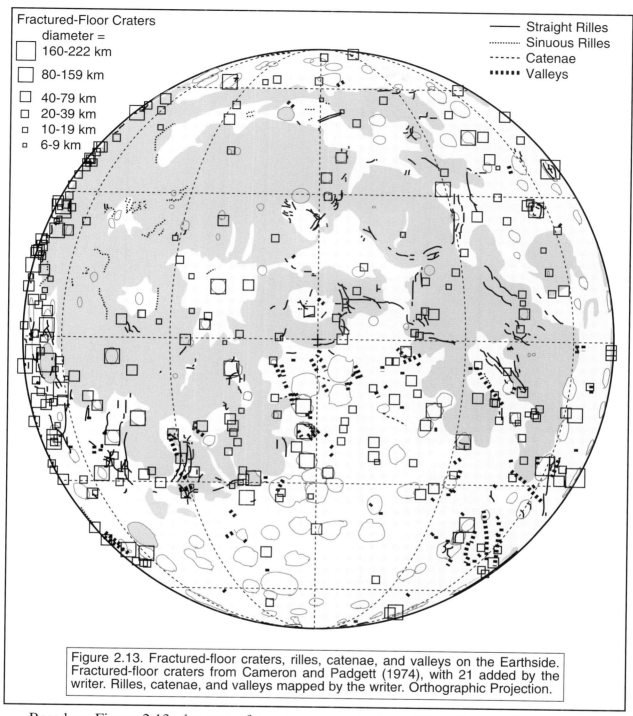

Figure 2.13. Fractured-floor craters, rilles, catenae, and valleys on the Earthside. Fractured-floor craters from Cameron and Padgett (1974), with 21 added by the writer. Rilles, catenae, and valleys mapped by the writer. Orthographic Projection.

Based on Figure 2.13, the area of greatest concentration of floor-fractured craters is in the Western and Northwestern Highlands near the western border of Oceanus Procellarum, even though their concentration is exaggerated by the figure's orthographic projection. The region of concentration includes Darwin (20°S) and extends 2300 km north to Xenophanes (57°N). The craters involved are mostly pre-Nectarian in age, whose floors contain rilles, scarps, and concentric ridges. It is best to observe this region with favorable westerly or northwesterly libration as well as with a low sun angle [see C075°-n/c, C080°-Fig. 1/n/c, Quadrant 2-n/nw, and C247°-n]. The northern portion of this region (32°N-60°N) is shown in Figure 2.14 (page 42); an area including several large rings. The rim of the adjoining Oceanus Procellarum is also noteworthy for a large number of mare-flooded craters. Clearly, volcanism and tectonic movement have been significant processes throughout this region.

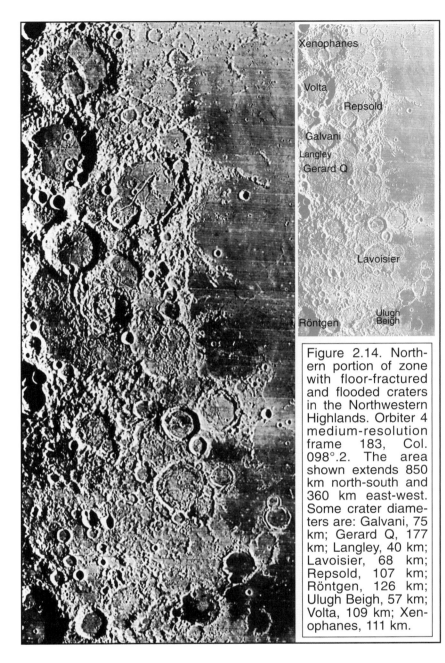

Figure 2.14. Northern portion of zone with floor-fractured and flooded craters in the Northwestern Highlands. Orbiter 4 medium-resolution frame 183, Col. 098°.2. The area shown extends 850 km north-south and 360 km east-west. Some crater diameters are: Galvani, 75 km; Gerard Q, 177 km; Langley, 40 km; Lavoisier, 68 km; Repsold, 107 km; Röntgen, 126 km; Ulugh Beigh, 57 km; Volta, 109 km; Xenophanes, 111 km.

Secondary Craters.
Large impacts can eject solid bodies from the excavation cavity, traveling many kilometers and impacting the surface at velocities of hundreds of meters per second. Clearly these impacting ejecta bodies will form their own craters, called *secondary craters.* "Secondaries" can be differentiated from "primaries" in several ways. In morphology, the former are the shallower, and have lower rims in relation to their diameters, and are also the more often elongated. Because debris is often ejected in clumps, secondary craters frequently form clusters, linear or curvilinear chains, and often distinctive V-shaped or "herringbone" patterns. The Copernicus secondaries lying in the Stadius area often are cited as examples; some of them are shown in Figure 2.15 (page 43).

Because of their low impact velocities, large secondaries are unusual. Nonetheless, several named craters are interpreted to be secondary craters: Müller (21.8 km), Auwers (23.4 km), Secchi (24.5 km), Ritchey (29.0 km), d'Arrest (30.0 km), Hypatia (33.6 km), Vogel (34.0 km), Schröter (34.5 km), and Zöllner (42.8 km).

Crater-Pits. For once the classical term appears still descriptive; *crater-pits* are rimless craters without ejecta. Impact craters are termed *exogenic* because they arise from collisions involving extralunar bodies. However, impacts cause craters with raised rims; it thus seems that crater-pits are *endogenic,* formed by processes originating within the Moon itself. More specifically, these depressions have been formed by surface collapse.

Crater-pits are thus often found along, or at the ends of, rilles, which also appear to be collapse features. One example is the "Cobrahead" bowl that forms the southeastern end of the Vallis Schröteri sinuous rille [C055°-n, C225°-n]. Sinuous rilles are usually interpreted as resulting from the collapse of lava tubes. Another form of surface collapse is caused by faulting, associated with straight rilles. The best-known example of the latter is the crater-pit Hyginus, which lies directly upon the Rima Hyginus. Hyginus' diameter is 9.7 km; its depth is 780 meters, but its rim does not protrude visibly above the outside surface. Hyginus does appear to be at the summit of a very gentle domical uplift, and crater-pits are often found on the summits of domes. The

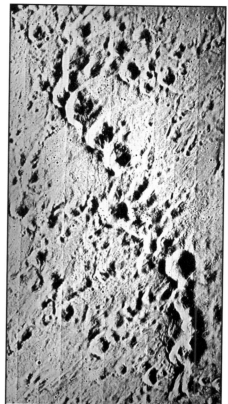

Figure 2.15. Chain of secondary craters originating from Copernicus; Stadius R (5.7 km) is 4 cm above the lower right corner. Orbiter 5 medium-resolution frame 142 (Col. 035°.1); this view extends 30 by 60 km. [C018°-Fig. 1 and C191°-Fig. 1]

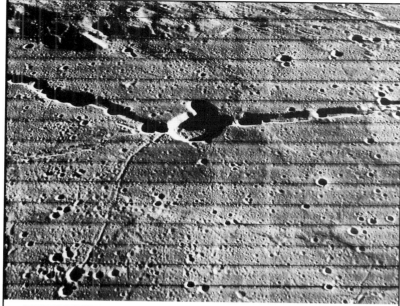

Figure 2.16. Looking north over the crater-pit Hyginus and the Rima Hyginus. Lunar Orbiter 3 medium-resolution frame 075 (Col. 000°.3)

crater-pit Hyginus, with its associated rille, is shown in Figure 2.16, above [see also C357°-Fig. 2/c and C174°-Fig. 2].

Other Crater Variants. Not all of the many forms of crater have been discussed, although some of the rarer examples are too small to be effectively viewed from Earth.

One form not found is summit craters as found on terrestrial stratovolcanoes; the Moon probably has none of the latter. However, volcanic calderas are found. Many selenologists (as lunar geologists were then called) once interpreted most craters as calderas. This is no longer so, but the craters Wolf [C191°-Fig. 3], in Mare Nubium; Daniell [C343°-Fig. 1], in Mare Serenitatis; and Kopff [Quadrant 3-w], adjoining Mare Orientale, may be calderas, along with the summit pits on many domes. The odd D-shaped depression Ina in Lacus Felicitatis is probably a caldera, but Ina is only 3 km in diameter and thus in a telescope resembles a typical crater-pit.

Collapse features take many forms besides circular features like Hyginus, such as racetrack-shaped ovals and arrowheads, the last exemplified by Krishna in Mare Serenitatis [C351°-n, C167°-n]. Krishna is also only 3 km across, so its unusual shape is apparent only in space-mission imagery.

Oval depressions are common among the smaller craters, and then are most often either collapse features or secondary craters. Still, elongated larger craters can also be found. Both craters of the Messier-Messier A pair are elongated [C119°-Fig. 2]. On a larger scale, Schiller is highly elongated [C048°-s]. Indeed, when seen in true projection, the Crisium Basin is elongated east-west. Oblique impacts probably explain the majority of the larger elongated craters, with multiple simultaneous impacts the cause of features like Schiller.

Indeed, the simultaneous impact of two associated bodies can explain the observed presence of more crater pairs than would be expected were impacts purely random. Naturally, this theory is viable only when the two craters are of similar age. Sabine and Ritter appear to fit this model [C154°-Fig. 1], as do Messier and Messier A, an unusual pair on two counts.

Rilles, Catenae, and Valleys

There is no single term for the several distinct forms of highly-elongated depression that are found on the Moon. The more prominent members of each type have been mapped in Figure 2.13 (page 41). Most of these features are shallow relative to their length, and even to their breadth, and are best seen under a low illumination angle.

Straight Rilles. Although called "straight", these features can form gentle arcs, branch, or even form complex networks. They are also called "clefts." Unfortunately, the Latin phrase for these features, *rima* (pl. *rimae*) is also used for the quite-different sinuous rilles.

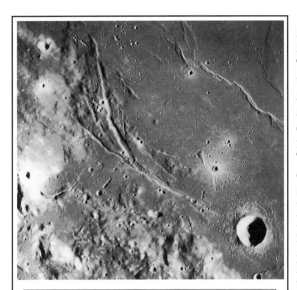

Figure 2.17. The Rimae Sulpicius Gallus. The 11.9-km crater Sulpicius Gallus is in the lower right, part of the Montes Haemus at the bottom, and Mare Serenitatis occupies most of the frame. Apollo-17 Mapping Camera frame 1817, covering an area 97 km on each side.

Figure 2.17, to the left, is centered on the Rimae Sulpicius Gallus on the southwest border of Mare Serenitatis. This rille system extends about 90 km. The rilles have flat floors lying between parallel walls. Thus, each component appears to be a *graben*, a depression formed by subsidence between two parallel faults (the Red Sea is a large-scale terrestrial example of a graben). Straight rilles range in width from kilometers down to below telescopic visibility. They are most prominent under a low Sun, but the direction of sunlight favors the visibility of north-south rilles over east-west ones. Wider and easily visible examples include the Rima Ariadaeus and Rima Hyginus [C357°-Fig. 2], which also shows the narrower Rimae Triesnecker and the Rimae Hippalus and Ramsden [C040°-Fig. 3]. Several straight rimae have a disconcerting habit of passing directly through older topographic features such as hills and even crater walls.

Straight rilles are concentrated in maria near the rims of basins, which they often parallel (Figure 2.13, page 41). A widely-held explanation for their formation is parallel faulting due to crustal tension, in turn caused by subsidence of the basin interior.

These features can be found elsewhere, however. Elaborate rille systems crisscross the floors of many fractured-floor craters, where floor subsidence creates the same effect at a small scale that basin subsidence does on a large scale (see Figure 2.12, page 40, and Figure 2.14, page 42). The largest example of a lunar graben is the Vallis Alpes, 180 km in length and so large that it is not even called a rima. This exceptional feature is found in the lunar highlands and is perpendicular, rather than parallel, to the rim of the Imbrian Basin [C167°-Fig. 1].

Sinuous Rilles. Like straight rilles, sinuous rilles usually have flat floors and parallel sides. Sinuous rilles, however, live up to their name in that they snake across relatively flat areas somewhat like meandering rivers. They look very much like the products of erosion, and indeed photogrammetry with the Apollo Mapping Camera photographs has shown that they consistently run downhill from a high end to a low end. They are of course now empty of whatever medium flowed along them, but the majority view is that it was low-viscosity lava. In fact, the sinuous rilles are probably a mixture of lava channels and collapsed lava tubes. Sinuous rilles can have straight segments, in which case faulting may also have been an agency. One reason for the popularity of the lava channel/tube theory is that numerous terrestrial analogies are found in lava fields in Iceland, Hawaii, and northeastern California. The lunar rilles are on a larger scale than the terrestrial ones, explained by the Moon's lower gravity and by assuming highly fluid lava.

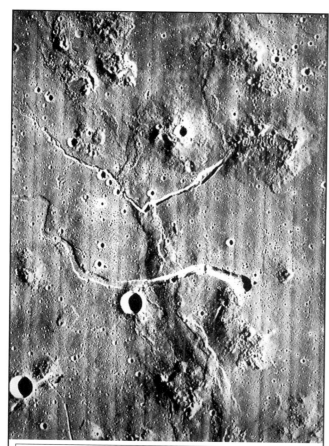

Figure 2.13, on page 41, shows the locations of the larger sinuous rilles on the lunar Earthside. Most of these distinctive features occur in maria, but some are found in the highlands or on crater floors. They sometimes occur singly, but often form complex branching systems. For example, an isolated sinuous rille, Rima Birt, lies northwest of Birt, beginning on top of a dome's summit crater, Birt E, then "flowing" about 50 km south to Birt F [C018°-Fig. 3]. Another single sinuous rille lies on the floor of the Vallis Alpes, narrow and difficult to see from Earth. At the base of the Apennine Scarp is the only sinuous rille visited by man, the Rima Hadley ("Hadley Rille"), site of the Apollo 15 Mission.

Along with a mixture of other types of volcanic features, the "Marius Hills" contains several sinuous rilles, as portrayed in Figure 2.18 (left). The two rilles shown are typical, originating in elongated crater-pits on what appear to be cinder cones and then meandering downhill, narrowing as they progress.

Figure 2.18. Unnamed sinuous rilles in the Marius Hills; the 3-km crater Galilaei E is in the lower left. Lunar Orbiter 5 medium-resolution frame 210 (Col. 070°.7). The area shown extends about 41 by 58 km.

Sinuous rilles are found on the floors of Posidonius, Humboldt, Pitatus, Hesiodus and other craters, but are far less common on crater floors than are straight rilles. The most prominent sinuous rille is the Vallis Schröteri, starting at the "Cobrahead" north of the crater Herodotus and trending successively north, northwest, and southwest for 160 km, with a maximum width of about 10 km but narrowing along its course [C055°-n, C225°-n].

Sinuous rilles are more common in the lunar western hemisphere than in the eastern. They are particularly concentrated in the Oceanus Procellarum north of Aristarchus, where they form an intricate system called the Rimae Aristarchus; see Figure 2.19 (left). This area shows frequent signs of tectonic movement, such as the Rupes Toscanelli scarp, as well as volcanism. The intricate, branching rilles possess straight portions along faults, as well as collapse features such as crater-pits. In addition to the Vallis Schröteri, the Rima Marius, Rima Brayley, Rima Diophantus, and Rima Delisle all lie within a few hundred kilometers distance.

Figure 2.19. The Rimae Aristarchus sinuous rille system. The 23-km diameter, 950-meter deep Imbrian-Period crater Krieger is at top of this 193-by-131-km frame. Apollo-15 Mapping Camera frame 2082. [C048°-Fig. 1]

Catenae. On the lunar Earthside and Marginal Zone are perhaps thirteen examples of *catenae* (sing. *catena*), of which twelve are so designated; the Catenae Abulfeda [C351°-Fig. 3, C161°-Fig. 2], Brigitte, Davy, Dziewulski, Humboldt [Quadrant 4-se], Krafft [C075°-Fig. 1], Littrow, Pierre, Sylvester [Quadrant 1-n, Quadrant 2-n], Taruntius, Timocharis, and Yuri, along with an unnamed example near the crater Müller [C174°-c]. The more prominent examples are plotted in Figure 2.13 (page 41).

Catenae are chains of craters; unlike secondary-crater chains, catenae craters follow straight lines or gentle curves and are sharply defined. Until recently, two theories competed to explain their formation. One is that they are chains of volcanic craters, the other that they are chains of secondaries. Indeed, these causes may apply to some of the catenae. Each explanation has problems, however. Some catenae are found in areas without other volcanic features; nor do they line up with obvious large craters of basins, and have no nearby similarly aligned secondary chains. Catenae craters appear fresh and, at least in some cases, like primary impact craters. Then, in 1994, Comet Shoemaker-Levy 9 impacted Jupiter, the different fragments impacting along a line. This suggested a third theory for lunar catenae; the collision of tidally-disrupted comets, which may fit at least those catenae that show no signs of volcanism and which, if they are supposed to be secondary chains, do not have an obvious source impact.

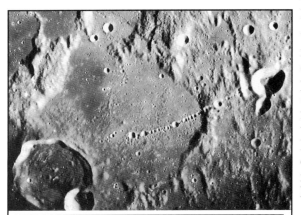

Figure 2.20. Catena Davy (near center). The extent of coverage is 122 by 88 km. Apollo-16 Mapping Camera frame 2197.

Two catenae are shown here. Figure 2.20 (upper left) is a photograph of Catena Davy, the most widely-known of these features. Considered of Copernican age, this crater chain is largely confined within the large ring Davy Y; the chain itself begins with the 3.1-km crater Davy C and extends east-northeast about 50 km, containing about 25 component craters. Catena Davy does not appear aligned with any large crater or basin, except perhaps the far-distant Orientale Basin.

The second example is the larger Catena Sylvester, near the north limb and visible from Earth only with favorable libration. Catena Sylvester extends for about 150 km, or 225 km if a poorly-defined northern portion is included. Its component craters have merged to a greater extent than those of Catena Davy. This catena is not clearly aligned with any basin; it is aligned with the rim, but not the center, of the farside crater Plaskett; Lucchitta (1978) considers both to be Imbrian in age. Catena Sylvester is shown in Figure 2.21 (left), which also shows a faint parallel structure to the west and another valley crossing the catena near its southern end. The intersecting valley is aligned with the Imbrian Basin, as is a parallel valley to its south. If the valley is a scar from the Imbrian event, Catena Sylvester must pre-date the Imbrian impact.

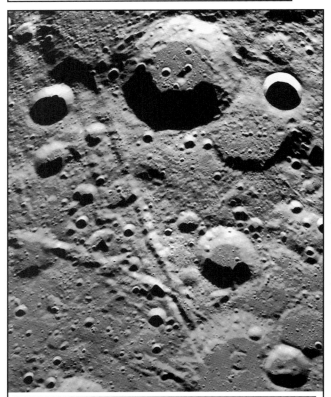

Figure 2.21. Catena Sylvester. A furrow about 150 km long. The 58-km crater Sylvester is at the top. USGS Clementine Digital Image Mosaic.

Valles. The IAU-approved term for these features, *vallis* (pl. *valles*), is unfortunate because all the linear depressions discussed up to this point could be called "valleys". However, as a lunar descriptor, *vallis* means a wide, approximately straight depression, formed of coalesced secondary craters. Indeed, these features are often called "secondary chains", "basin-secondary chains", or "radial valleys". They are broader than catenae, less straight, and have component craters that are more degraded.

Figure 2.13 (page 41) plots the major secondary chains, along with the other forms of "valley." This map gives the impression that the major secondary chains are radial to the Imbrian, Nectaris, Serenitatis, and Humorum Basins, which is partly correct. However, its orthographic projection obscures the prominent valleys radial to the Orientale Basin.

The Imbrium-radial valleys are the best placed for observation from the Earth. Those in the Central Highlands are shown in Figure 2.22, below. In that view, the Imbrian Basin itself is located off the figure to the upper left. It is clear that most of the "scars" in this area are radial to the Imbrian Basin. A few are not, however, and two prominent exceptions are the Catena Davy and the "Catena Müller", both mentioned in the previous section.

Curiously, none of the valleys radial to the Imbrian Basin are named, except for the Vallis Alpes, which is a graben and lies directly upon the basin rim. Similarly, Vallis Schröteri, already described, is obviously a sinuous rille, and not one of the secondary chains.

On the other hand, two valleys radial to the Nectaris Basin have names: Vallis Rheita, 500 km in length, and Vallis Snellius, of about the same length [C315°-s, C323°-s, C-124°-s]. Two or three other chains of Nectaris secondaries can also be identified in this area. The Vallis Palitzsch, adjoining Petavius on the east, appears radial to the Crisium Basin [C303°-Fig. 2, C112°-c], and the Vallis Capella is radial to the Serenitatis Basin [C135°-Fig. 1].

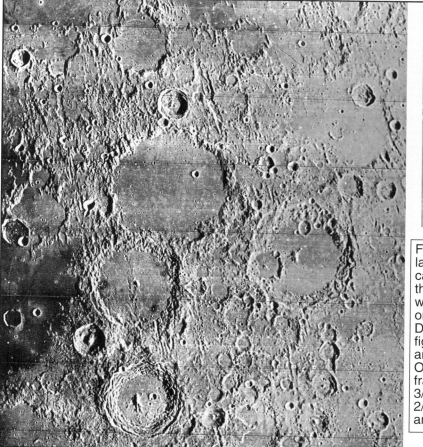
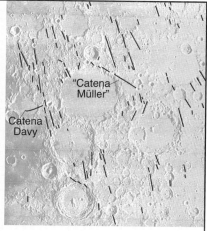

Figure 2.22. The Central Highlands, showing crater chains and catenae. The inset above shows the major crater chains, most of which are Imbrian Basin secondary chains, as well as Catena Davy and "Catena Müller". The figure extends 520 km east-west and 590 km north-south. Lunar Orbiter 4 medium-resolution frame 108. See also C357°-Fig. 3/c, C003°-Fig. 2/c, C009°-Fig. 2/c/s, C167°-c, C179°-Fig. 2/c, and C183°-c.

When seen in true projection, Mare Orientale has a magnificent collection of radial valleys. It was only in the 1960s, when earthbased photographs of the region were rectified by projection onto globes, that the nature of these radial valleys, and indeed of the Oriental Basin itself, became apparent. When viewed from Earth under favorable westerly libration, four prominent Orientale valleys can be seen near the limb: Valles Baade (160 km long), Bouvard (280 km), and Inghirami (140 km) lie south of Orientale [C080°-Fig. 1c/s, Quadrant 3-sw], while Vallis Bohr (300 km) lies to the north [Quadrant 2-nw].

Secondary chains were almost entirely obliterated in the maria, which were flooded after the basins, and hence the radial valleys, were formed. One possible mare example is the elongated depression Bullialdus W, radial to the Nubium Basin and extending northwest from Bullialdus; [C029°-Fig. 3, C201°-Fig. 2, C205°-Fig. 3]. Secondary chains are common in the highlands, however, when they are viewed under low lighting. Most of the highland radial valleys are anonymous or have only letter designations. Boscovich P, radial to Imbrium, is an example, just northwest of Boscovich [C351°-Fig. 3]. Another Imbrian scar nearby is the unnamed valley paralleling the northeast wall of Julius Caesar [C351°-Fig. 2].

Farther east, the Montes Taurus contains a number of Serenitatis and Crisium radial valleys. One evident Crisium-Basin radial valley lies north of Macrobius [at center of C131°-Fig. 1]. As with the Central Highlands, the terrain of the Eastern and Northeastern Highlands has been sculpted by ejecta from basin impacts.

Mountains

"Mountains" are here defined as units of highland material that are elevated above their surroundings, or as groups of such units. Most lunar mountain ranges are named after terrestrial ones. Individual peaks, if named at all, have been given the name of a terrestrial mountain or of a person; or a Greek letter following the name of a nearby feature. As with Roman letters, Greek-letter designations are no longer sanctioned by the International Astronomical Union, but are still used. "Official" IAU designations for these features begin with the modifiers *mons* for individual peaks, *montes* for mountain ranges, *promontorium* for promontory, and *rupes* for scarp. Perhaps intentionally, or perhaps because of their prominence, nearly all elevations that are formally named are indeed highlands units, rather than mare features. The writer will use the above Latin terms as generic feature descriptors, whether or not a particular feature has a proper name.

Whether isolated peaks or entire ranges, most mountains are either portions of the rims of impact basins, or formed by ejecta from those basins. Rings or partial rings of highlands surrounded by mare indicate the rims of submerged craters. Occasionally an elevated highlands structure has been created by tectonic uplift, such as is the case with crater central peaks. Elevations composed of highlands material that are attributable to volcanism exist, but are rare.

Several mountain ranges (montes), both large and small, and single peaks (mons) can be seen on Figure 2.23 (page 49). The area shown is the eastern Mare Imbrium and its rim.

Mons. The term *mons* is used here to mean isolated peaks, although when part of a name "mons" may indicate a peak within a range. Several isolated peaks appear in Figure 2.23, including Mons Pico (2400 meters high), Mons Piton (2200 meters), Pico β south of Pico, Piton γ south of Piton, and smaller unnamed "islands" of highlands in the mare. (Note: Except in areas photographed by the Apollo Mapping Cameras, the heights of lunar peaks were determined by measuring the lengths of their shadows and are quite uncertain; in many cases this writer has relied on elevations cited in Neison's *The Moon*, printed in 1876!) These peaks appear to be the surviving high points of basin interior rings that otherwise have been submerged by mare lava. Indeed, their pattern is the chief evidence for the existence and for the diameters of such inner rings. These inner-ring highlands fragments are most common in Mare Imbrium; Mare Crisium and Mare Humorum each have only a few examples. Isolated highlands units are found in Mare Orientale, Mare Frigoris, Mare Nectaris, Mare Nubium, Mare Serenitatis, Mare Tranquillitatis,

2. The Moon's Surface Features

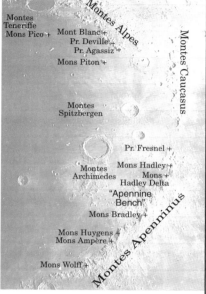

Figure 2.23. Mountain ranges (*montes*) and isolated peaks (*mons*) in the eastern Mare Imbrium. The coverage is approximately 800 by 1210 km. Lunar Orbiter 4 medium-resolution frame 114, colongitude 030°.7.

This area can be seen under varying lighting in these Section-II figures: C353°-n/c, C357°-Fig. 1/n/c, C003°-Fig. 1/n/c, C009°-Fig. 1/n/c, C018°-n/c, C161°-n/c, C167°-Fig. 2/n, C174°-Fig. 1/n/c, C183°-Fig. 1/n/c, and C191°-n/c.

In the inset above, "Pr." represents "Promontorium".

and Mare Fecunditatis, but cannot be said to form a coherent circular pattern within any of them.

To the right, Figure 2.24 gives a close view of an isolated highlands peak, Mons Vinogradov, in Mare Imbrium. Isolated peaks like this, when viewed under a low sun, appear steep. Actually, this feature is 1450 meters high and 25 km across, giving a mean slope of about 7 degrees. Mons Vinogradov appears more as an elevated land mass, with several distinct summits, than as a single peak.

Figure 2.24. Mons Vinogradov. This complex extends 115 km on each side. Crater Natasha in lower right (11.4 km diameter). Apollo-17 Mapping Camera frame 2732. [C036°-n, C040°-n, C201°-n, C205°-n, C213°-n]

To the southeast of Mons Vinogradov is a jumble of smaller peaks, some less than a kilometer across and under 100 meters high. The crater Natasha in the lower right has been flooded by mare lava. Adjoining Natasha on its upper right is a smaller crater whose rim has been broken. Similarly, referring back to Figure 2.14 (page 42), several arcs of highlands material can be seen in Oceanus Procellarum in the right-hand portion of the figure, representing craters whose rims have been more than half submerged.

Montes. Most of the Moon's named mountain ranges form portions of the rings of impact basins. When largely intact, in plan they take the form of circular arcs; in profile they are steepest facing the basin interior, and gentler facing outside. They thus appear to be blocks of tilted lunar crust, analogous to terrestrial block mountains like the Sierra Nevada in California. Some of the impact-ring mountain ranges are partly submerged in mare lavas and then appear as clusters of highland peaks, rather than as continuous ranges.

The main rim of the Imbrian Basin is marked on all sides by mountain ranges. In the eastern portion, shown in Figure 2.23 (page 49), they consist of (moving counterclockwise around the rim): the Montes Apenninus, Montes Caucasus, and Montes Alpes. West of the Montes Alpes, the rim is marked by a highlands strip, rather than a well-defined range, which is interrupted by Sinus Iridum, itself bounded on the north by the Montes Jura [C036°-n]. The southwest end of this highlands strip (shown in Figure 2.6, page 35 as the "Jura-Alpes-Caucasus Highlands") terminates where Oceanus Procellarum and Mare Imbrium meet. About 200 km farther south, the Imbrian rim reappears as the Montes Harbinger [C040°-Fig. 1, C048°-Fig. 1]. The remaining mountain range defining the basin rim is the Montes Carpatus on its southern boundary [C024°-c, C029°-c, C201°-Fig. 1]. These Imbrian-rim mountains exceed 5000 meters in height for the higher peaks of the Montes Apenninus, almost 6000 meters for some Montes Caucasus peaks near Calippus, then dropping to perhaps 2000-3000 meters in the Montes Alpes, except for 3600-meter Mont Blanc and 3700-meter Plato λ. West of Plato the highland peaks are in the 1000-2000 meter range, but rise to about 4000 meters in the Montes Jura. The Montes Harbinger peaks rise 700-1900 meters, while those for the Montes Carpatus attain about 2000 meters elevation above the mare.

Within Mare Imbrium, some small ranges and single peaks define the innermost ring: the Montes Spitzbergen (800-1700 meters), Mons Pico (2400 meters), Montes Teneriffe (2400 meters), Montes Recta (1900 meters) [C024°-n], and Mons La Hire (1800 meters) [C029°-n; see also Fig. 2.25 on page 52]. A ring intermediate between the innermost and the main rim may be marked by Mons Delisle (1200 meters) [C040°-Fig. 1] and the Montes Archimedes (1200 meters high at most), but when the markers are so far apart the identity of the rings becomes uncertain.

The rim of the Imbrian Basin has been sketched out in detail because it is well-marked and well-presented to our view. Were the Orientale Basin nearer the center of the disk it would be at least as good an example as the Imbrian Basin of circum-basin mountain ranges. Both of its ringwalls are visible from the Earth near full phase when there is a westerly libration. An inner basin ring, about 620 km in diameter, is marked by the Montes Rook, about 800-900 km in length. However, the most prominent ring, 930 km across, is marked by the Montes Cordillera, which extend 600 km (according to the *Gazetteer of Planetary Nomenclature 1994*; 1500 km in Rükl, 1990) [C080°-c, Quadrant 3-w]. We unfortunately have no accurate altitudes for their impressive peaks; Neison (1876) quotes heights of 5000-7000 meters for the Cordillera, while Wilkins and Moore (1955) give 5500 meters for the Montes Rook (both sources interchange the two names, as compared with modern usage). It is not clear how these elevations were obtained, but these peaks can be seen, and presumably measured, as limb irregularities when the libration is right [C237°-c/s, C247°-c].

All the remaining well-defined circular basins have fringing mountains, although only a few of the ranges are named. The Montes Haemus mark the southwest portion of the main ring of the Serenitatis Basin [C351°-c, C161°-c]. Likewise, the Montes Pyrenaeus form the eastern boundary of the main ring of the Nectaris Basin [C323°-c, C135°-Fig. 1]. The Humorum Basin [C048°-c, C055°-Fig. 2, C225°-Fig. 2] and the Crisium Basin [C303°-Fig. 1, C112°-n, C119°-c] also have mountain ranges on portions of their main rings, but the ranges are unnamed. The Montes Taurus [C323°-n, C135°-n] lie between the Serenitatis and Crisium Basins, but constitute an area of irregular uplands, rather than a basin rim. On the other hand, the little Montes Riphaeus, 800 meters high, may well mark the northwest rim of the possible Cognitum Basin [C029°-Fig. 2].

Rupes. This term denotes fault scarps, where tectonic movement has altered the relative elevations of the terrain on either side. Like the straight rilles (which are formed by two parallel faults), rupes follow straight lines, a series of straight line segments, or form gentle curves. Rupes may be found in either maria or highlands. When they bound a basin, their steeper face overlooks the basin interior. Rupes are usually named for a nearby feature.

Seven such faults, large enough to been seen from Earth, have been recognized by naming them. Many more smaller unnamed rupes can be found wherever there has been tectonic motion, while some large features could be called "rupes" but are not, such as the "Apennine Scarp".

By far the most prominent rupes is Rupes Altai, which defines the southwestern main rim of the Nectaris Basin and lies entirely within highlands. Over 400 km in length, its steep inner rim averages 1200-1800 meters above the basin interior, with one peak near Fermat rising 3400 meters [C334°-s, C343°-c/s, C143°-s, C148°-s, C154°-s]. As with all rupes, its appearance depends on whether it is seen under morning or afternoon lighting. In the lunar morning, the Rupes Altai appear as a bright line; it casts a shadow only under afternoon illumination.

Rupes Kelvin bounds part of the southeast main rim of the Humorum Basin. Extending about 80 km, the scarp rises a maximum of 1500 meters above the mare [C036°-s; C213°-Fig. 1, bottom right]. In a similar manner, the 1200-meter high Rupes Mercator occupies about 90 km of the southwest rim of the Nubium Basin [C024°-s; C201°-Fig. 1, bottom right].

Like the previous rupes, Rupes Liebig parallels a basin rim, in its case the western rim of the Humorum Basin, but it lies within the rim [C055°-Fig. 2, lower right; C225°-Fig. 2, below center]. The rupes extends approximately 180 km, rising some 600 meters above the mare interior.

The remaining named scarps all fall within maria. The best known is Rupes Recta (the "Straight Wall"), east of Birt in the eastern Mare Nubium [C018°-Fig. 2, C191°-Fig. 2]. This scarp faces west and runs for 130 km, rising about 270-300 meters above the mare. Located in northeastern Mare Tranquillitatis, Rupes Cauchy runs 120 km northwest-southeast, paralleling the Rima Cauchy to its north [C328°-c; C143°-Fig. 1, left center]. Facing southwest, it rises some 200-300 meters. Finally, the Rupes Toscanelli lies in the area of volcanism that contains the Rimae Aristarchus [C048°-Fig. 1, 2.5 cm above Aristarchus]. Rupes Toscanelli starts at the crater of that name and runs south about 70 km; the scarp faces west and is 200-300 meters high.

Highlands Volcanism. This term refers to volcanism involving light-toned highlands material, whether it is found in highlands or maria. Admittedly, highlands volcanism is rarer than mare volcanism, and also less obvious if it has taken place in a highlands area. Nonetheless, a number of elevations attributable to highlands volcanism have been identified.

One fairly extensive area that illustrates highlands volcanism is the "Apennine Bench formation" (Figure 2.23, page 49), along with some smaller units to its northeast.

Some areas within the highlands have been interpreted as volcanic, such as plains units like the widespread Cayley Formation, but these have also been interpreted as ejecta blankets. In addition, the highlands contain a few unusual elevated areas that may be volcanic. One is the "Kant Plateau", about 40 km in length, adjoining Zöllner on the southwest [C154°-c; C161°-Fig. 2]. Another, smaller, distinctive plateau is southeast of Dembowski [C357°-Fig. 2/c; C167°-c].

Most domes are composed of mare materials and are discussed in the next section. A few highland domes have been identified, however. A group near the southwestern end of the Jura-Alpes-Caucasus Highlands includes Mons Gruithuisen Gamma, 1200 meters high with a 18-by-24-km base; Mons Gruithuisen Delta, about 1500 meters high and measuring 14 by 33 km [C048°-n, C213°-n]; and, to their northwest, the similar formation Mairan T (*sic.*; *not* a crater, perhaps a "crater cone") [C225°-Fig. 1, west of Mairan], 7 km wide and 900 meters high. Southeast of Mairan T are two more such features. Another is Mons Hansteen in southwest Oceanus Procellarum [C055°-c, C225°-c], and perhaps the pitted and fissured Darney χ plateau in Mare Cognitum [C029°-Fig. 2, lower right; C205°-Fig. 2, center bottom]. Several of these seven features resemble the rhyolite domes found in the Mono Craters in California.

Mare Features

Maria. Craters and highland elevations within the maria have already been discussed, but the maria also contain distinctive landforms composed of mare material. The first such type of feature is the maria themselves. It is surprising how often the literature ignores the maria as a landform type. Visible to the naked eye and occupying about one-third of the Earthside hemisphere, they are perhaps too large, and considered as simply the background for other features. When seen under a high sun, the maria surfaces may look flat and featureless. Even then their varying tones and color indicate a complex history. Crater counts show that the ages of the maria vary, and vary within the same mare. About three-quarters of the area of the present-day maria were flooded by highly fluid basaltic lava in the Upper Imbrian Epoch, but in four or more separate stages. Mare Vaporum and sizeable portions of Oceanus Procellarum and Maria Imbrium, Frigoris, and Smythii flooded in the Eratosthenian Period. The lavas of the portion of Oceanus Procellarum near Lichtenberg have been dated to the Copernican Period. Naturally, these more recent flows must be merely the latest layers overlaying a series of previous flows.

Figure 2.25. Apollo-15 Mapping Camera frame 1535 looking north over Mare Imbrium. Promontorium Laplace is on the horizon, about 600 km distant. In the foreground is Mons La Hire. Near the horizon are the twin craters Helicon (24 km; left) and Le Verrier (21 km; right). Below Helicon is the 11-km crater Carlini. The Dorsum Zirkel extends 190 km from the lower right to the upper left; to its left is the 150-km Dorsum Heim. [C029°-n]

Under a low sun angle, maria surfaces clearly are not flat; extensive areas show gentle slopes, while adjacent areas lie at different levels. Besides these general characteristics of the mare surfaces themselves, grazing illumination reveals differences in crater density and low-lying features such as flow fronts, ridges, and domes.

A representative area in central Mare Imbrium is shown in Figure 2.25 (left). Its more obvious features consist of highlands mountains, craters, and mare ridges. A closer look indicates that the lower right portion of this area is higher than the upper left portion, at least near the zone of contact. In the foreground, several different layers are apparent, differing in elevation. Following the principle of superimposition, the higher layers should be the more recent, and their lower crater density confirms this. The bottom and lower right represent an Eratosthenian flow, while the rest of the area is Imbrian in age.

One aspect of the maria that is possible to study from earthbased observations is the thickness of the shallower portions of the mare flows. This is done by inferring the depth of flooding from pre-mare craters that have been partly submerged [DeHon, 1974, 1979; DeHon and Waskom, 1976]. Low lighting is preferred for identifying and measuring these "ghost craters". Extensive mare areas, especially near their margins, are less than 500 meters deep, although flow thicknesses exceed 1500 meters in a few areas.

Another study possible from the Earth is the location of lobate (rounded-edged) *lava-flow fronts* under low lighting. Our best knowledge of these low features comes from Apollo 15-17 photographs, with low-sun views confined largely to Mare Imbrium and Mare Nubium; Figure 2.25 (p. 54) is an example of one of these photographs. Some low flow scarps have also been found in Oceanus Procellarum, Mare Cognitum, Mare Vaporum, and Sinus Medii [Schaber *et al.*, 1976]. The thicknesses of individual flows range from 1-96 meters, mostly less than 15 meters, as measured from the Apollo photography [Taylor, 1982, p. 270]. The question is whether sinuous flow scarps of similar heights would be telescopically visible in other areas. With grazing illumination, a 15-meter scarp would have a maximum shadow length of 7.2 km, while one 96 meters high could cast an 18-km shadow. Naturally these shadows would be cast upon a landscape itself faintly illuminated, but given adequate exposure of the CCD image, detection of lobate flow scarps appears possible. Indeed, one example can be seen in the atlas section; C357°-Fig. 1 (lower right), where the Eratosthenian flow of Mare Serenitatis has overlapped the earlier Imbrian flow. The mare on the east is higher than that on the west, so the scarp casts a shadow in C357°-Fig. 1. In C174°-Fig. 2, with afternoon lighting, it appears as a bright line.

Mare Ridges. Most maria contain complex networks of *wrinkle ridges*. Some are straight, others sinuous, and others may outline circular or oval areas. The major mare ridges are mapped in Figure 2.26 (page 54). The figure omits many short ridges, and many that are very low and visible only very close to the terminator. There is a tendency for ridges to be concentric with basin rims, particularly in Mare Imbrium, Mare Serenitatis, Mare Crisium, Mare Humorum, and Mare Nectaris. They cross the centers of several basins; apparently more often north-south than east-west, but this is a bias caused by the east-west direction of sunlight near the terminator. Ridges are even found on some crater floors.

Most mare ridges are unnamed; those that are have the prefix *dorsum* (pl. *dorsa*), followed by a person's name (e.g., Dorsum Zirkel in Fig. 2.25). A large ridge may extend for 100-300 km, be perhaps 5-10 km in width, and rise 100-200 meters. Their slopes are thus of the order of a few degrees, making these features best studied under low lighting. Probably the best known such feature is the famous "Serpentine Ridge" in eastern Mare Serenitatis, now named Dorsa Smirnov in its northern part and Dorsa Lister in its southern. Together the two dorsa extend over 400 km and rise to 300-350 meters above the mare [C334°-Fig. 1, C154°-n/c].

The prevalent theory of ridge formation is that they are tectonic features caused by compression of the lunar crust. Supportive of this is the observation that ridges are often paralleled by rilles (see Fig. 2.17, page 44), both being caused by subsidence of basin interiors. One other theory is that they are extrusive, caused by lava flows from fissures, and it may be that some individual ridges were caused this way.

Lunar Domes. "Classical" lunar *domes* are low swellings, chiefly in the maria and composed of maria materials. Figure 2.26 (page 54) plots the distribution of these features as catalogued by the Association of Lunar and Planetary Observers [A.L.P.O.; Jamieson and Phillips, 1992]. This map includes objects which have not yet been confirmed but excludes steeper specimens and those composed of highlands material, although most lunar geologists term these features "domes" as well [Smith, 1973]. Most domes are unnamed, except for A.L.P.O. catalog numbers, although a few have Greek-letter designations similar to those of peaks. One "Dome Complex" has been named, Mons Rümker in Oceanus Procellarum [C061°-Fig. 1]. The older literature does not recognize these features as a class, and even Rümker was described by Neison (1876, p. 267) as "... a kind of low plateau ... [rising] from the union of a number of ridges."

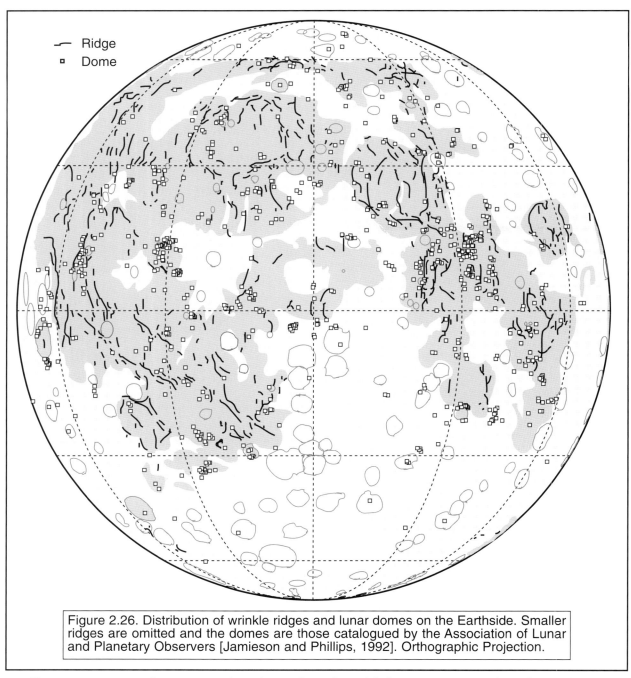

Figure 2.26. Distribution of wrinkle ridges and lunar domes on the Earthside. Smaller ridges are omitted and the domes are those catalogued by the Association of Lunar and Planetary Observers [Jamieson and Phillips, 1992]. Orthographic Projection.

Domes appear to cluster near the edges of maria, with heavy concentrations in north-central Mare Tranquillitatis, the Hortensius-Milichius-Tobias Mayer area west of Copernicus, the Marius Hills, and several lesser groupings. A few have been reported on crater floors.

"Classical domes" are roughly circular and have slopes under five degrees [Westfall, 1964]. These features typically are 5-20 km across and at most a few hundred meters high. They frequently have summit craters, sometimes with rilles extending from them, and most likely are shield volcanos, built up by extruded lava, resembling on a smaller scale Mauna Loa in Hawaii. Figure 2.27 (page 55) shows two examples near Hortensius, while Figure 2.28 (page 55) shows more of the many domes in the Hortensius-Milichius-Tobias Mayer cluster.

Some extremely low swellings, such as that centered on Hyginus, may represent *laccoliths*, which are subsurface lava intrusions that have elevated the surface above them. Study of terrain very near the terminator may reveal other examples of these subtle features [Lucchitta, 1976;

2. The Moon's Surface Features

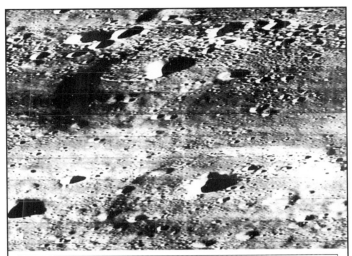

Figure 2.27. Two shield-type domes, each about 5 km in diameter. In the foreground is Hortensius σ, with A.L.P.O. Catalog -450+137 behind it. Oblique view looking north; Orbiter-3 high-resolution frame 123-2, colongitude 035°.5. [C029°-Fig. 1].

Figure 2.28. Part of the Hortensius (lower right) - Milichius (center) - Tobias Mayer (top) dome field. Coverage is 225 by 300 km. Orbiter 4 medium-resolution frame 126, colongitude 043°.0. [C029°-Fig. 1, C205°-Fig. 1]

Westfall, 1966]. The steeper features, with slopes up to 15-20 degrees, such as many of those in the Marius Hills, may be *cinder cones*, cones surmounted with depressions, which the older literature calls "crater cones."

The Moon: A Unique Analogy

Two reasons to observe the Moon are that it contains types of features unique to itself, but also has features that are analogous to those found on other planetary bodies. First, what the Moon (and indeed the Earth) has that is not found on other worlds that have been visited by space missions is widespread basaltic lava flows; the maria. In addition to these maria, of course, are their associated features, particularly sinuous rilles, mare ridges, and mare domes.

Second, the Moon's proximity allows us to observe conveniently (and economically) types of features that *can* be found elsewhere, and thus help our understanding of general planetary processes. Impact craters are found on all the bodies in the Solar System that have solid surfaces whose features are not gradually obliterated by endogenic processes. The larger satellites and the planets Mercury and Mars even exhibit multiring basins. Secondary craters, catenae, and valles are found elsewhere as well. The planet Mercury is the Solar System's closest analogy to our Moon, followed by Mars.

In addition, other bodies show signs of tectonic activity. Fault scarps are found on Mercury, Venus, the Earth, and Mars, for example. Naturally, each world has its own "style" and exhibits features that have no lunar analogy. Even those that have types of features like those of the Moon will display subtle differences.

But planetary science would lose much of its interest were worlds too similar. The Moon's chief attraction is that it provides a most accessible variation on a planetary theme.

Section II. Atlas

Arrangement of the Atlas

The CCD image mosaics in this section show the Moon at 47 different phases, arranged in ascending order of colongitude (291°-357°, 003°-247°) from waxing crescent to waning crescent. North is at the top in all views. Each phase begins with a page that describes selected features best seen under that lighting. The top margin of the descriptive page lists the telescopic, lighting, and topocentric libration conditions of the mosaic. Accompanying the descriptive text are one or more large-scale CCD images of selected features. Following this page is the CCD image mosaic, shown in two or three sections arranged in north-to-south order, with the latitude range and scale given on each section. For example, the view at colongitude 067°.2 is titled simply "Colongitude 067°." Its descriptive page includes enlarged views of two selected lunar areas then visible. Its three mosaic pages are labeled "067°-N" (north), "067°-C" (central), and "067°-S" (south).

Each mosaic section includes an inset that gives the names of the formations that appear in the image. In addition, on pages 269-288 is an overall index of named formations, divided into craters and non-crater features. The index gives features' longitudes and latitudes and the colongitude mosaics on which they appear. For craters, diameters and depths are also given On pages 59 and 60 are two lunar index maps that show the position of the terminator for each mosaic; the first map for the waxing Moon (Figure II.1) and the second for the waning Moon (Figure II.2). In addition to colongitude, the effect of varying selenocentric solar declination has been taken into account in plotting the terminator positions, so that the terminator arcs do not intersect neatly at the poles, as would be the case if they always followed meridians.

Making the Mosaics

Chapter 1 has described the general process of obtaining individual CCD images. However, one image does not make a mosaic, so it is pertinent here to describe how images intended for mosaics are acquired, processed, and then assembled.

A series of images that are to be mosaiced must be as uniform as possible; this means no changes in the exposure time, camera orientation, optical system, or filter (or lack of filter) can occur while the series is being exposed. Any cloudiness, even thin clouds, is usually fatal because the Moon's brightness will change between images. Indeed, if the Moon is at a low altitude, the series must be completed quickly because the transparency even of the clear atmosphere will change rapidly. Another reason not to dawdle excessively is that the colongitude changes by a tenth of a degree every 11-12 minutes.

The writer's procedure is to begin each series at the northern limb, then to expose frames successively southward along the terminator until he reaches the southern limb. The exposure time is adjusted by trial and error so that features 8°-10° from the terminator are on the point of "saturation" (overexposure), to ensure continuity with the previous and subsequent mosaics. Naturally, it is important to be sure that each frame overlaps its neighbors; preferably by 10-20 percent. Depending on the telescope used, the amplification, if any, and the lunar phase and distance, the terminator strip will constitute approximately 12-30 images, and take 20-40 minutes to acquire. Because of seeing changes, the writer frequently exposes several (or many!) frames of an area before he finds one worth being saved to disk.

The lunar images are initially stored on the disk drive of a Macintosh IIsi computer that is dedicated to acquiring CCD images. At the end of an observing session, the images acquired are transferred to floppy disks and read into the disk drive to the writer's "indoor" computer, currently a Macintosh 8600/250. This more powerful computer is used to process the images, which begins with performing a linear "stretch" of their contrast with the software supplied by SpectraSource, the manufacturer of the CCD camera. First, the DN values (brightness levels in a range 0-65,520) of the darkest and the brightest pixels in each frame are determined. Then, the contrast of each frame is "stretched" so that its darkest pixel is set to a DN (brightness value) of 0. A brightness range is calculated so that the frame with the highest initial DN for its brightest pixel has its brightest pixel set to a DN of 65,520 (the camera's highest possible DN). This ensures that the maximum stretching for each frame is achieved consistent with every frame's contrast being exaggerated by a uniform amount.

At this point, the images are converted to PICT-format files using the Macintosh "screen copy" function. This allows them to be read by the Photoshop image processing program (the writer began with Version 2.5 and is now using Version 5.0). All of the frames are placed on one Photoshop image. Then a non-linear contrast stretch is used to selectively exaggerate brightness near the terminator. Next, each of the individual frames is selected and moved to fit the previous frame, again proceeding in a north-to-south order. The successive frames are fitted by eye, matching detail along their edges; with aerial photographs, this would be called an "uncontrolled mosaic." (An experiment in measuring the positions in ten selected mosaics indicated that this "eyeballing" process created a median scale error of just 0.4 percent, more than sufficient for pictorial purposes, if not for precise measurement.)

Despite exercising all the care possible, there may remain visible breaks between occasional frames. For example, feature positions may disagree slightly because of short-term differences in seeing. Also, brightness levels may differ due to differences in the exposure time of the camera's mechanical shutter, to short-term variations in atmospheric transparency, or to both. Small-scale differences at the frame edge can be obscured by judicious use of Photoshop's "smudge" tool; inter-frame brightness differences are reduced by using a slight amount of further contrast stretching on individual troublesome frames; fortunately this is rarely necessary.

Once the frames are all assembled, and edge-to-edge differences reduced as much as possible, a black fill is used for the areas behind the terminator or beyond the limb. This prepares the mosaic for unsharp masking, which must be employed to the entire mosaic in a single step because the process can create "artifacts" (bands of dark or light shading) along the margin of the area of application. The writer typically uses high levels of unsharp masking which creates "grain" in areas of uniform shading; this is removed using the program's "despeckle" function.

The processing of all images in a mosaic must be identical in terms of contrast stretching and sharpening, just as for their initial exposure. This means that all the adjustments employed are compromises intended to give the best results for the mosaic as a whole. Were the individual frames exposed and enhanced singly, they often would appear better-exposed and more clearly defined than they do in the mosaic. Unfortunately, were such a procedure followed, it would result in obvious breaks between frames in the resulting mosaic.

Although the writer tries to expose the frames so that their narrow dimension is aligned with lunar north-south, the mosaic may show some inclination to the lunar axis. If this misalignment exceeds one degree, the writer uses Photoshop's "rotate" command to correct. At this point the mosaic is essentially completed. However, the scales of those mosaics used in this atlas have been adjusted, and each mosaic divided into two or three segments, in order better to fit the pages.

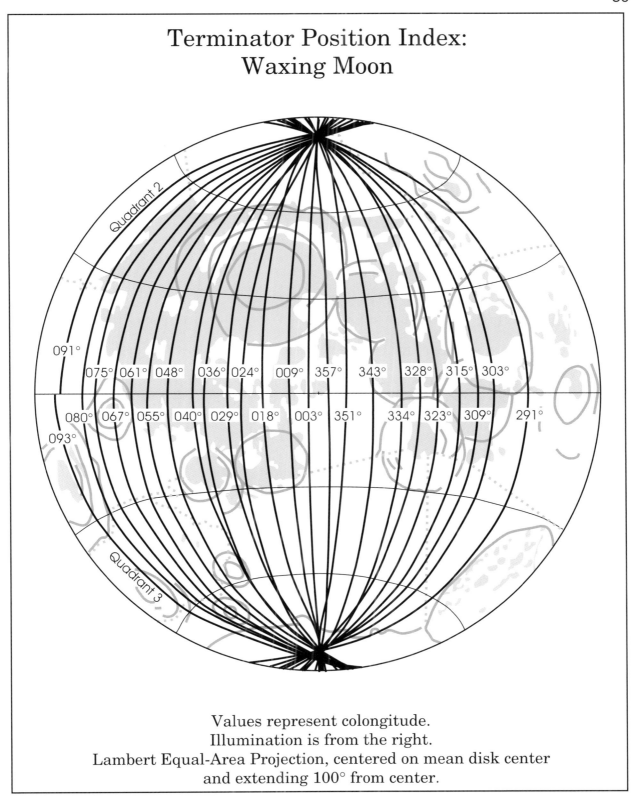

Figure II.1. Terminator position index for colongitude mosaics, waxing Moon.

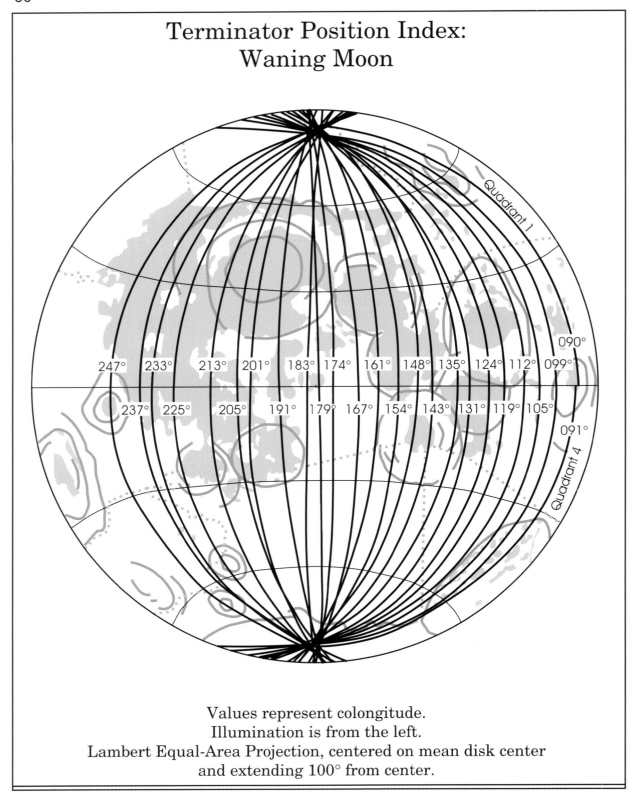

Figure II.2. Terminator position index for colongitude mosaics, waning Moon.

Colongitude 291°

1995 APR 02, 03h00m UT. 28-cm Sch.-Cass., f/10, 0.30 s, W58 (green) Filter.
Colongitude = 290°.5, Solar latitude = 0°.6N, Librations = 1°.8E/0°.2N. 1.50 km/pixel.

Spotting young Moons, as young as possible, is an observational challenge. The current world record, set in January, 1996, is 12.1 hours, and this value may well drop. Figure 1 records a Moon at an age slightly less than 24 hours.

One of the most interesting aspects of a young Moon (or indeed an old one) is the earthlight that illuminates the Moon's night hemisphere, as shown in Figure 2. Because we are located on the source of this illumination, this is the only time when we can see features under exactly frontal lighting; technically, when the *phase angle* is zero. This condition is never possible with direct sunlight because the Moon would then be undergoing a total eclipse. Thus it is unfortunate that earthlight is so faint, about 1/10,000 as bright as full sunlight. Nonetheless, the great dynamic range of the CCD camera might show useful earthlit detail with a fast optical system.

Very young Moons are not useful for the observation of terminator detail. One problem is that we are then seeing the shadowed faces of the Moon's relief features. Another problem is that we must then observe the Moon either in a twilight sky; or if in a dark sky, at a very low elevation and thus most likely under conditions of poor seeing. A useful experiment would be to image young Moons high in the sky, in full daylight, using an infrared filter to darken the sky.

The Moon needs to be about two days old before one can see useful terminator detail, as in the mosaic on the following two pages. The young Moon is best observed shortly after a springtime sunset, as is the case here. Even so, observing conditions are difficult; yet this is our only opportunity to see features near the east limb under morning lighting. Another constraint is that these limb features are best seen with an easterly libration.

Although the area exposed is chiefly highlands, it includes four small maria. Mare Humboldtianum and Mare Smythii occupy impact basins; Mare Marginis does not and is more irregular; and Mare Australe is probably within a very old basin, containing a complex group of merged flooded craters.

The Eastern Highlands clearly contains a variety of craters. Among these, Gauss, Neper, Kästner, and Humboldt are particularly impressive even under morning lighting, and will be more so with an evening Sun, shortly after full phase.

1. Young Moon. (Film Photograph) 1989 APR 07, 03h19.2m UT. 400-mm, f/6.3. 1 sec on Kodachrome 200 Film. Lunar age 23h46m. Lunar north to right.

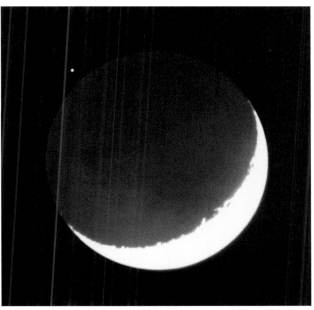

2. Earthlit Moon and Aldebaran. (Film Photograph.) 1980 APR 18, 04h33m UT. 10-cm Refractor, f/10. 10 sec on Ektachrome 400 Film. Lunar north to upper right.

291°–N

90°N to 1°S

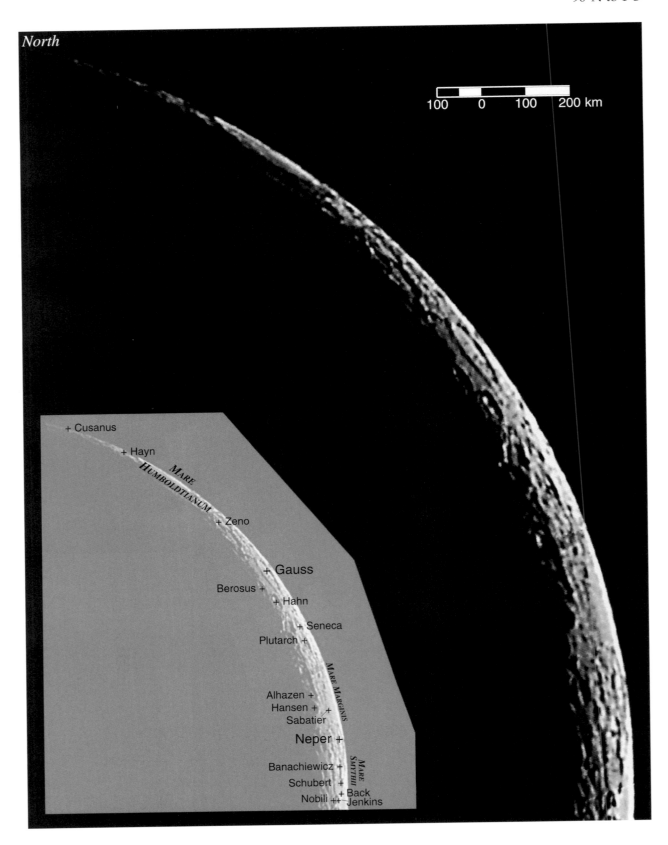

291°–S

5°N to 90°S

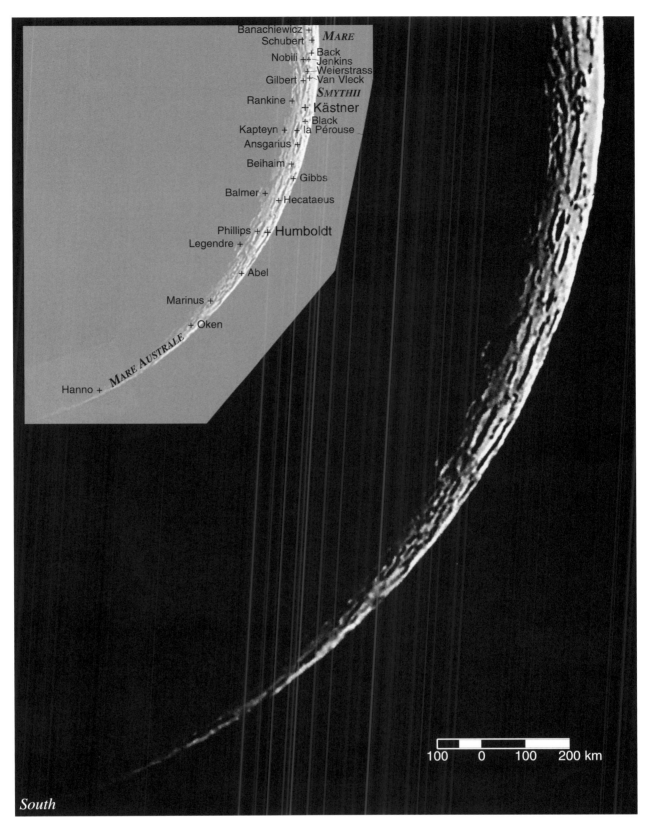

63

Colongitude 303°

1995 APR 03, 04h09m UT. 28-cm Sch.-Cass., f/10, 0.25 s, W58 (green) Filter.
Colongitude = 303°.3, Solar latitude = 0°.6N, Librations = 0°.7E/1°.7N. 1.48 km/pixel.

As the crescent Moon draws away from the Sun in the evening sky the visibility of detail rapidly improves. The higher altitude of the Moon improves atmospheric seeing and lunar features are now better illuminated. Also, the terminator is farther from the east limb, so that features are not as foreshortened as earlier.

The Northeastern Highlands are much better exposed at this phase than in the previous view, with Messala currently the most prominent crater in the area.

The Sun is now rising on Mare Crisium (Figure 1), which lies inside the first impact basin to be clearly seen in each lunation. Foreshortening makes the mare appear elongated north-south, in actuality it is stretched east-west. Within the mare is an approximately concentric system of ridges, of which only the eastern portion is visible at this phase.

South of Mare Crisium, near the equator, are the irregular Maria Undarum and Spumans. To their south is a striking north-south chain of four craters (Figure 2 shows the northern three); in north-south order, Langrenus, Vendelinus, Petavius, and Furnerius. Petavius is distinctive with its circumferential valleys, central peak and floor rille. Their apparent linear arrangement suggests that these four depressions are associated in some way. However, they appear to be of different ages. Also, in a less distorted projection, they are not quite so precisely aligned or so regularly spaced as they appear from Earth

To the south of Furnerius we can now see the crater-saturated condition of the Southeastern Highlands.

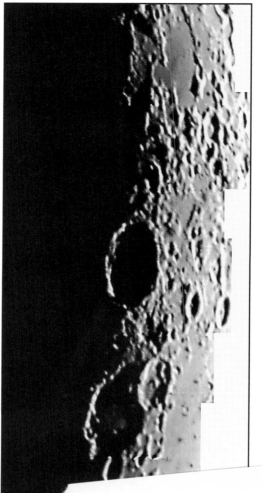

1. **Mare Crisium (left).** 1996 JAN 23, 01h58m UT. 28-cm Sch.-Cass., f/21, 0.10 s. Col. = 299°7, Solar Lat. = 1°.5N, Librations = 5°.0E/4°.0S.

2. **Eastern Crater Chain (right).** 1996 JAN 23, 02h14m UT. 28-cm Sch.-Cass., f/21, 0.10 s. Col. = 299°.9, Solar Lat. = 1°.5N, Librations = 5°.0E/4°.0S.

303°–N

90°N to 20°N

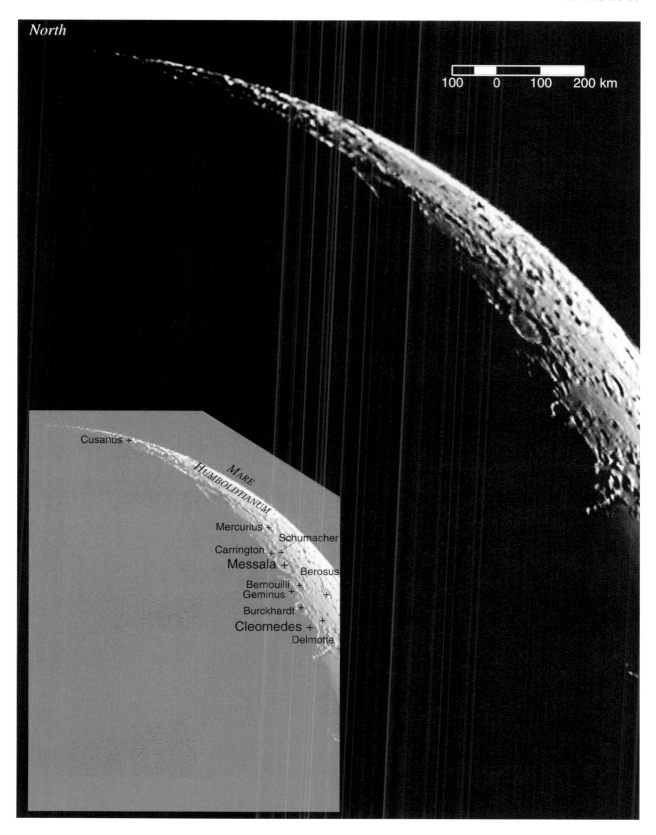

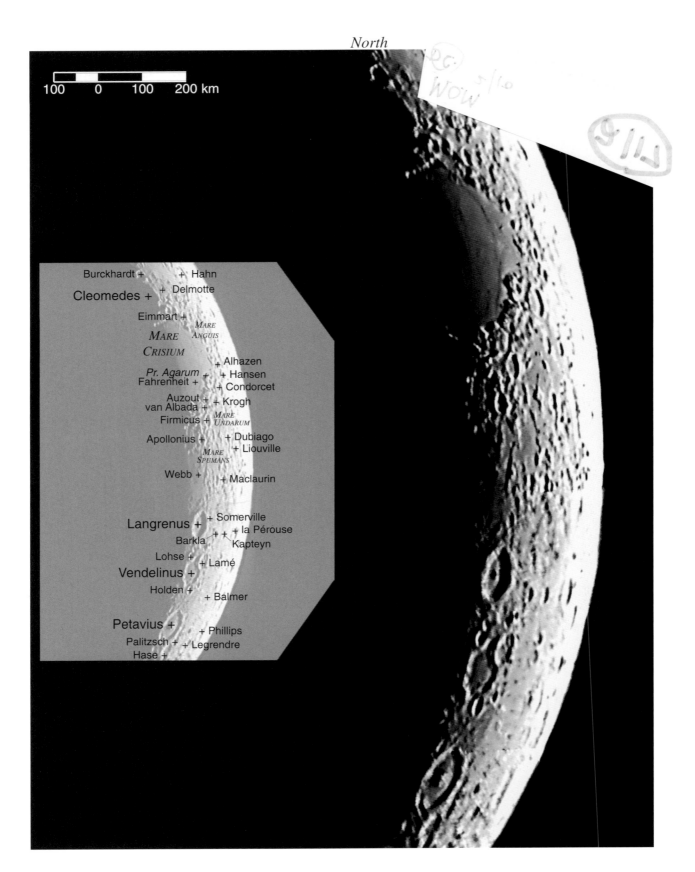

303°–S

3°N to 88°S

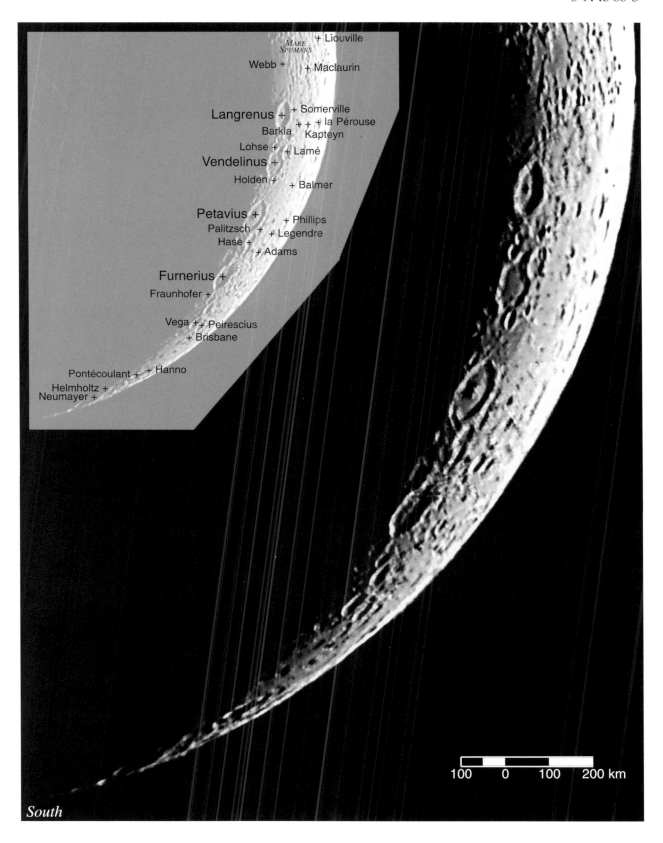

Colongitude 309°

1995 May 03, 03h32m UT. 28-cm Sch.-Cass., f/10, 0.15 s, W58 (green) Filter.
Colongitude = 308°.9, Solar latitude = 0°.2S, Librations = 2°.4W/5°.2N. 1.50 km/pixel.

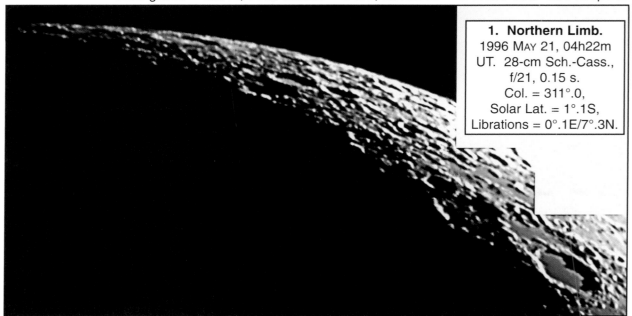

1. Northern Limb.
1996 May 21, 04h22m UT. 28-cm Sch.-Cass., f/21, 0.15 s.
Col. = 311°.0,
Solar Lat. = 1°.1S,
Librations = 0°.1E/7°.3N.

It is now approximately three days after New Moon, with the phase sufficiently large to show detail near the north or south limb; depending on the sign of the libration in latitude. In the case of the image on the following three pages, as well as in Figure 1, there is a strong northerly libration. This situation shows favorably several features near the north and northwest limbs, such as Nansen, Petermann, Cusanus, de la Rue, and Endymion, and even lets us peek beyond the mean limb in this area.

The increased phase has also brought into view a considerably greater extent of the lowlands. Mare Crisium is now completely sunlit, but the solar elevation remains low enough to highlight the concentric ridge system in its western portion, shown in Figure 2. This lighting also brings out the half-flooded pre-mare crater Yerkes (bottom center in Figure 2) and the two "capes" just to its west, Promitorium Olivium (north) and Promitorium Lavinium (south). In 1953 a (presumably) natural bridge was reported under sunset lighting as spanning the gap between the two promontories. This feature was later shown to be a lighting illusion, demonstrating the need to examine the Moon's features under all conditions of lighting before attempting to interpret their nature.

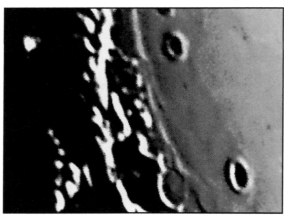

2. Western Mare Crisium.
1996 May 21, 04h31m UT. 28-cm Sch.-Cass., f/21, 0.15 s. Col. = 311°.0, Solar Lat. = 1°.1S, Librations = 0°.1E/7°.3N.

West of the Langrenus-Vendelinus-Petavius crater chain, the surface of Mare Fecunditatis is now illuminated. This is the first large mare visible as the moon waxes, stretching over 1000 km north-south. Elongated and irregular in outline, this mare's ridges form a pattern more haphazard than concentric. Besides the frequent ridges, the mare surface contains single craters, a crater cluster northwest of Langrenus, and isolated patches of highlands material. The actual pre-Nectarian mare impact basin is interpreted as occupying only the northern two-thirds of the mare, with the southern portion formed by lava that broke through the basin rim.

309°–N

90°N to 28°N

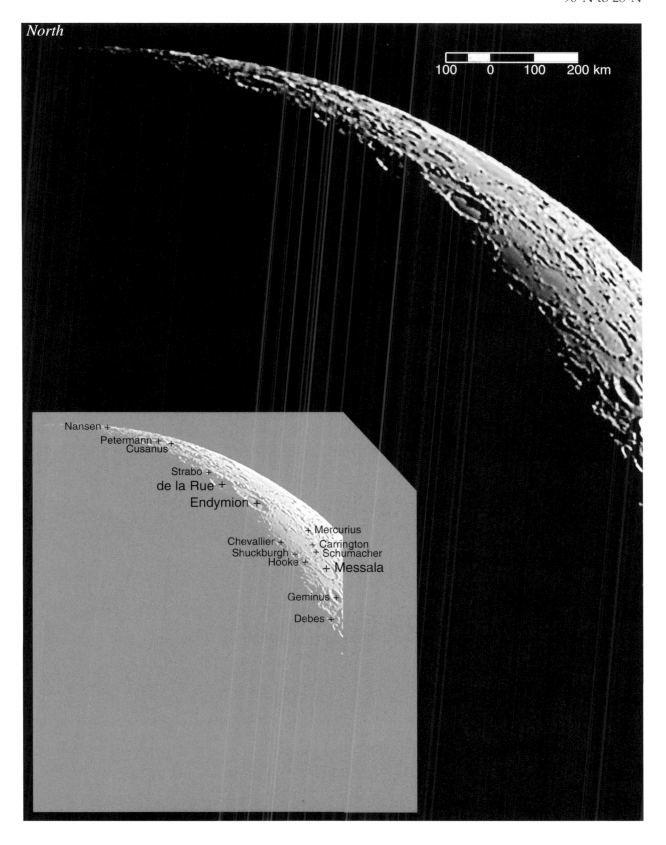

309°–C
40°N to 24°S

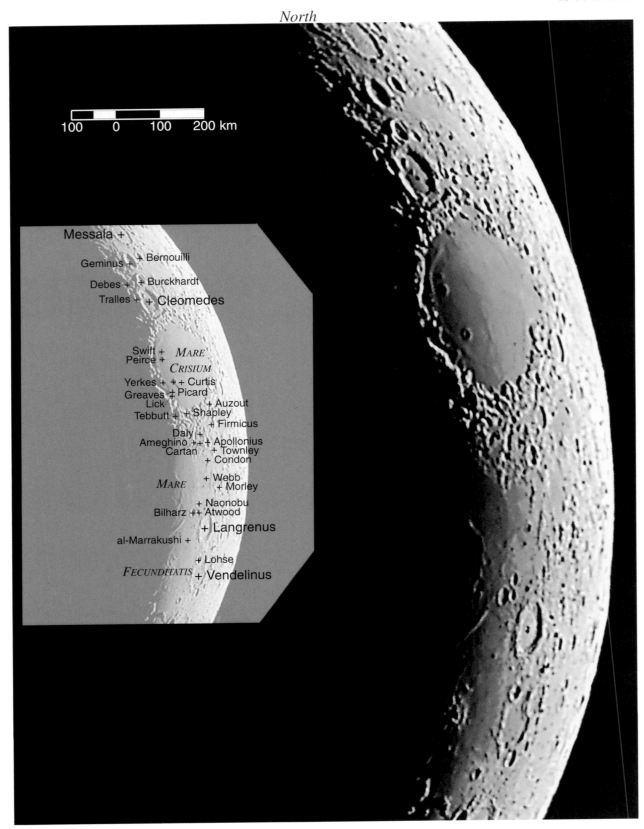

309°–S
3°N to 85°S

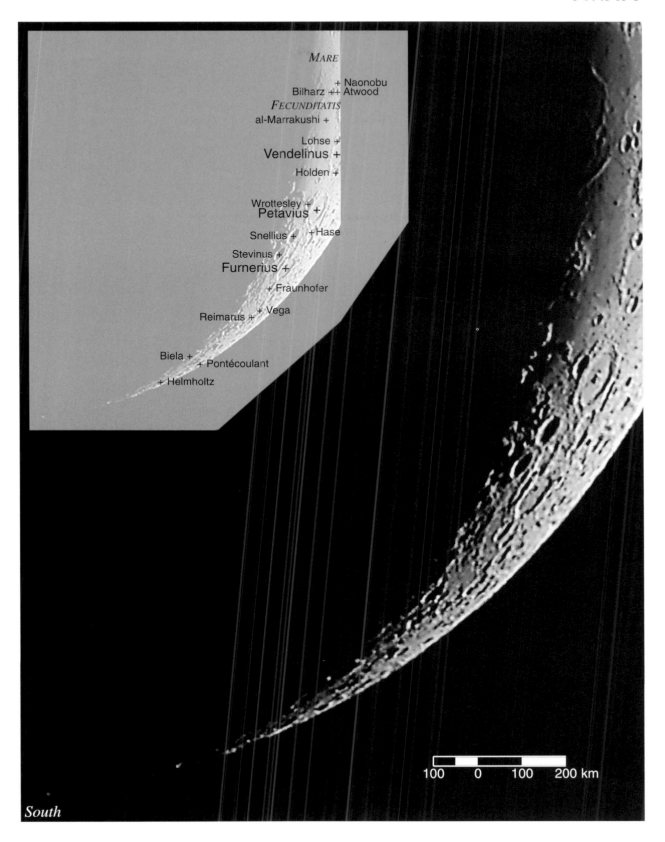

Colongitude 315°

1995 Apr 04, 03h22m UT. 28-cm Sch.-Cass., f/10, 0.20 s, W58 (green) Filter.
Colongitude = 315°.1, Solar latitude = 0°.6N, Librations = 0°.6W/3°.1N. 1.49 km/pixel.

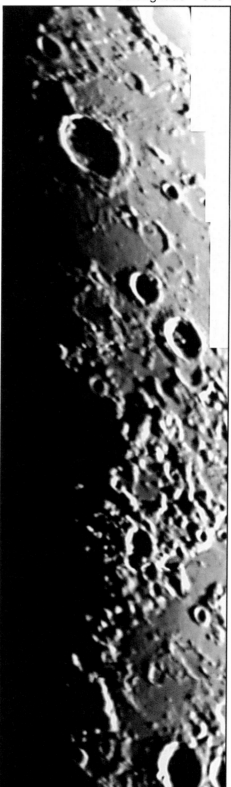

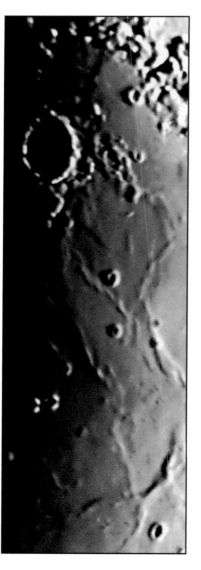

By the time the Moon reaches an age of four days, several interesting features are perched on the terminator. In the Northern Hemisphere one of the more prominent is the crater Atlas (Figure 1, near top), standing out from the unusually flat highlands region around it. South of Atlas the terrain becomes more rugged in the area containing the craters Cepheus, Franklin, and Macrobius (Figure 1, lower portion). Note the furrowed terrain north of Macrobius.

More of Mare Fecunditatis is sunlit now, revealing in its northern portion the low crater Taruntius (Figure 2, upper left); when the Sun rises higher, Taruntius will show its concentric, double-walled nature.

Slightly farther south, Messier and Messier A are on the terminator; a pair of similar, but not quite identical craters (Figure 2, below center). In the Taruntius-Messier area the low Sun brings out the complex ridge system of western Mare Fecunditatis.

Goclenius is on the eastern "shore" of the peninsula between Mare Nectaris and Mare Fecunditatis, here called the Pyrenaeus Highlands.

Just northwest of Snellius, the Vallis Snellius has appeared, and to its south is a similar but larger feature, Vallis Rheita, immediately southeast of the crater of that name. Three smaller valleys, all unnamed, lie farther south, north of the Steinheil-Watts pair. These valleys point toward Mare Nectaris, whose existence we can already infer from its widespread ejecta, although the mare itself is still in the night hemisphere.

1. Atlas to Macrobius (left). 1997 Mar 13, 03h30m UT. 28-cm Sch.-Cass., f/21, 0.15 s. Col. = 318°.6, Solar Lat. = 0°.2N, Librations = 6°.0E/5°.5N.

2. Western Mare Fecunditatis (right). 1997 Feb 11, 02h29m UT. 28-cm Sch.-Cass., f/21, 0.25 s. Col. = 312°.9, Solar Lat. = 0°.9N, Librations = 4°.8E/1°.7N.

315°–N

90°N to 8°N

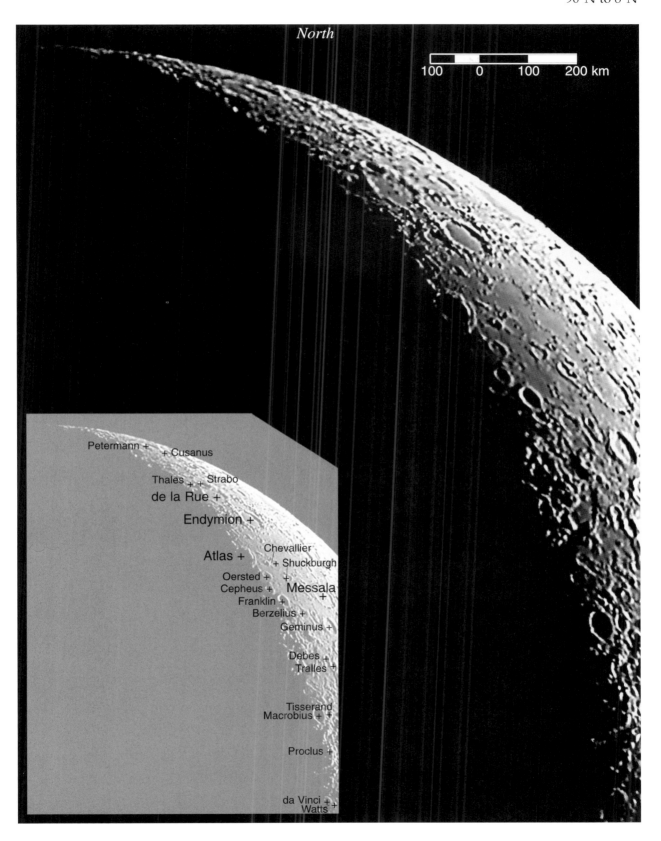

315°–C

32°N to 25°S

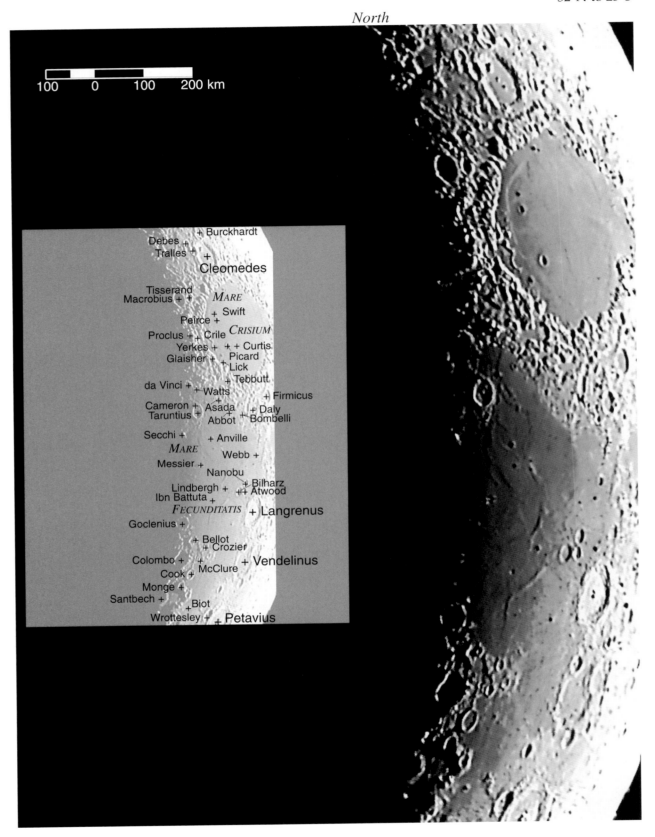

315°–S
1°S to 87°S

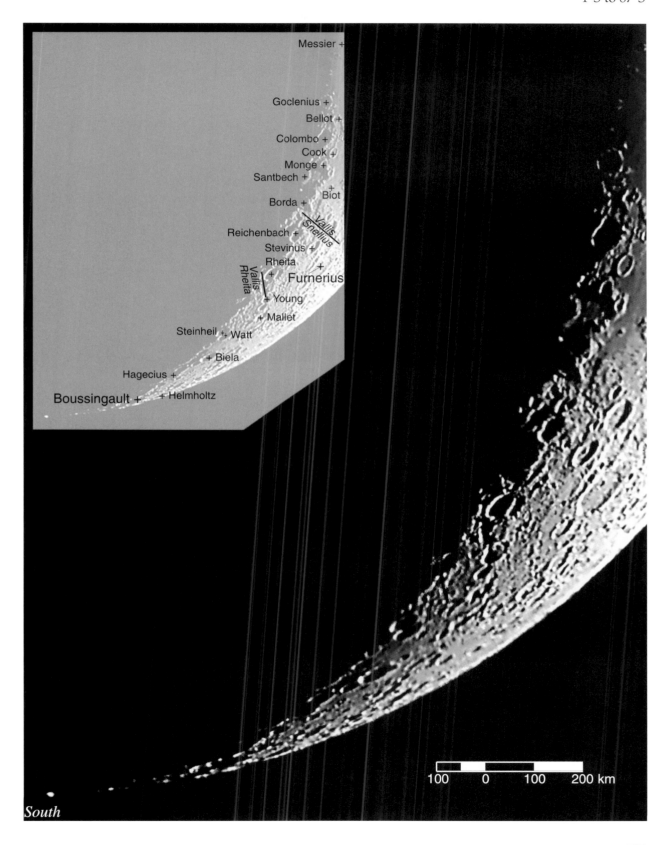

Colongitude 323°

1993 Dec 18, 01h56m UT. 28-cm Sch.-Cass., f/10, 0.07 s.
Colongitude = 322°.8, Solar latitude = 0°.6N, Librations = 4°.7E/6°.0S. 1.98 km/pixel.

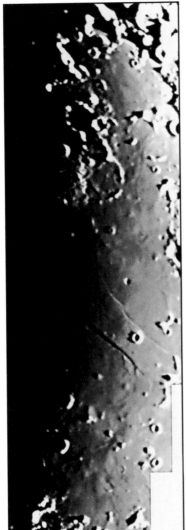

The western movement of the terminator has now reached the eastern portions of three maria not seen before this phase: Mare Frigoris, Mare Tranquillitatis, and Mare Nectaris.

Eastern Mare Frigoris, between the Northern Highlands and Atlas, contains several ridges, chiefly oriented approximately north-south. Atlas itself is now clearly only the eastern component of the Atlas-Hercules crater pair. The flat area south of Atlas is a good example of highlands plains material; probably a volcanic flow, but of the light tone associated with highlands material, rather than mare.

Counting its northeast extension, Sinus Amoris, Mare Tranquillitatis extends over 700 km along the terminator from 24°N to the Equator (Figure 1). Its surface holds a variety of formations, from fresh craters to several flooded rings. A good example of a partial, flooded ring is Maraldi D, straddling the terminator above center in Figure 1. The fresh crater Cauchy can be identified by the prominent rille to its north, Rima Cauchy, while to its south is a fault scarp, Rupes Cauchy. South of the scarp are two domes; Cauchy ω on the east and Cauchy τ on the west (Figure 1, below center).

Mare Nectaris is the southernmost mare visible, and the rising Sun now reveals its multiring nature (Figure 2). A inner ring is defined by a ridge system; the next ring out is marked in part by the Montes Pyrenaeus, which run north-south through the Pyrenaeus Highlands between Mare Nectaris and Mare Fecunditatis.

South of Mare Nectaris, the Southern and Southeastern Highlands stretch to the limb. Within them, Vallis Rheita is now completely visible. Slightly farther south the most prominent crater near the terminator is the ancient ring Janssen (Figure 3). Besides being battered by subsequent impacts, Janssen is noteworthy for its hexagonal outline; its north wall is particularly well-defined by a fault scarp. Within Janssen, linear patterns of hills marking more fault scarps suggest that the crater is really a small multiring basin.

The terrain south of Janssen is so heavily cratered that many large formations have not even been named, such as the Boussingault E-B-C chain just northwest of that crater. Among the prominent named features are Vlacq and Rosenberger, along with Demonax, Scott, and Amundsen farther south.

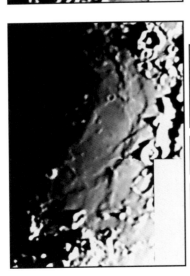

1. Eastern Mare Tranquillitatis (upper left). 1997 Feb 12, 02h30m UT. 28-cm Sch.-Cass., f/21, 0.20 s. Col. = 325°.1, Solar Lat. = 0°.9N, Librations = 5°.8E/3°.4N.

2. Eastern Mare Nectaris (lower left). (Same data as Figure 1.)

3. Janssen (lower right). 1997 Feb 12, 02h44m UT. 28-cm Sch.-Cass., f/21, 0.10 s. Col. = 325°.2, Solar Lat. = 0°.9N, Librations = 5°.8E/3°.3N.

323°–N

84°N to 2°N

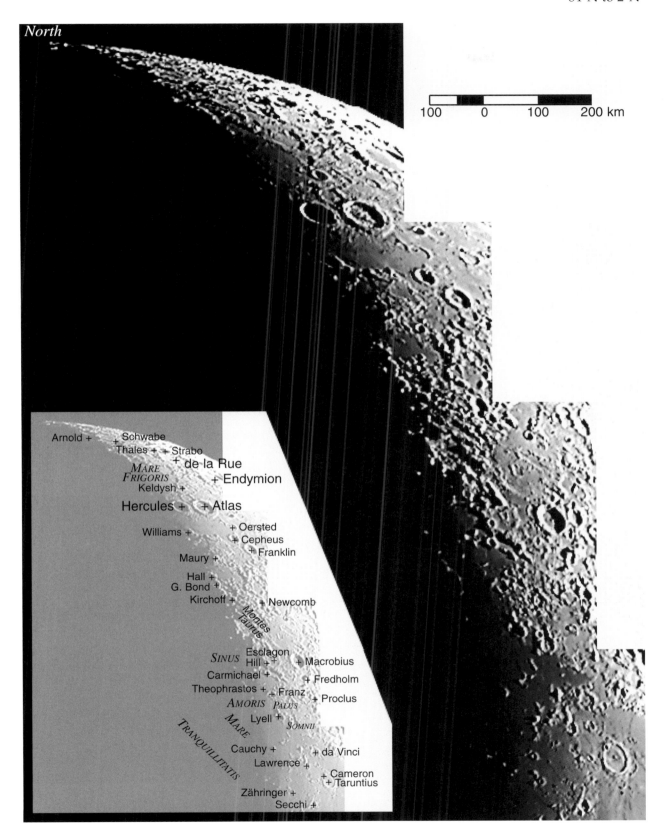

323°–C

18°N to 32°S

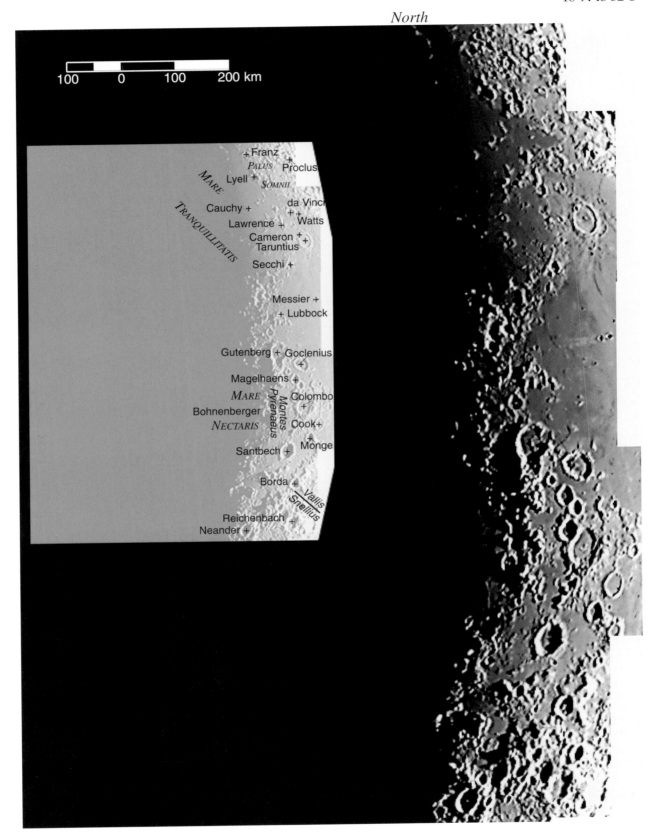

323°–S

16°S to 90°S

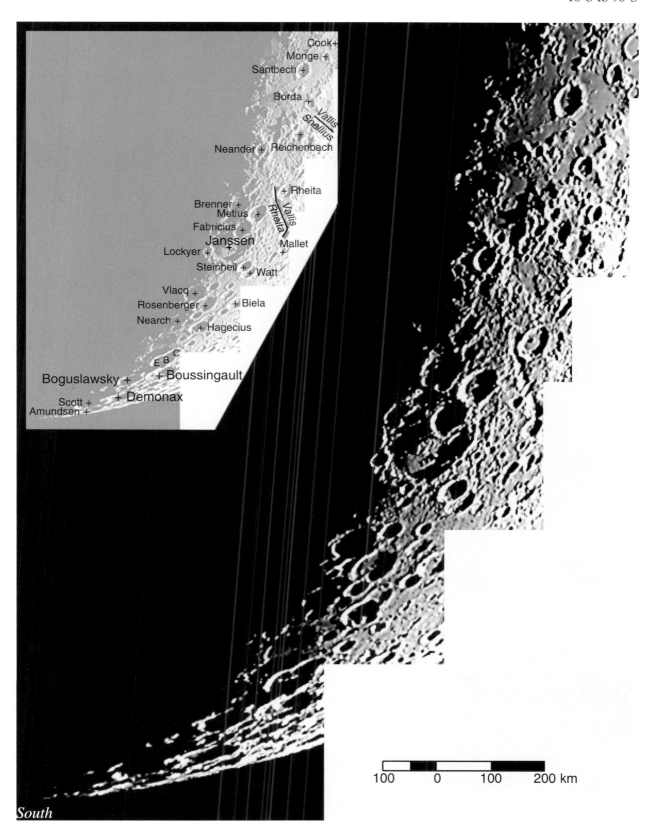

Colongitude 328°

1995 JUN 03, 04h26m UT. 28-cm Sch.-Cass., f/10, 0.38 s, W58 (green) Filter.
Colongitude = 328°.0, Solar latitude = 0°.9S, Librations = 6°.1W/7°.4N. 1.48 km/pixel.

By now, the Moon appears about one-quarter sunlit and large areas of both maria and highlands are visible. The image on the following pages has a strong northerly libration, favoring the Northern Highlands, which contain unusually large numbers of ancient flooded craters. One such is right on the terminator; Gärtner, on the northern shore of Mare Frigoris.

Mare Frigoris and its southern extensions, Lacus Mortis and Palus Somniorum, are now highlighted (Figure 1). Mare Frigoris' system of "wrinkle ridges" is clearly visible. There is no obvious boundary between Mare Frigoris and Lacus Mortis nor between Lacus Mortis and Palus Somniorum, both irregular lowland units too small to be termed maria.

There is, however, a clear highlands division between Palus Somniorum and Mare Tranquillitatis. The central portion of that mare is now sunlit, showing isolated hills, domes, and small craters (Figure 2).

South of this portion of Mare Tranquillitatis the Pyrenaeus Highlands between Maria Tranquillitatis and Nectaris contain several prominent north-south ejecta valleys.

The roughly-circular form of Mare Nectaris is now evident, with its southern "bay" formed by the crater Fracastorius, whose interior has been flooded, and its north wall disrupted, by lava flows originating in the mare to its north (Figure 3).

Straddling the terminator in the highlands south of Fracastorius is Piccolomini, whose central peak is just catching the sun. Beyond this crater are more valleys, shown by their alignments to be gouged by ejecta from Mare Imbrium. Near the southern limb, to the southwest of Janssen, the large crater immediately west of Vlacq and Rosenberger is Hommel, one of the many large craters between Janssen and the southern limb.

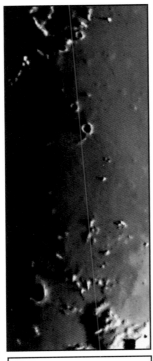

1. Mare Frigoris-Lacus Mortis-Palus Somniorum. 1997 FEB 12, 02h53m UT. 28-cm Sch.-Cass., f/21, 0.20 s. Col. = 325°.3, Solar Lat. = 0°.9N, Librations = 5°.8E/3°.3N.

2. Central Mare Tranquillitatis. 1993 Nov 19, 02h18m UT. 28-cm Sch.-Cass., f/21, 0.40 s. Col. = 330°.3, Solar Lat. = 0°.2S, Librations = 6°.3E/5°.2S.

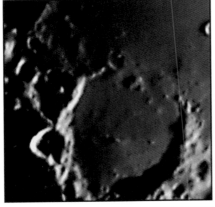

3. Fracastorius. 1993 Nov 19, 01h52m UT. 28-cm Sch.-Cass., f/21, 0.25 s. Col. = 330°.1, Solar Lat. = 0°.2S, Librations = 6°.4E/5°.1S.

328°–N

90°N to 18°N

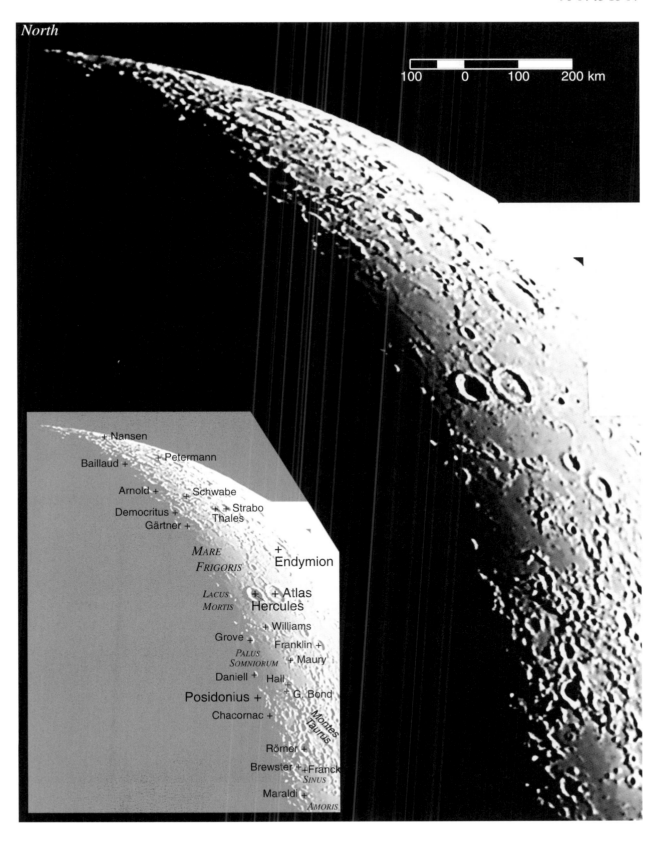

328°–C
34°N to 17°S

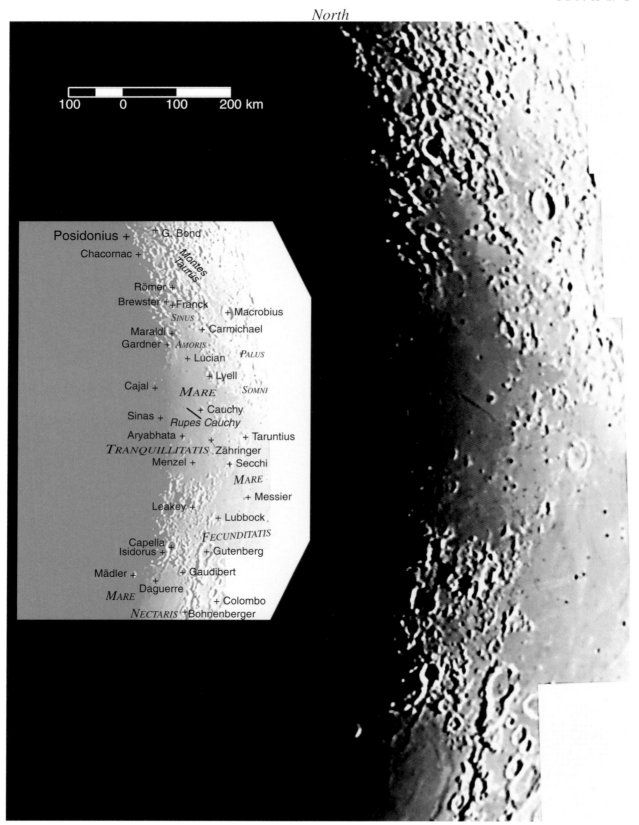

82

328°–S

1°S to 83°S

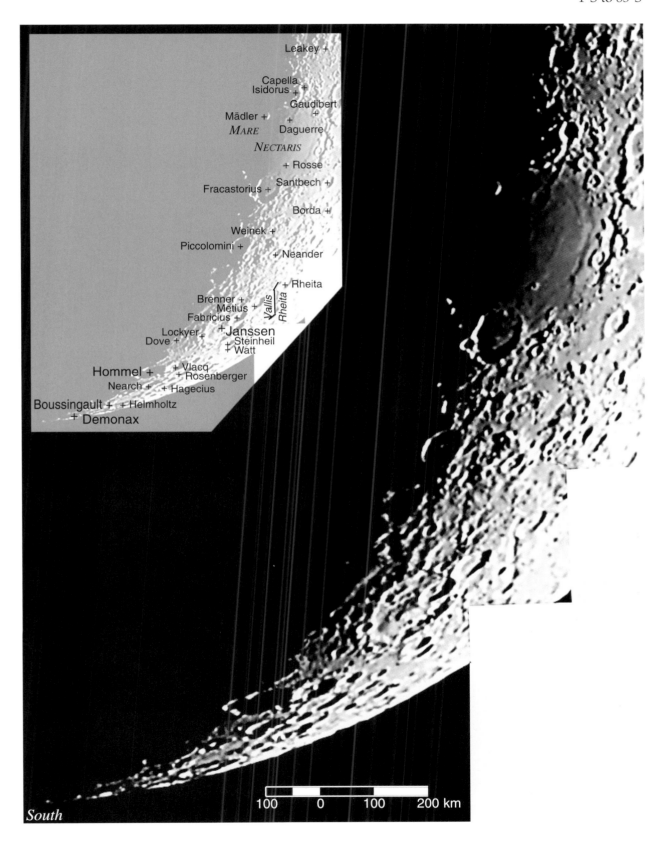

Colongitude 334°

1995 MAR 07, 02h56m UT. 28-cm Sch.-Cass., f/21, 0.10 s.
Colongitude = 334°.0, Solar latitude = 1°.2N, Librations = 0°.5E/2°.0N. 0.67 km/pixel.

The mosaic on the following pages is at a higher resolution than the previous mosaics; it thus shows finer detail than they did, but is confined to a narrower strip along the terminator.

On the north, more of Mare Frigoris is now visible, making clear its east-west elongation even when allowing for foreshortening. Mare Frigoris clearly is not a typical mare basin; perhaps it represents flooding between the main and outermost rims of the Imbrian Basin.

The polygonal outline of Lacus Mortis is now clear, being roughly centered on the crater Bürg. Lacus Somniorum, however, is now clearly the northeast extension of a mare visible for the first time at this phase, Mare Serenitatis. Just becoming sunlit in the mosaic is the great "Serpentine Ridge" (now named Dorsa Smirnov for its north-south portion, with its southwest extension named Dorsa Lister), one of the Moon's most prominent ridges, made more obvious by being aligned with the terminator. In Figure 1, with the Sun about 3 degrees higher than in the mosaic, this ridge system is more clearly shown.

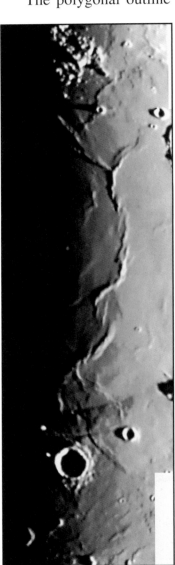

1. "Serpentine Ridge" (Dorsa Smirnov and Lister). 1996 JAN 26, 02h44m UT. 28-cm Sch.-Cass., f/21, 0.25 s. Col. = 336°.7, Solar Lat. = 1°.5N, Librations = 6°.8E/0°.4N.

The eastern rim of Mare Serenitatis begins with the concentric-ringed crater Posidonius on its north, and extends south to the cape Mons Argaeus on the south. Between this cape and Plinius, Mare Serenitatis is linked with Mare Tranquillitatis.

The terminator now falls within western Mare Tranquillitatis, where even a small change in colongitude highlights different features. In the mosaic on the following pages (Col. 334°.0), Carrel and the hills nearby are near the terminator, with only a few mare ridges visible. However, Figure 2 (Col. 336°.6), shows a complex system of mare ridges, many forming, or associated with, one of the Moon's best examples of a "ghost ring," Lamont.

South of the equator, the elongated, possibly secondary, crater Torricelli occupies the strait between Mare Tranquillitatis and Mare Nectaris; this inter-maria zone has been named Sinus Asperitatis.

Mare Nectaris is now fully visible, with the prominent crater Theophilus on its northwestern shore. By this phase, it appears that this mare has a partial outer rim on its southwest, comprising the scarp of the Rupes Altai, now starting to catch the Sun. Indeed, there are signs of another partial rim between Fracastorius and the Rupes Altai.

The great extent of the Southern Highlands is increasingly evident. Within them, Hommel is now free of the terminator; its interior is now illuminated, revealing peaks and craters on its walls and floor. Farther south, the large craters Mutus and Manzinus have replaced Hommel on the terminator in this densely cratered region.

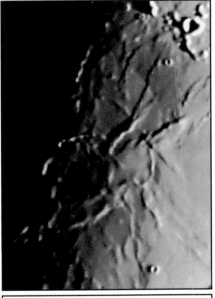

2. Lamont. 1996 JAN 26, 02h38m UT. 28-cm Sch.-Cass., f/21, 0.25 s. Col. = 336°.6, Solar Lat. = 1°.5N, Librations = 6°.8E/0°.4N.

334°–N

90°N to 20°N

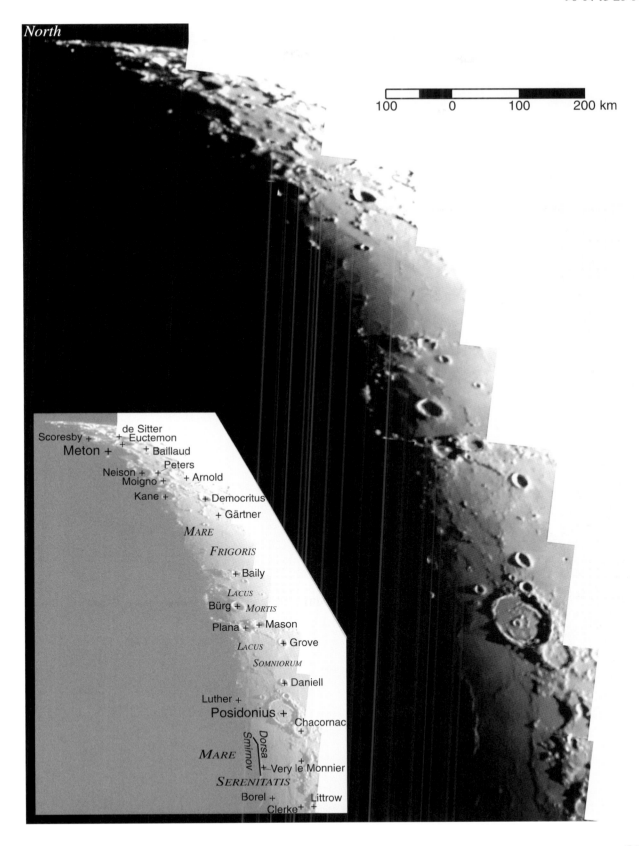

334°–C

23°N to 18°S

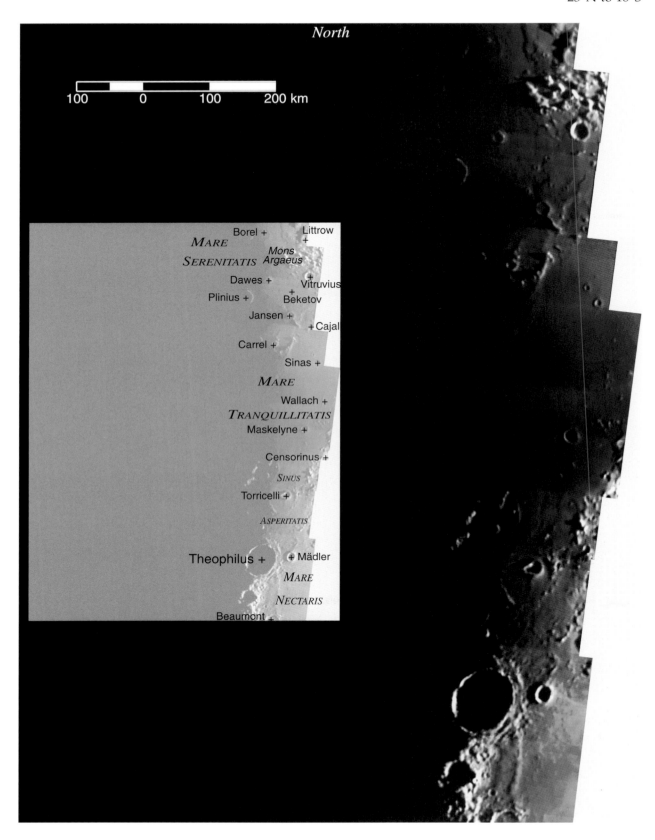

86

334°–S
14°S to 88°S

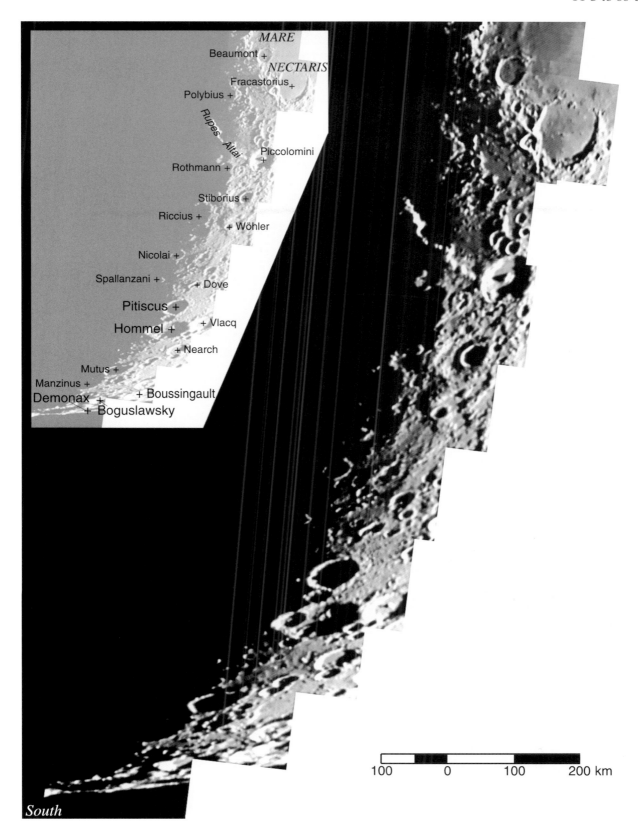

87

Colongitude 343°

1994 DEC 09, 01h44m UT. 28-cm Schmidt-Cassegrain, f/10, 0.10 s, W58 (green) Filter.
Colongitude = 342°.9, Solar latitude = 0°.8N, Librations = 7.4°E/5°.3S. 1.42 km/pixel.

By this phase, the terminator falls across several highland and mare units. The image mosaic for this lighting has a libration that favors viewing the Southern Hemisphere but tilts the northern regions away from us. Nonetheless, several of the Northern Highland's flooded rings are clearly seen, including Meton, Euctemon, Neison, and Arnold. Meton is particularly interesting, being composed of a group of approximately five coalesced craters.

On the south side of Mare Frigoris lie the Jura-Alpes-Caucasus Highlands, which include the prominent pair of craters Aristoteles and Eudoxus, now directly on the terminator, with their ejecta blankets well shown.

Low relief in the Mare Serenitatis is highlighted in this mosaic, particularly the ridges in its interior. On the northeast mare rim, the concentric scarp inside the crater Posidonius is now sunlit (Figure 1).

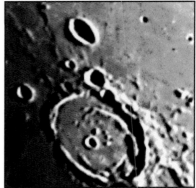

1. Posidonius. 1994 MAY 16, 04h12m UT. 28-cm Sch.-Cass., f/21, 0.25 s. Col. = 338°.4, Sol. Lat. = 0°.0N, Librations = 7°.7W/6°.8N.

Western Mare Tranquillitatis has several interesting low-relief features, such as the domes Arago α and β and the complex ridge systems that extend to the northwest and southwest of Lamont (Figure 2). Note also the east-west Rima Hypatia near the mare's southwestern edge.

Sinus Asperitatis and Mare Nectaris are the next mare units to the south, and the terminator crosses the Central Highlands to their west. Lying on the highlands margin, the crater trio of Theophilus, Cyrillus and Catharina are now fully visible (Figure 3). To their south, the Rupes Altai marks the southwest edge of the Nectaris Basin.

The Southern Highlands extend from Mare Nectaris to the south limb. In the jumble of craters near the terminator here are many low damaged rings that are now prominent, but will become harder to see under a higher sun.

Some better-defined craters in this area are Zagut, Rabbi Levi, Maurolycus, Mutus, and Manzinus. Nearer to the south limb, the southerly libration favors our view of Boguslawsky and Demonax, as well as the craters Scott and Amundsen in the marginal zone.

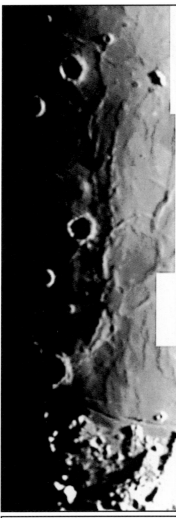

2. W. Mare Tranquillitatis. 1994 MAY 16, 04h18m UT. 28-cm Sch.-Cass., f/21, 0.50 s. Col. = 338°.5, Sol. Lat. = 0°.0N, Librations = 7°.7W/6°.8N.

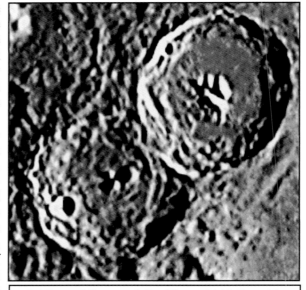

3. Theophilus and Cyrillus. 1993 DEC 20, 01h49m UT. 28-cm Sch.-Cass., f/21, 0.10 s. Col. = 347°.1, Sol. Lat. = 0°.7N, Librations = 3°.2E/5°.7S.

343°–N

85°N to 14°N

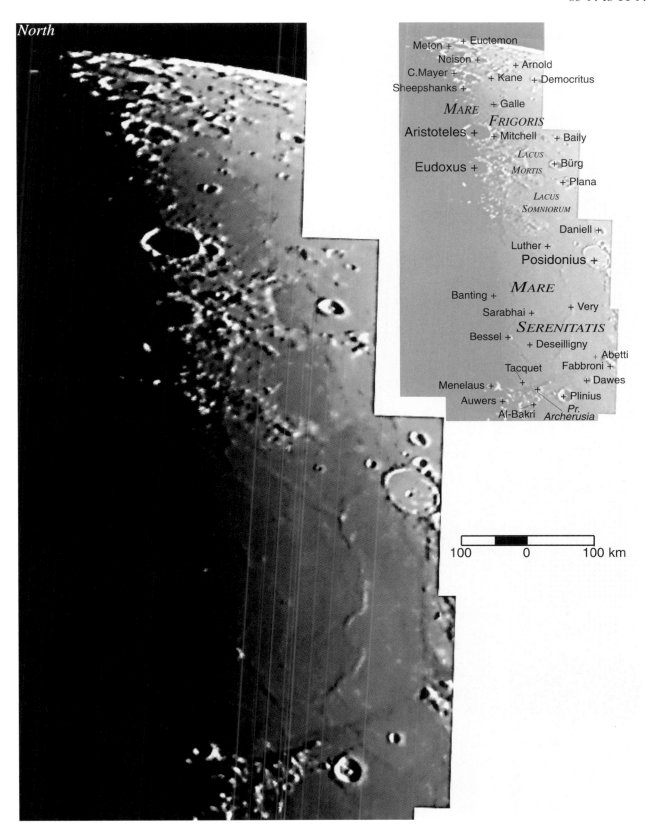

343°–C

16°N to 26°S

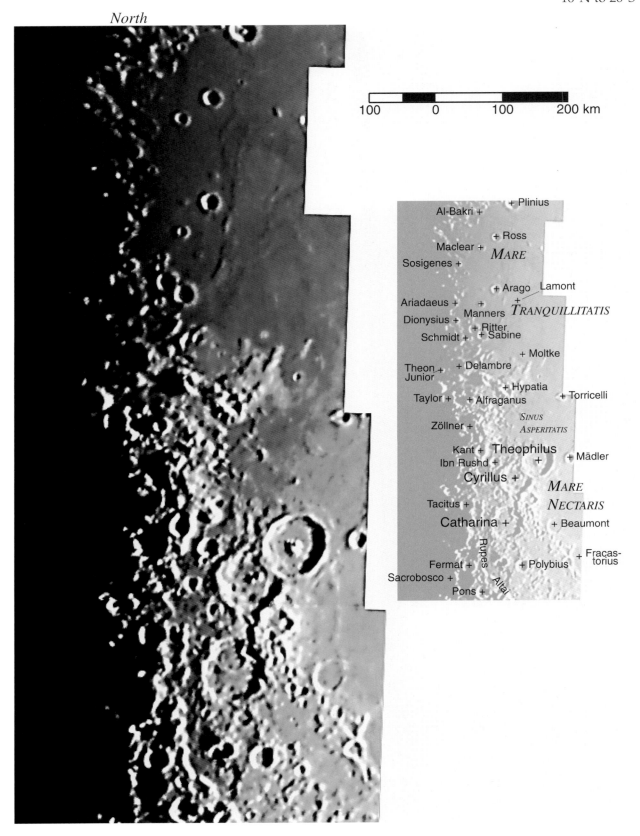

343°–S
24°S to 90°S

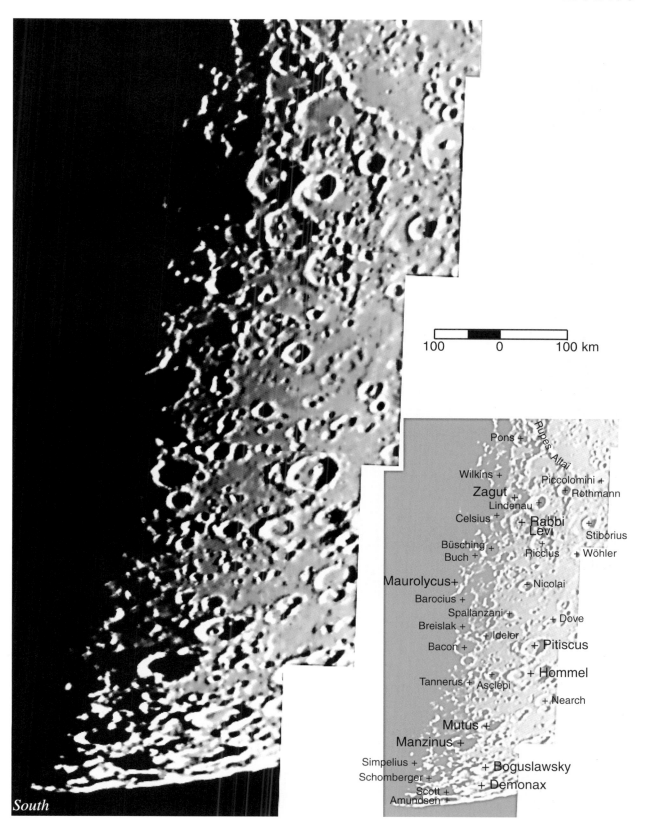

Colongitude 351°

1994 JUL 15, 03h54m UT. 36-cm Sch.-Cass., f/11, 0.05 s, Infrared Filter.
Colongitude = 351°.4, Solar latitude = 1°.3S, Librations = 3°.1W/5°.3N. 1.32 km/pixel.

The Moon is now nearing quarter phase, the best time to observe the terminator as the direction of sunlight is nearly perpendicular to our line of sight and lunar shadows are especially prominent.

The view on the next pages and in Figure 1 have a northerly libration, causing the Northern Highland's flooded craters, such as Meton and Barrow, to show up well.

All the eastern maria are now sunlit, as well as the eastern portions of Maria Frigoris and Vaporum. The ridges of western Mare Serenitatis form part of a chain that encloses most of the mare's interior. A prominent dome (the "Valentine Dome") is located just northeast of the "strait" on the west edge of the mare, connecting Mare Serenitatis with Mare Imbrium. This area contains the crater Linné, once reported as having disappeared, and the collapse feature Krishna.

The area of southwest Mare Tranquillitatis (Figure 2) contains several ridges, the Arago domes and the Ariadaeus Rille (see also Figure 3), and is crowded with valleys radiating from the Mare Imbrium Basin.

South of Mare Vaporum, highlands extend to the southern limb (Figure 3), with well-formed craters, such as the Godin-Agrippa pair, among flooded partial rings like Tempel and Lade. Abulfeda is now prominent. Playfair and Apianus lie on the terminator. East of Apianus, the heavily damaged crater Poisson is almost lost in the rugged terrain; its southeastern neighbor, Gemma Frisius, is also a ruin, but stands out more clearly.

Next to the south is the giant crater Maurolycus, an unusual example of a large crater overlapping a smaller. The large "notch" in the terminator to its west is the yet larger crater Stöfler.

1. Northern Highlands. 1996 FEB 26, 02h55m UT. 28-cm Sch.-Cass., f/21, 0.08 s. Col. = 353°.9, Solar Lat. = 1°.0N, Librations = 4°.3E/5°.6N.

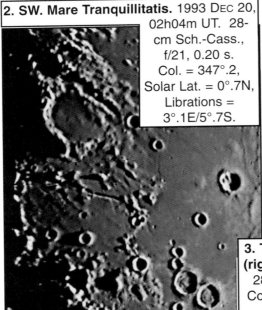

2. SW. Mare Tranquillitatis. 1993 DEC 20, 02h04m UT. 28-cm Sch.-Cass., f/21, 0.20 s. Col. = 347°.2, Solar Lat. = 0°.7N, Librations = 3°.1E/5°.7S.

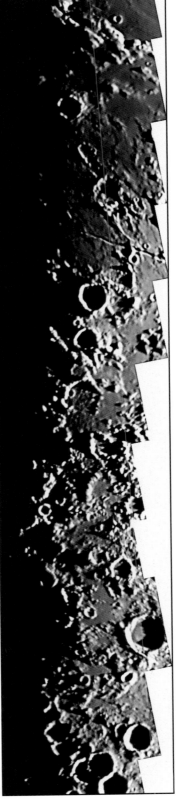

3. Terminator from 22°N to 22°S (right). 1995 Nov 29, 02h24m UT. 28-cm Sch.-Cass., f/21, 0.10 s. Col. = 351°.2, Solar Lat. = 1°.0N, Librations = 6°.3E/ 4°.4S.

351°-N

90°N to 24°N

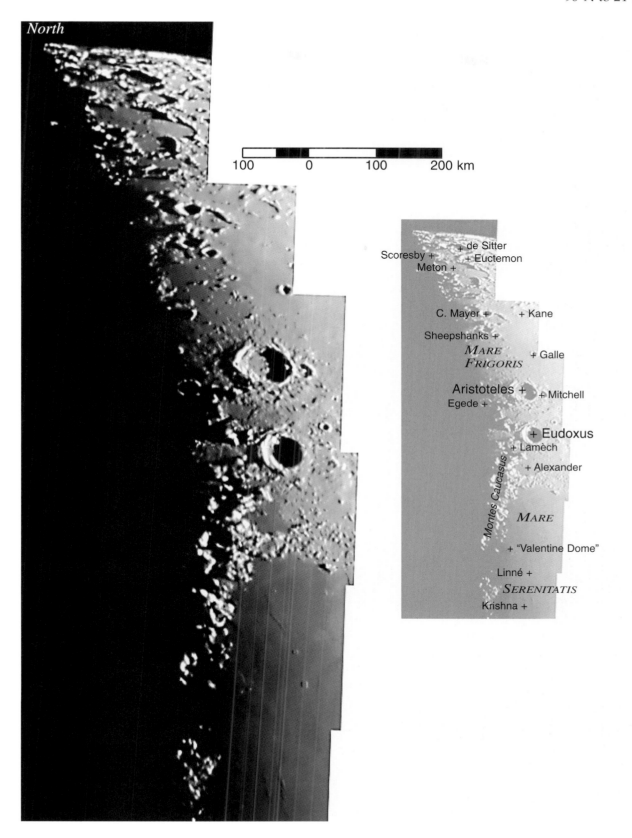

351°–C

24°N to 16°S

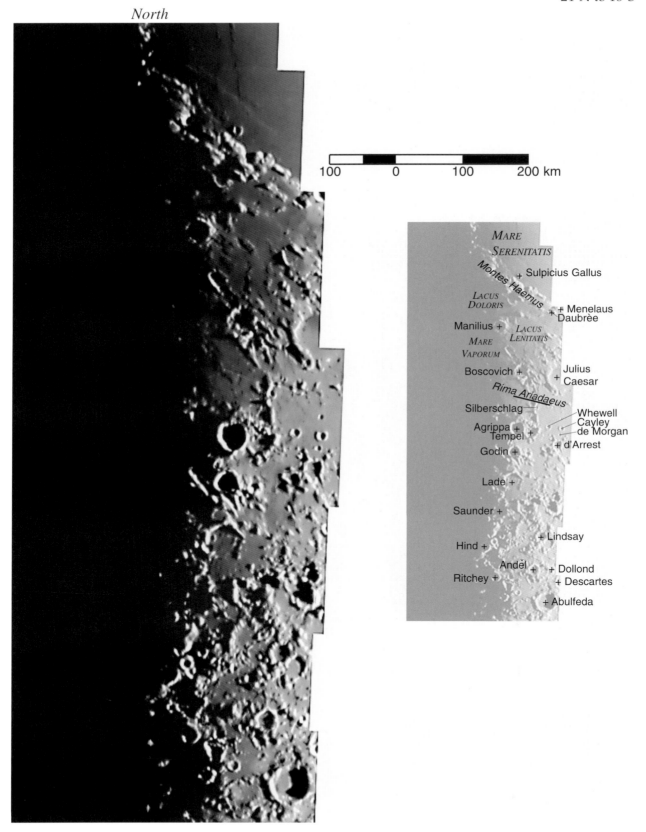

351°–S

14°S to 85°S

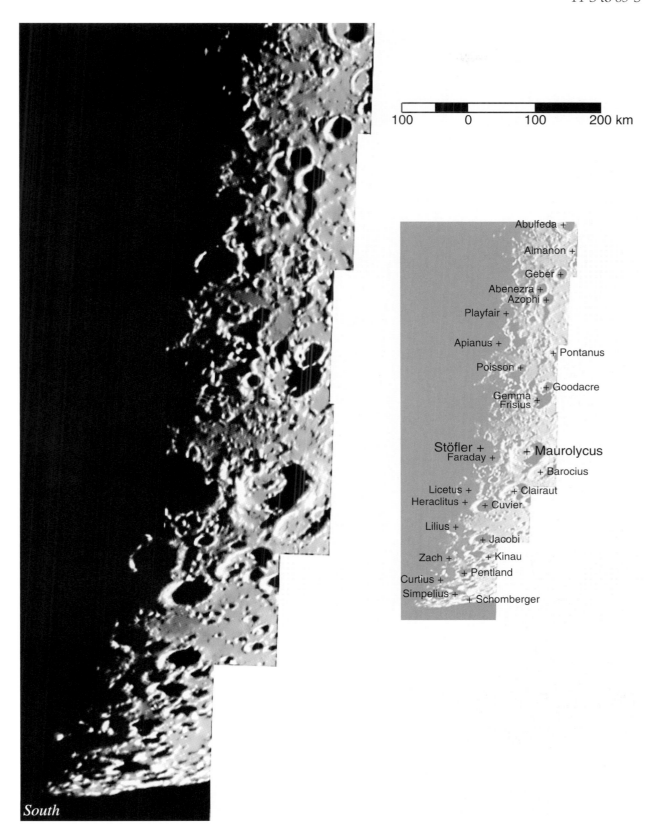

95

Colongitude 357°

1994 Apr 18, 04h26m UT. 28-cm Sch.-Cass., f/10, 0.07 s.
Colongitude = 356°.8, Solar latitude = 0°.7N, Librations = 8°.4W/6°.0N. 1.96 km/pixel.

At First Quarter, the present phase, glancing rays of sunlight are streaming through the passes of the Montes Caucasus, illuminating eastern Mare Imbrium and four craters within it: Cassini, Theaetetus, Aristillus and Autolycus (Figure 1).

Northwest of the Caucasus, the Montes Alpes is just becoming visible, along with the Vallis Alpes that cuts through them. To their north, yet more of Mare Frigoris and the Northern Highlands have appeared. On the Mare Frigoris-Northern Highlands boundary the flooded polygonal crater W. Bond is especially striking; its apparently rough floor will appear smooth when the Sun rises higher.

The "strait" connecting Maria Serenitatis and Imbrium contains the fronts of several lava flows, their morphology suggesting that the Serenitatis lava flow was later than that of the Imbrian Basin. Southwest of this strait the curve of the Montes Apenninus is just starting to be defined by the rising Sun.

In the highland plains zone between Mare Vaporum and Sinus Medii the Rima Hyginus stands out, one of the Moon's best-known rilles (Figure 2). This region abounds in rilles, including the Rima Ariadaeus east of Hyginus and Rimae Triesnecker to the south.

The Central Highlands contain the Godin-Agrippa crater pair, but currently their most impressive feature is another north-south pair, Hipparchus and Albategnius (Figure 3).

There is no definite boundary between the Central and the Southern Highlands, and both highland areas are saturated with craters. Two medium-sized examples are the pair Aliacensis and Werner, now straddling the terminator. To their south, the rim of Stöfler is now entirely illuminated, and the Sun is beginning to light the crater floor.

Even though only its rim is sunlit, it is clear that there is an unusual linear feature south of Stöfler; the elongated, parallel-sided depression Heraclitus, with one peak of its central ridge already sunlit.

Farther south, it is easy to become lost in a jumble of craters of all sizes. One guide to making one's way within this area is an east-west chain of four craters slightly south of Heraclitus, of which Lilius and Jacobi are the two largest components. Another landmark in this region is the sharply defined crater Manzinus, lying on a direct line between Jacobi and the southeast limb. Manzinus is at latitude 68° south; due to its northerly libration, features farther south are difficult to distinguish on the mosaic for this phase.

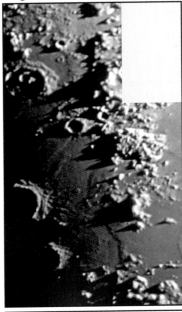

1. Montes Caucasus.
1994 Apr 18, 03h37m UT. 28-cm Sch.-Cass., f/21, 0.20 s. Col. = 356°.4, Solar Lat. = 0°.7N, Librations = 8°.3W/6°.1N.

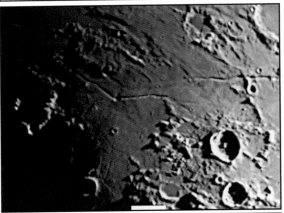

2. Hyginus Area. 1994 Apr 18, 03h52m UT. 28-cm Sch.-Cass., f/21, 0.30 s. Col. = 356°.5, Solar Lat. = 0°.7N, Librations = 8°.3W/6°.1N.

3. Hipparchus and Albategnius.
1994 Apr 18, 04h04m UT. 28-cm Sch.-Cass., f/21, 0.30 s. Col. = 356°.6, Solar Lat. = 0°.7N, Librations = 8°.3W/6°.1N.

357°–N

90°N to 24°N

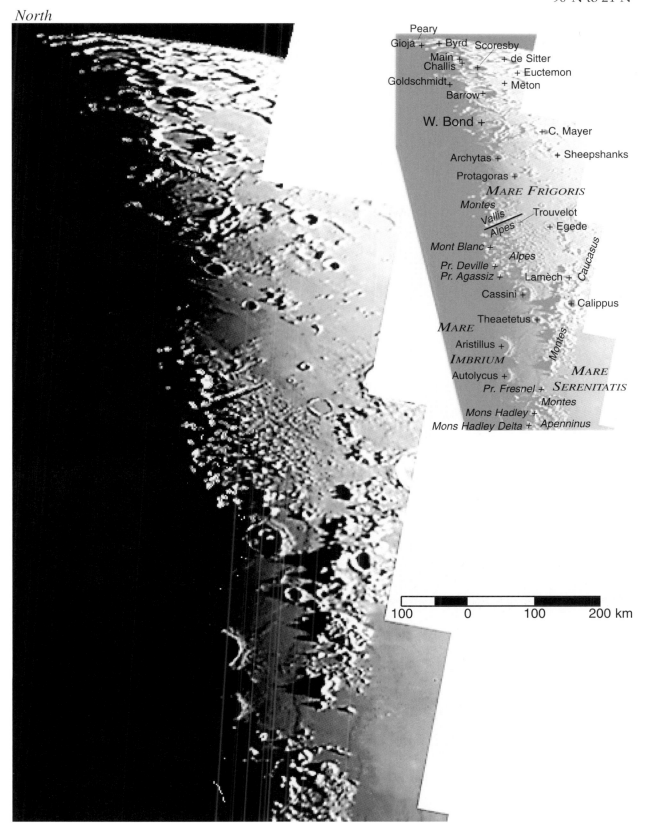

357°–C

26°N to 14°S

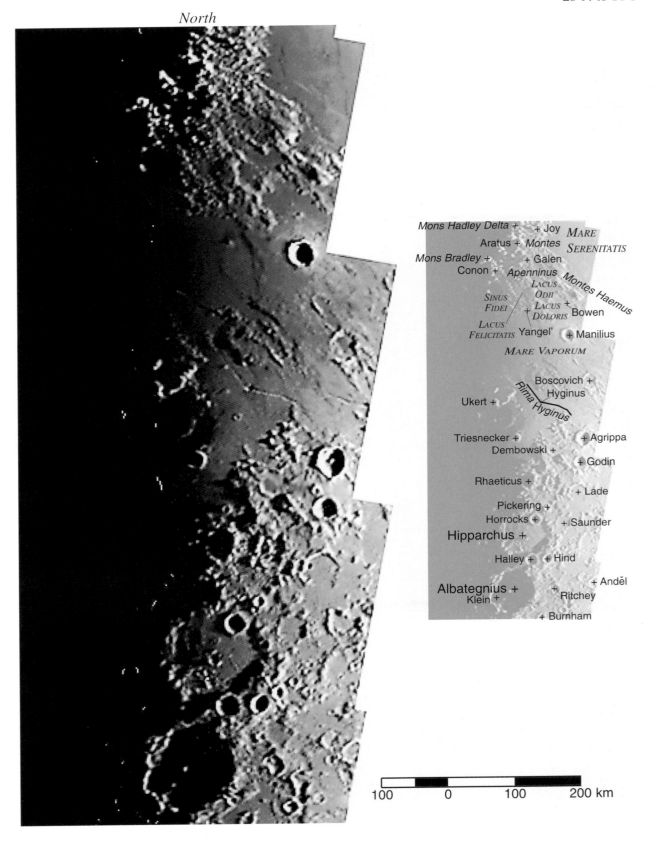

357°–S

13°S to 84°S

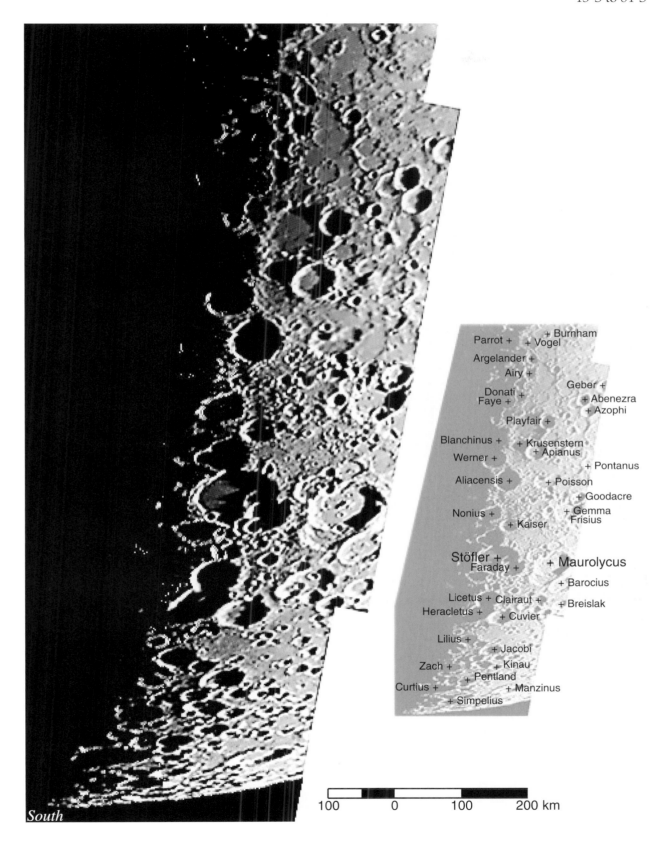

99

Colongitude 003°

1994 MAY 18, 04h46m UT. 28-cm Sch.-Cass., f/10, 0.05 s.
Colongitude = 003°.2, Solar latitude = 0°.0S, Librations = 7°.7W/7°.7N. 1.92 km/pixel.

This terminator mosaic has a northerly libration so favorable that the lunar North Pole's location is marked on the index chart. The terminator moves slowly near the poles, so the lighting in the Northern Highlands is not very different from that in the previous view. The floor and walls of W. Bond are now fully sunlit, however, revealing this depression's polygonal shape.

South of W. Bond, several ridges in Mare Frigoris are near the terminator, which then crosses the Montes Alpes. The great graben that forms the Vallis Alpes is now completely visible.

Several isolated peaks, exemplified by Mons Piton, and a chain of peaks, the Montes Spitzbergen, mark an inner ring of the Imbrian Basin (Figure 1). The ejecta patterns surrounding Autolycus, and especially Aristillus, are best seen at this phase. Meanwhile, the Sun is rising on the east wall of Archimedes to their southwest.

Between Archimedes and the Montes Apenninus is an area of complex relief known as the "Apennine Bench," often cited as a rare example of volcanism involving highlands material.

South of the Montes Apenninus, the Sun is rising on the low ridges in Sinus Aestuum. To their south is the mean center of the disk, within Sinus Medii (Figure 2). The glancing sunlight on the floor of Ptolemaeus shows the many "saucers" that dot its interior, unlike the surface of Sinus Medii to the north. To Ptolemaeus' south, Alphonsus' central peak is just catching the Sun.

Farther south, the large craters Purbach, Regiomontanus, and Walter form a distinctive chain. The two large "notches" jutting from the terminator south of them are caused by the craters Orontius and Maginus.

Finally, the apparent giant depression straddling the terminator farther south is really due to a group of medium-sized craters centered on Moretus, as will become clear later when the Sun rises higher.

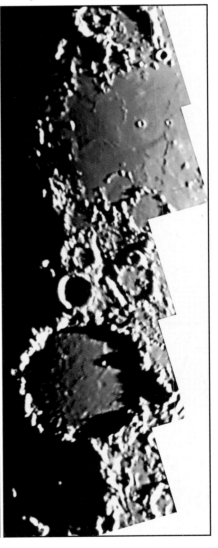

1. Eastern Mare Imbrium (left). 1995 Nov 30, 02h22m UT. 28-cm Sch.-Cass., f/21, 0.10 s. Col. = 003°.4, Solar Lat. = 1°.0N, Librations = 6°.8E/3°.2S.

2. Sinus Medii - Ptolemaeus - Alphonsus (right). 1995 Nov 30, 02h08m UT. 28-cm Sch.-Cass., f/21, 0.10 s. Col. = 003°.2, Solar Lat. = 1°.0N, Librations = 6°.9E/3°.2S.

003°–N

90°N to 26°N

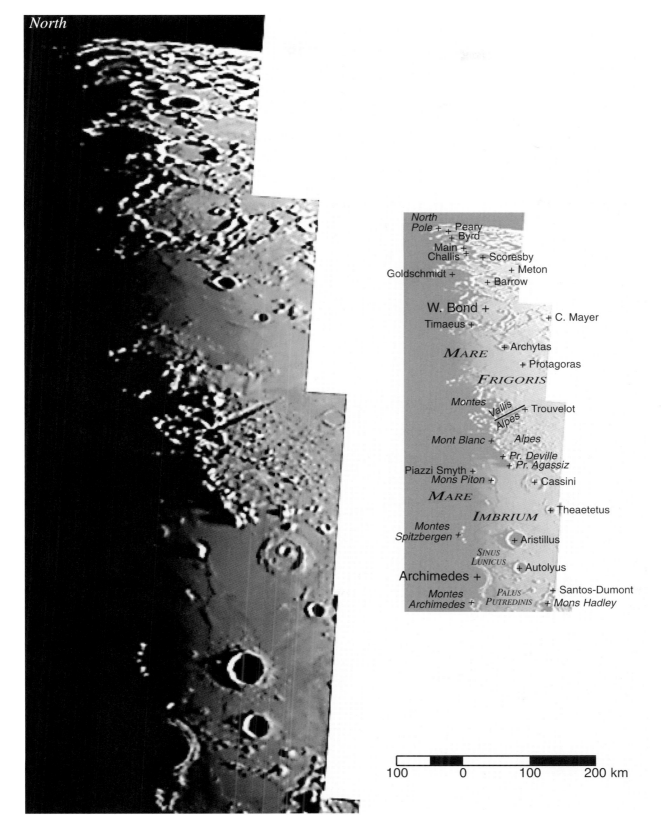

003°–C
28°N to 13°S

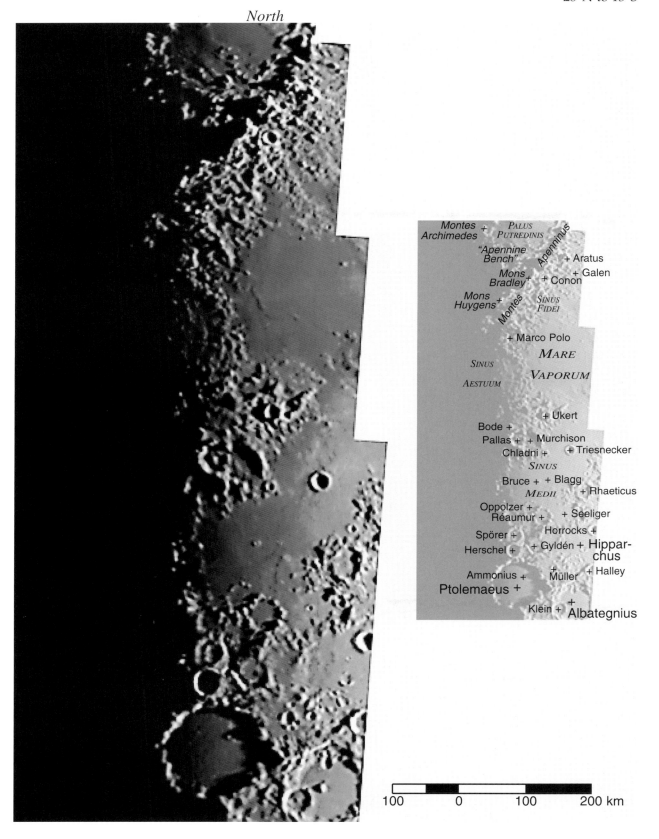

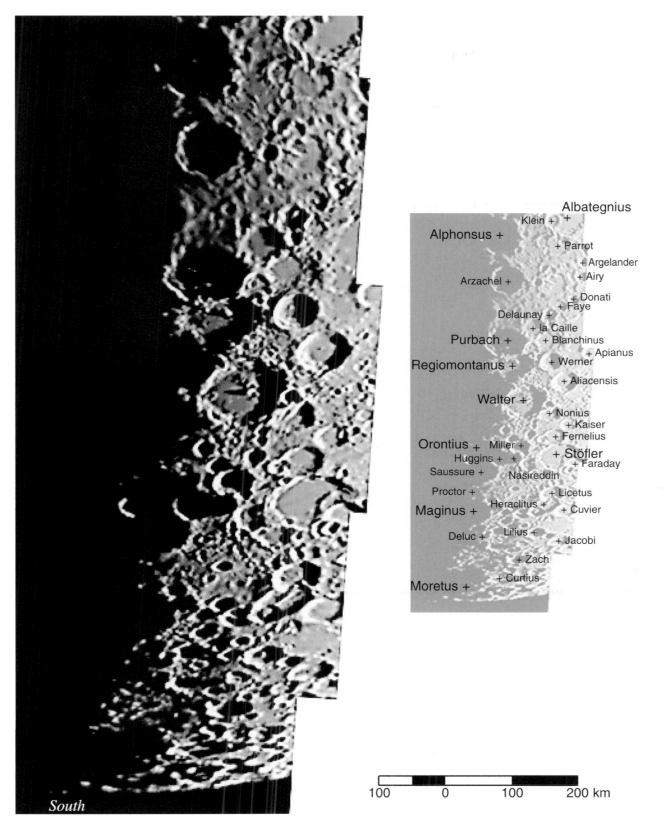

Colongitude 009°

1994 Apr 19, 04h58m UT. 28-cm Sch.-Cass., f/10, 0.05 s.
Colongitude = 009°.3, Solar latitude = 0°.7N, Librations = 8°.6W/6°.9N. 1.93 km/pixel.

This is another view that favors the Northern Highlands, where Goldschmidt can be added to the list of large flooded depressions that characterize this area.

An unnamed scarp has appeared in Mare Frigoris southwest of Timaeus. To its southwest the large crater Plato is on the terminator, its flooded interior still in shadow. South of Plato, the Montes Teneriffe are starting to appear; with Mons Pico, the Montes Spitzbergen, and the Montes Archimedes (Figure 1), they trace out part of the inner ring of the Imbrian Basin. Note that Archimedes' flooded floor is now sunlit; Plato will be seen to resemble this crater when the sunlight reaches its floor.

With the Sun rising on Eratosthenes, at the southwest tip of the Montes Apenninus, the entire chain is now visible for the first time, as is the Sinus Aestuum with its complex system of wrinkle ridges.

South of Sinus Medii is a roughly north-south chain of interesting craters (Figure 2). The northernmost, Flammarion, has a floor that contains several domes, visible only under very low lighting. Ptolemaeus is clearly an approximate hexagon, and its floor "saucers" are starting to disappear. Just south of Ptolemaeus, Alphonsus' interior is sunlit, revealing its prominent central peak and north-south medial ridge. Alphonsus is another crater with a hexagonal outline.

South of Alphonsus is the deep crater Arzachel with a terraced inner wall and a high central peak. To its southwest a large low ring, informally called the "Birt Basin," straddles the terminator where Birt juts out of shadow. A famous north-south fault scarp named the Rupes Recta ("Straight Wall") splits the Basin in two.

A large "ruined" feature, Deslandres (Figure 3) was not even formally named until the 1960s. Perhaps its late recognition was due to its irregular outline and low, damaged walls.

At this phase a wedge of sunlight shows part of the floor of Maginus, a large deep feature, whose irregular outline and poor contrast make it hard to identify under high lighting. Southwest of Maginus, the terminator juts eastward; a "bulge" that is sometimes visible to the naked eye at a slightly later phase. It is made by the giant crater Clavius, called the Moon's largest in older books, although we now know several that are larger yet.

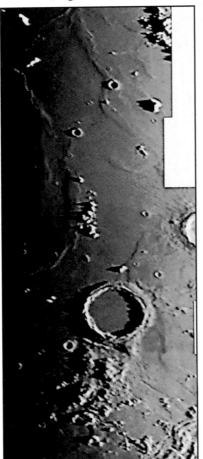

1. Eastern Mare Imbrium.
1994 Apr 19, 03h34m UT. 28-cm Sch.-Cass., f/21, 0.20 s. Col. = 008°.6, Solar Lat. = 0°.7N, Librations = 8°.3W/7°.0N.

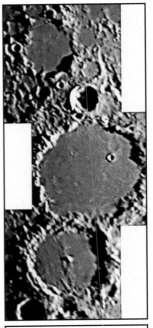

2. Ptolemaeus Area. 1994 Apr 19, 04h05m UT. 28-cm Sch.-Cass., f/21, 0.25 s. Col. = 008°.8, Solar Lat. = 0°.7N, Librations = 8°.5W/7°.0N.

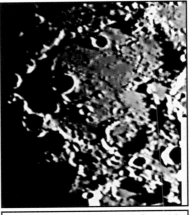

3. Deslandres. 1994 Apr 19, 04h22m UT. 28-cm Sch.-Cass., f/21, 0.25 s. Col. = 009°.0, Solar Lat. = 0°.7N, Librations = 8°.5W/7°.0N.

009°–N

90°N to 25°N

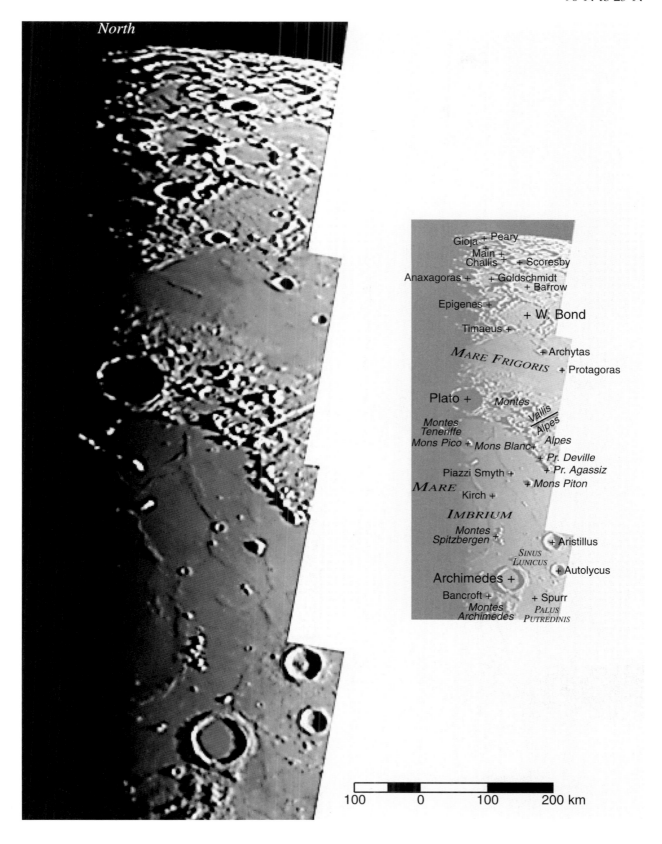

105

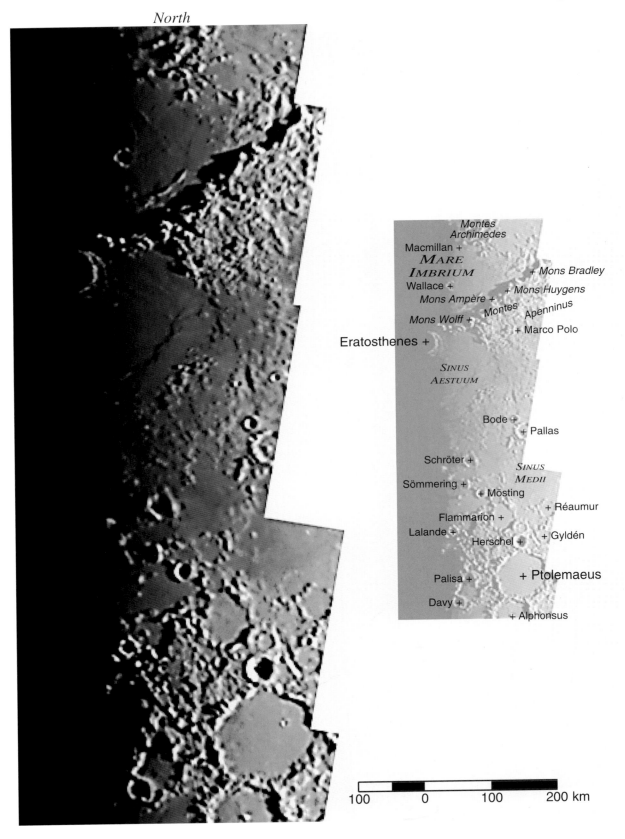

009°–C
27°N to 14°S

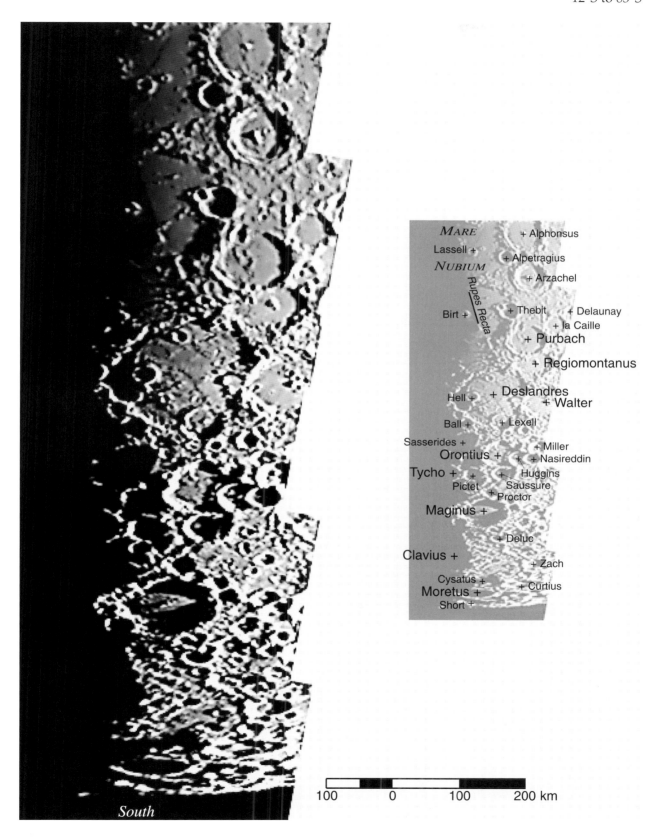

Colongitude 018°

1995 Feb 09, 03h06m UT. 28-cm Sch.-Cass., f/10, 0.10 s, W58 (green) Filter.
Colongitude = 017°.5, Solar latitude = 1°.5N, Librations = 0°.3W/4°.0N. 1.47 km/pixel.

The phase is gibbous now, with Fontenelle and the Montes Recti squarely on the terminator. South of the latter stretches central Mare Imbrium; smooth except for several wrinkle ridges.

South of Mare Imbrium the surface is anything but smooth. First, the Montes Carpatus are starting to appear. Second, chains of secondary craters are highlighted west of the Stadius ghost ring (Figure 1). They are associated with Copernicus, a spectacular crater whose eastern wall is now illuminated.

The zone south of Copernicus is called Mare Insularum, and this area contains crater ejecta, wrinkle ridges, isolated patches of highlands material, and a cluster of four large flooded craters: Fra Mauro, Bonpland, Parry, and Guericke. This cluster marks the approximate northern boundary of Mare Nubium, within which the "Birt Basin" is now entirely visible (Figure 2).

The Southern Highlands begin with Deslandres, and Pitatus to its west; the latter one of the best examples of a concentric-walled feature. Deeper into the highlands the central peak of Tycho is catching the sunlight. Maginus' floor is now completely visible; just to its south the inside of Clavius remains dark but the walls of its interior craters emerge from the shadow. It takes almost two terrestrial days for the interior of Clavius to become fully sunlit.

South of Clavius, the highlands become particularly rugged, with Moretus (Figure 3) one of the most prominent craters in these high latitudes.

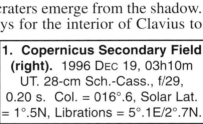

1. Copernicus Secondary Field (right). 1996 Dec 19, 03h10m UT. 28-cm Sch.-Cass., f/29, 0.20 s. Col. = 016°.6, Solar Lat. = 1°.5N, Librations = 5°.1E/2°.7N.

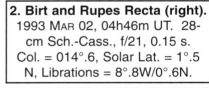

2. Birt and Rupes Recta (right). 1993 Mar 02, 04h46m UT. 28-cm Sch.-Cass., f/21, 0.15 s. Col. = 014°.6, Solar Lat. = 1°.5 N, Librations = 8°.8W/0°.6N.

3. Moretus (below). 1993 Nov 23, 03h51m UT. 28-cm Sch.-Cass., f/21, 0.07 s. Col. = 019°.7, Solar Lat. = 0°.1S, Librations = 2°.8E/6°.6S.

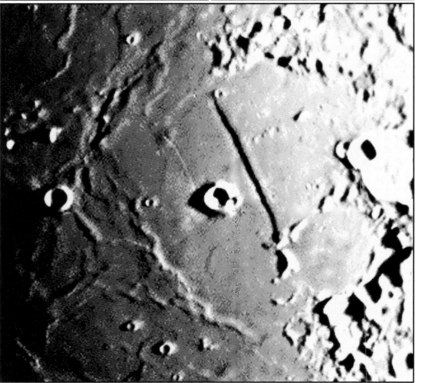

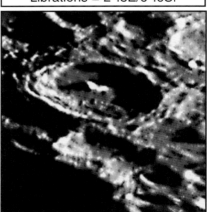

018°–N
90°N to 21°N

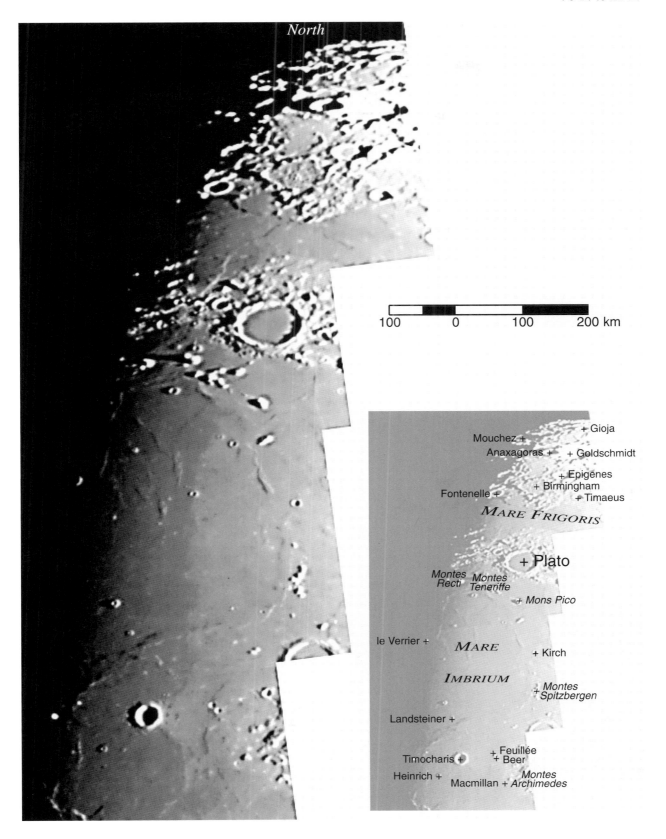

018°–C

23°N to 17°S

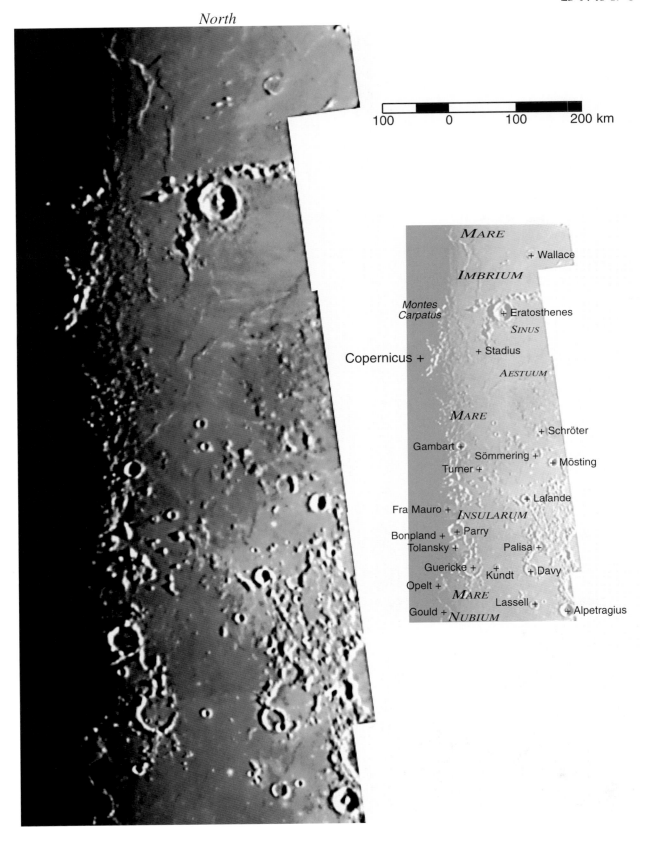

018°–S

15°S to 86°S

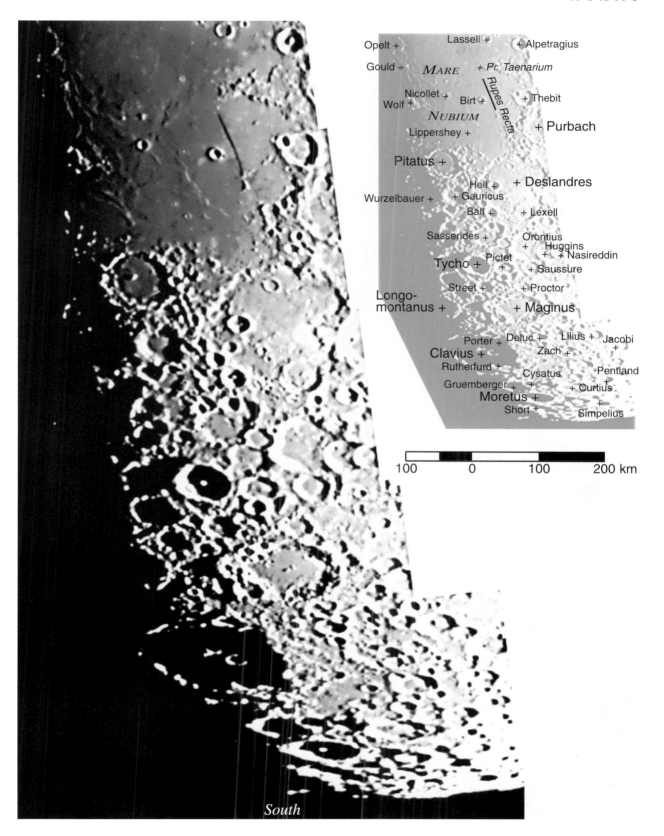

111

Colongitude 024°

1995 JUL 07, 05h36m UT. 28-cm Sch.-Cass., f/10, 0.20 s. W58 (green) Filter.
Colongitude = 024°.2, Solar latitude = 1°.4S, Librations = 6°.4W/1°.0N. 1.39 km/pixel.

As the Moon's phase advances, its *photometric function*, by which the longitude difference between the terminator and an object helps to determine its brightness, increases contrast near the poles, now making it necessary there to overexpose ("saturate") areas away from the terminator.

Hills and ridges have appeared in the western Mare Frigoris. In Mare Imbrium, the Montes Recti are now fully visible. To their south is the crater pair le Verrier-Helicon. In the southwestern Mare Imbrium are several wrinkle-ridge systems and isolated highlands-materials hills, especially near the crater Lambert, which adjoins a ghost ring to its south (Figure 1, top).

The Montes Carpatus are well illuminated now, forming the southern part of the main ring of the Imbrian Basin. South of these highlands is an excellent example of a fresh, large crater; Copernicus (Figure 1, below center). This prominent crater has a multiple central elevation, a terraced inner wall; and extensive ejecta ridges, mounds, and secondary craters. Copernicus is located in the northern portion of the vaguely defined Mare Insularum, which contains several medium-size craters such as Reinhold and Gambart, along with extensive areas of low hills and plateaus.

The flooded-crater complex that contains Fra Mauro, Parry, Bonpland, and Guericke is the next prominent feature southward (Figure 2). The Imbrian Basin ejecta in this area is the type feature for the "Fra Mauro Formation." To its southwest in Mare Nubium the flooded crater Lubiniezky falls on the terminator. More prominent, though, is its neighbor Bullialdus; a smaller version of Copernicus with terraces, a multiple central peak, and radial ejecta ridges.

The southwest outer rim of the Humorum Basin is defined by the Rupes Mercator scarp; beyond that a portion of the Palus Epidemiarum is beginning to appear.

In the highlands several craters are emerging from darkness, such as Wurzelbauer, which displays a concentric structure similar to its neighbor Pitatus. The largest feature on the terminator is Longomontanus, with an off-center peak on its flat floor. The yet-larger feature Clavius is finally fully illuminated, showing that its walls and floor are littered by craters of all sizes. In the heavily cratered area south of Clavius are several features visible for the first time in this lunation, including Scheiner and Blancanus, the crater pair Klaproth and Casatus, and the deep multiple formation Newton.

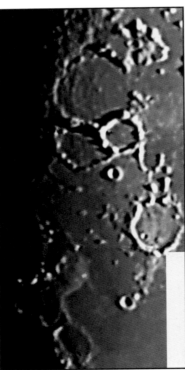

2. Fra Mauro - Opelt.
1995 OCT 03, 06h46m UT.
28-cm Sch.-Cass., f/21,
0.10 s. Col. = 019°.4,
Solar Lat. = 0°.5S,
Librations = 2°.0E/6°.2S.

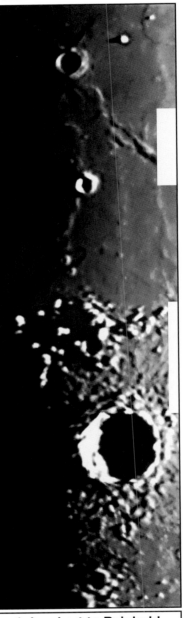

1. Lambert to Reinhold,
1995 MAY 09, 03h46m UT.
28-cm Sch.-Cass., f/21,
0.10 s. Col. = 022°.4,
Solar Lat. = 0°.3S,
Librations = 7°.3W/7°.0N.

024°–N

90°N to 19°N

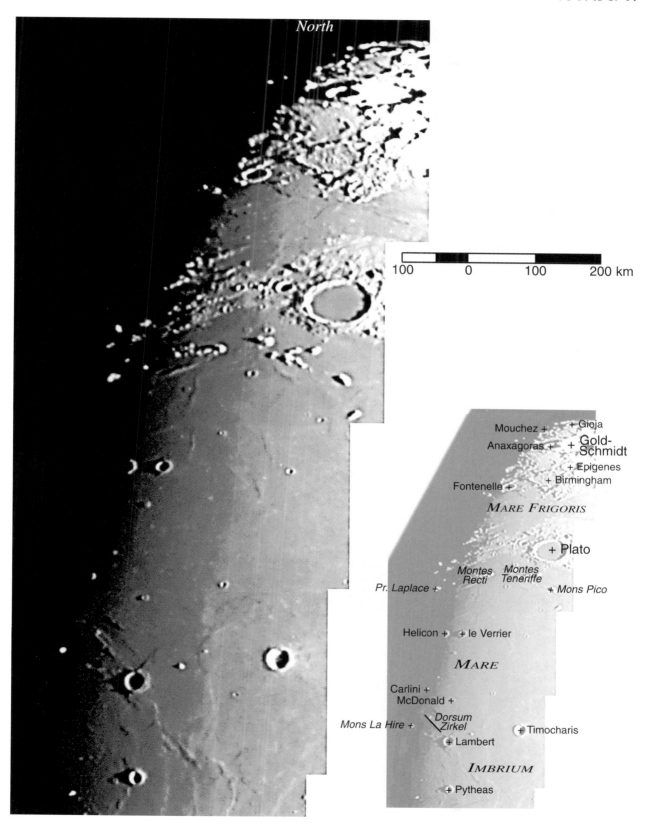

113

024°–C

21°N to 19°S

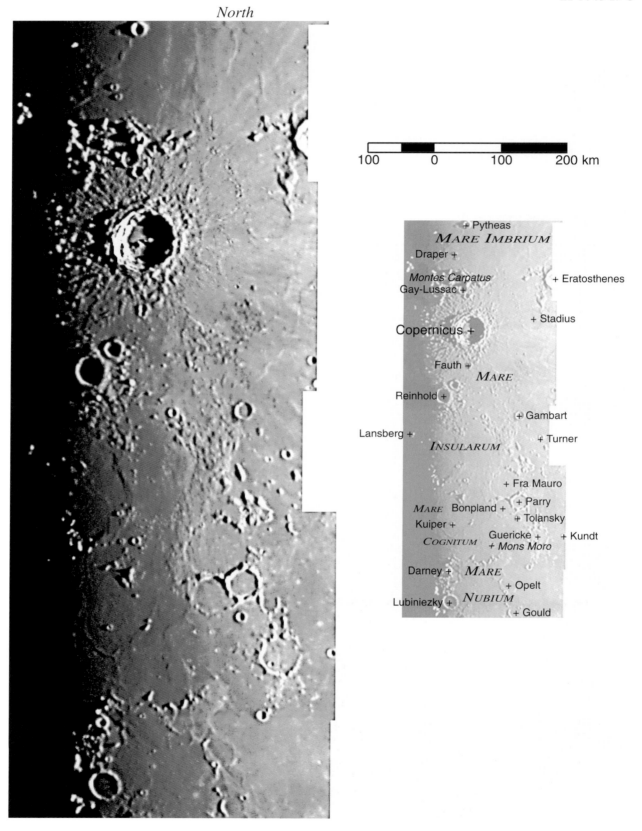

024°–S

17°S to 89°S

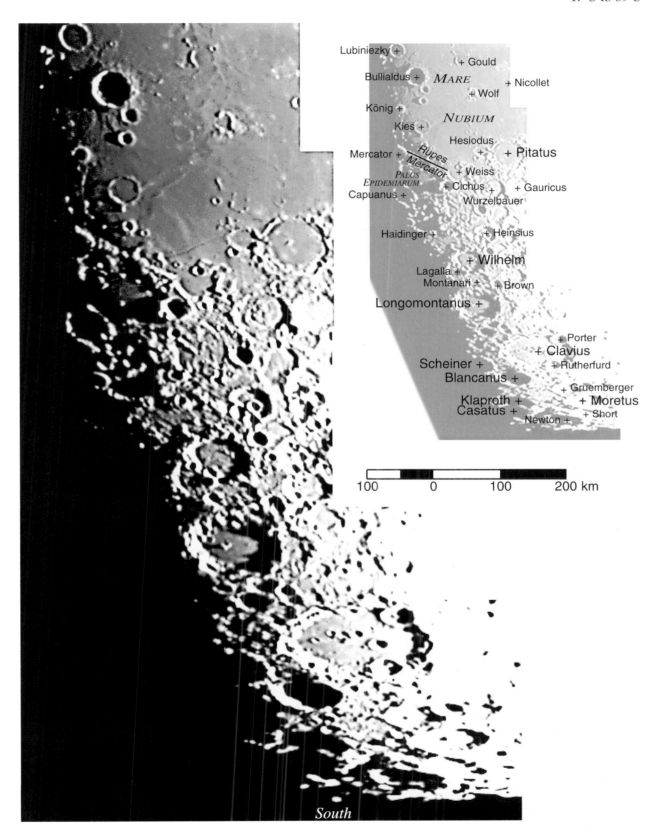

115

Colongitude 029°

1995 Apr 10, 04h00m UT. 28-cm Sch.-Cass., f/10, 0.10 s, W58 (green) Filter.
Colongitude = 028°.6, Solar latitude = 0°.5N, Librations = 6°.3W/7°.7N. 1.43 km/pixel.

Another mosaic with a highly favorable northerly libration, this view displays near-polar features like Gioja, Byrd, Peary, and Mouchez.

On the northern shore of Mare Imbrium the Montes Jura are emerging from darkness, defining the northern rim of a Clavius-size crater, the Sinus Iridum, half destroyed by the formation of Mare Imbrium. The southern rim of the Sinus is outlined by wrinkle ridges.

The area west of Copernicus deserves attention (Figure 1). A complex mixture of highlands, primary and secondary craters, and mare material, this zone contains an unusually large number of domes. Several are in the vicinity of the crater Hortensius (in the lower left of Figure 1), with two domes north of the crater and a cluster of four to its northeast. Many more domes will appear in the region farther to the northwest of Hortensius as the Sun rises higher.

Forming the boundary between Oceanus Procellarum (to the west) and Mare Cognitum (to the east), the arc of the Montes Riphaeus may represent an isolated remnant of the northwestern rim of an ancient crater or basin (Figure 2).

Bullialdus' interior can be clearly seen now, with prominent wall terraces and a central peak (Figure 3). A broad, shallow trough extends northwest of the crater, interrupted by hills and a "causeway" of lowlands material. South of Bullialdus, note the flooded ring Kies, with a prominent dome (Kies π) to its west.

Most of Palus Epidemiarum is sunlit now, including the western segment of Rima Hesiodus, extending from Hesiodus to just north of Capuanus. Just to the south, the small mare unit Lacus Timoris can be seen.

South of the Lacus, highlands extend to the southern limb, within which the craters Scheiner, Klaproth, and Casatus are emerging from the lunar night.

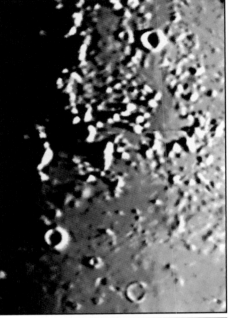

1. Hortensius Domes (upper right). 1995 Apr 10, 04h30m UT. 28-cm Sch.-Cass., f/21, 0.10 s. Col. = 028°.9, Solar Lat. = 0°.5N, Librations = 6°.5W/7°.7N.

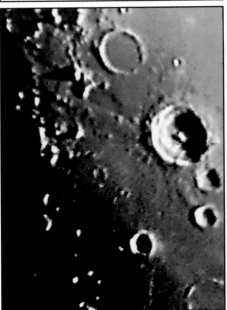

3. Bullialdus (above). 1994 Nov 13, 04h56m UT. 28-cm Sch.-Cass., f/21, 0.05 s. Col. = 028°.2, Solar Lat. = 0°.2N, Librations = 5°.8E/4°.5S.

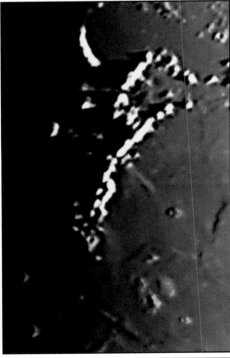

2. Montes Riphaeus (above). 1995 Apr 10, 04h36m UT. 28-cm Sch.-Cass., f/21, 0.10 s. Col. = 029°.0, Solar Lat. = 0°.5N, Librations = 6°.5W/7°.7N.

029°–N

90°N to 25°N

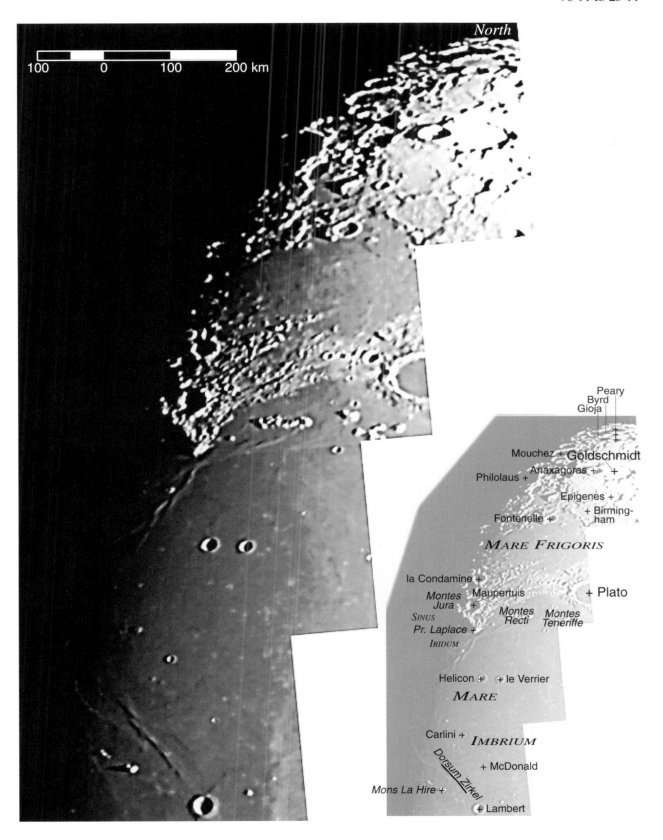

117

029°–C

27°N to 14°S

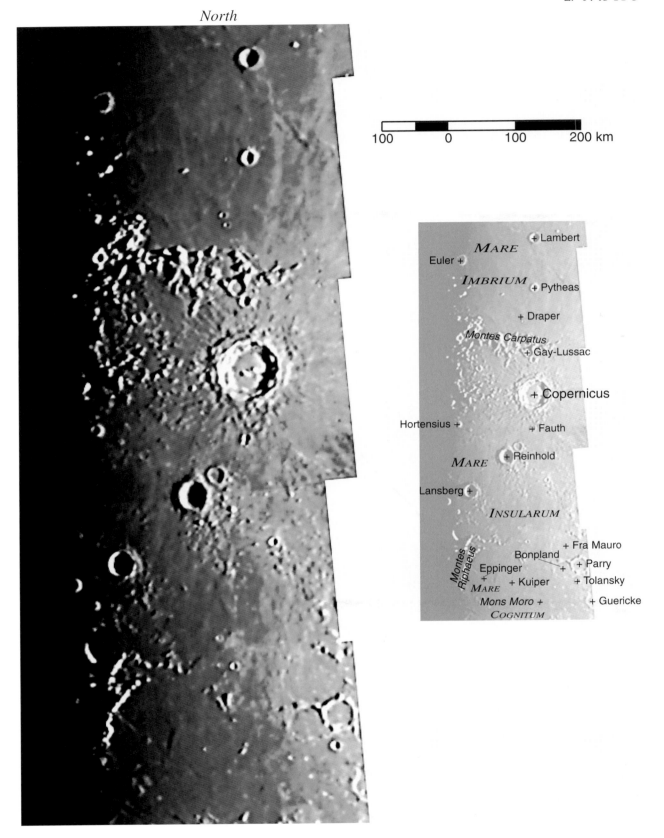

029°–S

11°S to 82°S

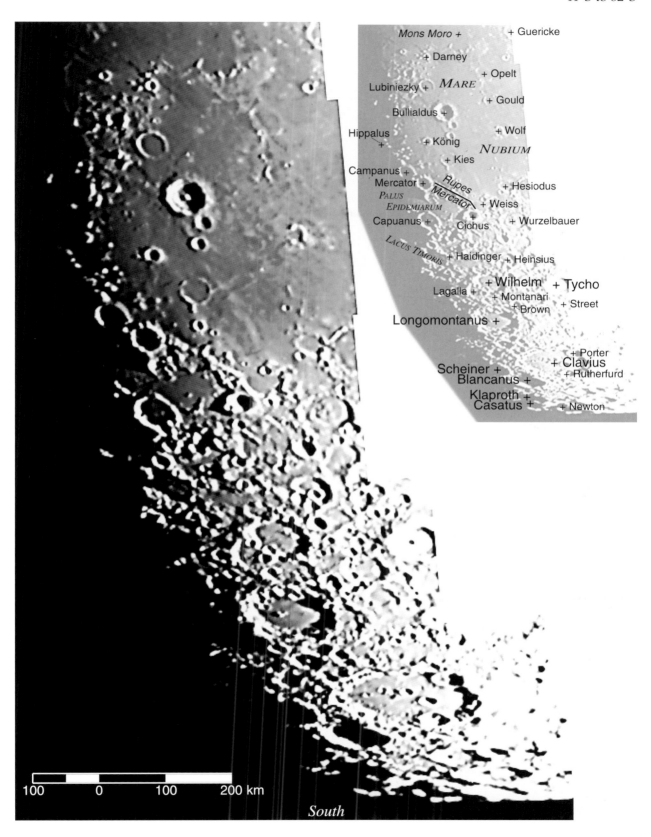

Colongitude 036°

1995 Jul 08, 04h40m UT. 28-cm Sch.-Cass., f/10, 0.10 s, W58 (green) Filter.
Colongitude = 035°.9, Solar latitude = 1°.4S, Librations = 5°.2W/0°.6S. 1.34 km/pixel.

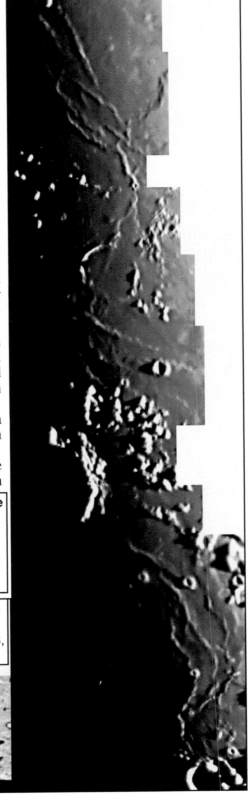

With the remainder of the Montes Jura sunlit all the way to their southern termination at Promontorium Heraclides, the semicircular form of Sinus Iridum is now evident.

Along the terminator in Mare Imbrium lies the crater pair Delisle-Diophantus with the isolated Mons Delisle. To their south is another isolated peak, Mons Vinogradov. Between Oceanus Procellarum and Mare Insularum are numerous scattered hills with dome fields in the area of Tobias Mayer, Milichius, and Hortensius; the dome fields are visible before Full Phase only in the colongitude range between the previous mosaic and this one (029°-036°).

The southeastern Oceanus Procellarum contains a complex wrinkle-ridge system west of the Montes Riphaeus (Figure 1, upper portion), along with low hills and an unnamed ghost ring northwest of Herigonius.

Scattered highlands divide Oceanus Procellarum and Mare Nubium from Mare Humorum (Figure 1, below center). Within this hilly terrain the Sun is rising on the east wall of the large crater Gassendi. South of it, the circular form of the Mare Humorum Basin is revealed by a concentric system of wrinkle ridges (Figure 1, lower portion).

Near the terminator in the Southern Highlands is an unusual triple crater; the largest component is Hainzel (south); the other two portions are Hainzel A (north) and Hainzel C (east). To this triplet's southwest, Hainzel in turn adjoins the irregular and damaged crater Mee.

South of Mee, the terminator outlines the east wall of a highlands basin, the Schiller-Zucchius Basin, best seen when the Sun is a few degrees higher than in this view.

The area near the south limb can be seen well if the libration is favorable (Figure 2). In the rugged terrain south of Casatus and Moretus are several large craters: Newton, Schomberger, Demonax, Scott, Malapert, and Amundsen; the last two just five degrees from the South Pole.

1. Oceanus Procellarum - Mare Humorum. 1996 May 28, 05h01m UT. 36-cm Sch.-Cass., f/23, 0.10 s. Col. = 036°.6, Solar Lat. = 1°.2S, Librations = 6°.4W/2°.0N.

2. South Polar Region. 1993 Nov 24, 07h24m UT. 28-cm Sch.-Cass., f/21, 0.05 s. Col. = 033°.7, Solar Lat. = 0°.0S, Librations = 1°.0E/5°.4S.

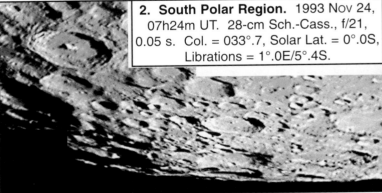

036°–N

89°N to 14°N

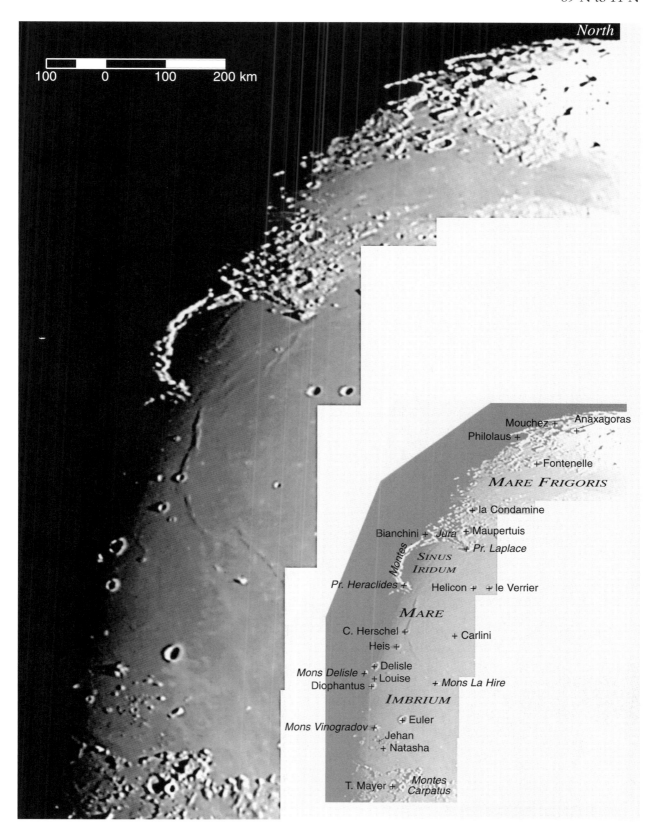

036°–C

23°N to 23°S

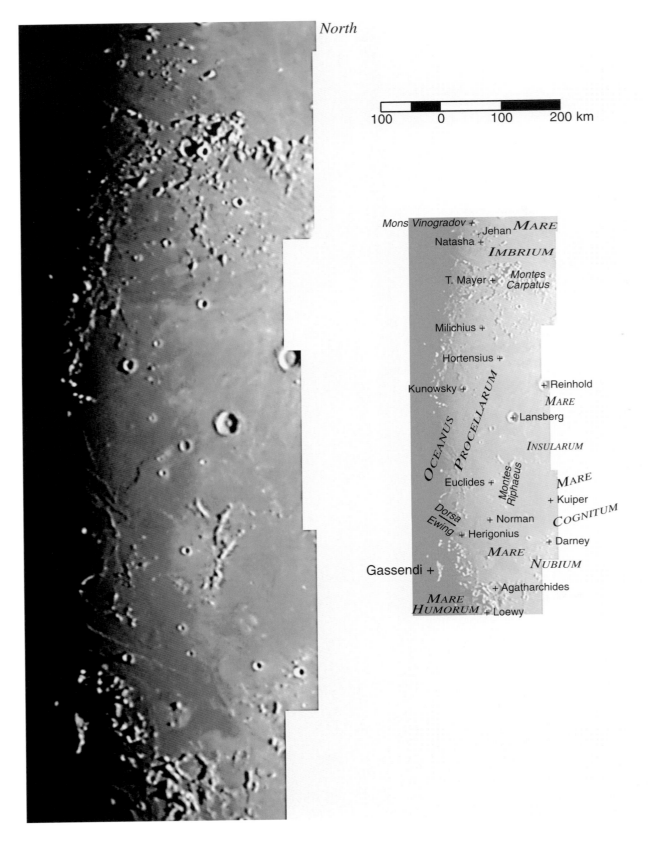

036°–S

14°S to 90°S

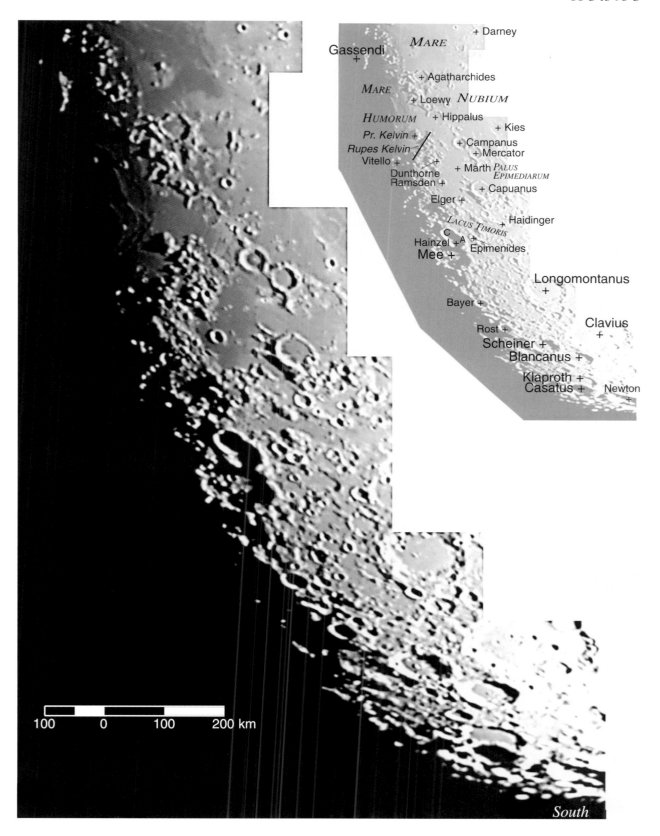

Colongitude 040°

1994 Nov 14, 03h20m UT. 28-cm Sch.-Cass., f/10, 0.50 s, 550-nm narrow-band filter.
Colongitude = 039°.6, Solar latitude = 0°.2N, Librations = 5°.6E/3°.4S. 1.45 km/pixel.

The terminator has reached the western edge of Mare Frigoris, containing the crater Harpalus. Southwest of Sinus Iridum, the rising Sun is also spotlighting two unusual highland domes, Mons Gruithuisen Gamma and Mons Gruithuisen Delta (Figure 1, upper right corner).

On the Oceanus Procellarum - Mare Imbrium boundary, the Montes Harbinger define part of the outer ring of the Imbrian Basin; their name derives from the fact that they become sunlit shortly before the brilliant crater Aristarchus.

Farther south in Oceanus Procellarum is another, more extensive, area of scattered hills containing the craters Kepler and Encke (Figure 2). The former is the center of a ray system, while the latter lies on the northeast rim of the ghost crater Encke T (Figure 2, lower left).

The entire circuit of Gassendi's rim is sunlit at this phase, and its multiple central peak is also beginning to appear. Adjoining the crater on the south is Mare Humorum, with concentric wrinkle ridges in its interior. On the south edge of the mare are two concentric-walled craters, Vitello and Doppelmayer, both near the flooded ring Puiseux.

With the Palus Epidemiarum fully sunlit, the Rimae Hippalus rille system can be seen as concentric to the Humorum Basin (Figure 3, upper portion). To its south, another rille system surrounds Ramsden in the Palus Epidemiarum (Figure 3, lower portion).

In the Southern Highlands the interiors of Hainzel and Mee are now sunlit, with the outline of the elongated crater Schiller protruding from the terminator. What appears to be a large crater just south of Schiller, west of the crater Rost, is actually formed by the inner and outer rings of the Schiller-Zucchius Basin.

The terrain south of Schiller is mountainous and foreshortened, particularly limbward of Casatus and Newton. The high mountains along the south limb, once known as the "Leibnitz Mountains," are the earthside portion of the rim of the Moon's largest impact basin, the Farside-Aitken Basin.

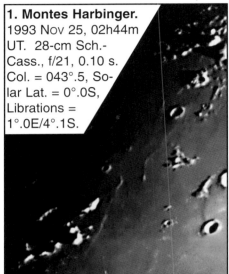

1. Montes Harbinger.
1993 Nov 25, 02h44m UT. 28-cm Sch.-Cass., f/21, 0.10 s. Col. = 043°.5, Solar Lat. = 0°.0S, Librations = 1°.0E/4°.1S.

2. Kepler and Encke.
1993 Mar 04, 03h07m UT. 28-cm Sch.-Cass., f/21, 0.15 s. Col. = 038°.1, Solar Lat. = 1°.5N, Librations = 6°.7W/3°.7N.

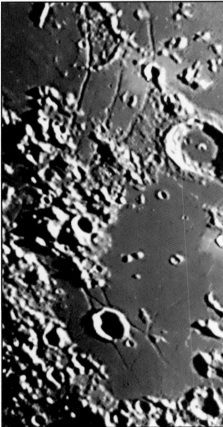

3. Rimae Hippalus and Ramsden.
1993 Mar 04, 03h18m UT. 28-cm Sch.-Cass., f/21, 0.15 s. Col. = 038°.2, Solar Lat. = 1°.5N, Librations = 6°.7W/3°.7N.

040°–N
87°N to 4°N

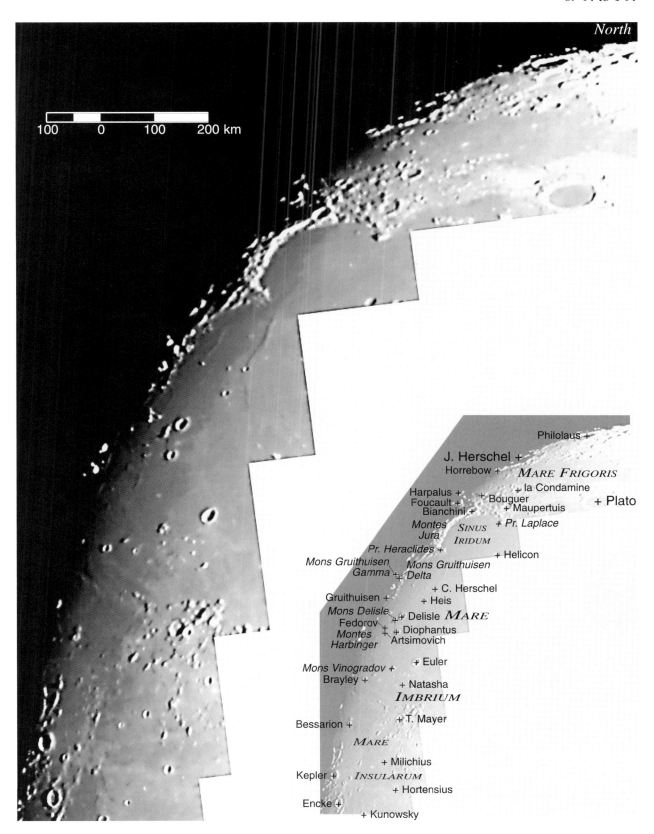

040°–C
20°N to 31°S

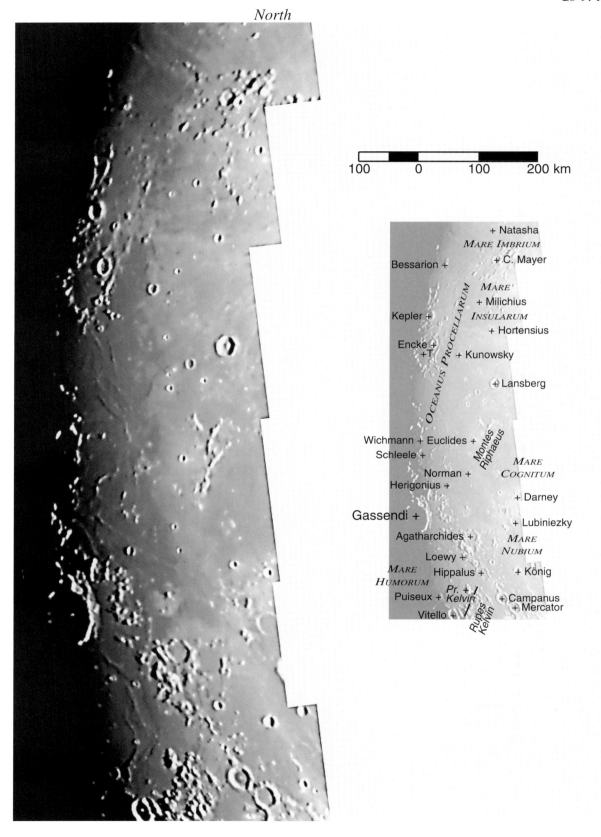

040°–S

14°S to 90°S

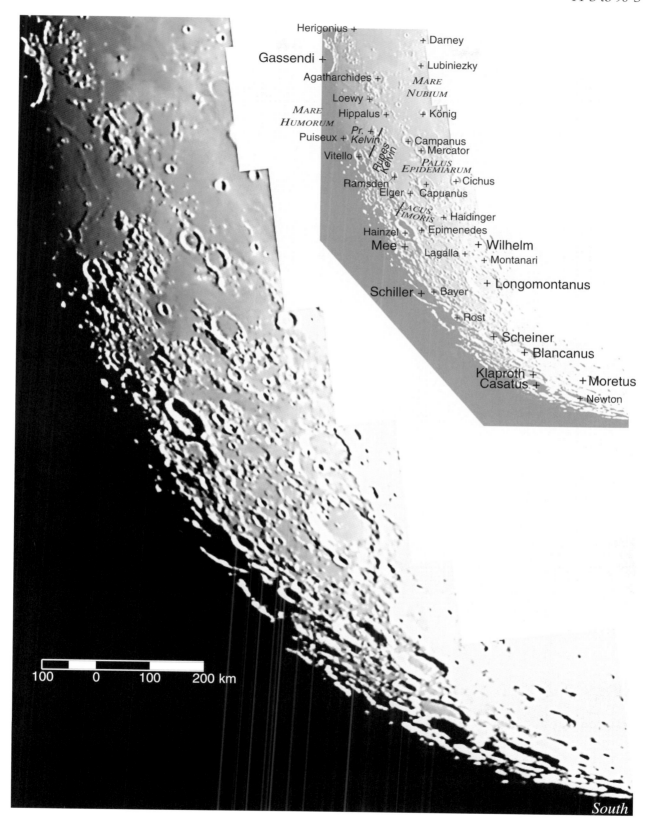

Colongitude 048°

1995 Sep 06, 03h22m UT. 28-cm Sch.-Cass., f/10, 0.08 sec, W58 (green) Filter.
Colongitude = 048°.2, Solar latitude = 1°.1S, Librations = 2°.4E/5°.7S. 1.36 km/pixel.

With the disk more than 90 percent sunlit, the same exposure time will no longer correctly image the terminator region in both low and high latitudes. The polar regions can best be seen later in the near-Full polar views.

The floor of J. Herschel is now sunlit, along with the crater Harpalus, marking the boundary between Sinus Roris and Mare Frigoris; the latter extends for over 90° of longitude. Likewise, the terminator has passed west of the Jura-Alpes-Caucasus Highlands.

In Oceanus Procellarum on the terminator is the Aristarchus Plateau, interpreted as thick deposits of volcanic material (Figure 1). Aristarchus itself is possibly the Moon's brightest crater, and is the site of numerous LTP reports. Between the plateau and the Montes Harbinger is a region of much past volcanic activity, flooding the crater Prinz and creating several sinuous rilles.

Farther south in Oceanus Procellarum is a broken-walled flooded ring, Maestlin R, just southwest of Maestlin. A similar feature, Flamsteed P, lies immediately north of Flamsteed.

On the southwest shoreline of Oceanus Procellarum, Letronne's northern wall and floor have been covered by mare lava (Figure 2). A short distance south, the rille system on Gassendi's floor is clearly visible, while Mare Humorum lava flows have breached the crater's southwest wall. Also, with Mare Humorum entirely sunlit, the concentric scarps and rilles on its western shore can be seen.

In the Southwestern Highlands, the Schiller-Zucchius Basin displays two concentric, if partial, rings. Schiller itself, consisting of two merged craters, lies between the two rings. Immediately to the south of the Schiller-Zucchius Basin are three similar craters, Zucchius, Bettinus, and Kircher.

Nearer to the pole, the floors of the deep craters Casatus and Newton can be well seen if the libration is favorable (Figure 3).

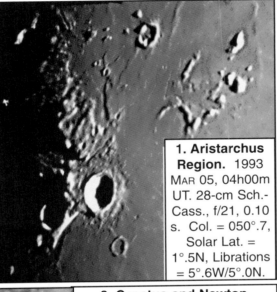

1. Aristarchus Region. 1993 Mar 05, 04h00m UT. 28-cm Sch.-Cass., f/21, 0.10 s. Col. = 050°.7, Solar Lat. = 1°.5N, Librations = 5°.6W/5°.0N.

2. Letronne and Gassendi. 1993 Jun 01, 04h42m UT. 28-cm Sch.-Cass., f/21, 0.10 s. Col. = 044°.5, Solar Lat. = 0°.0N, Librations = 2°.1E/5°.7N.

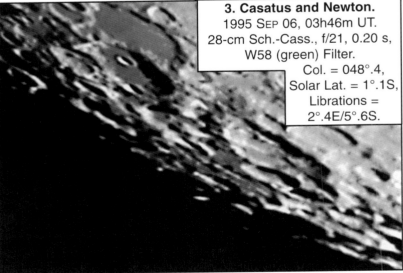

3. Casatus and Newton. 1995 Sep 06, 03h46m UT. 28-cm Sch.-Cass., f/21, 0.20 s, W58 (green) Filter. Col. = 048°.4, Solar Lat. = 1°.1S, Librations = 2°.4E/5°.6S.

048°–N

84°N to 14°N

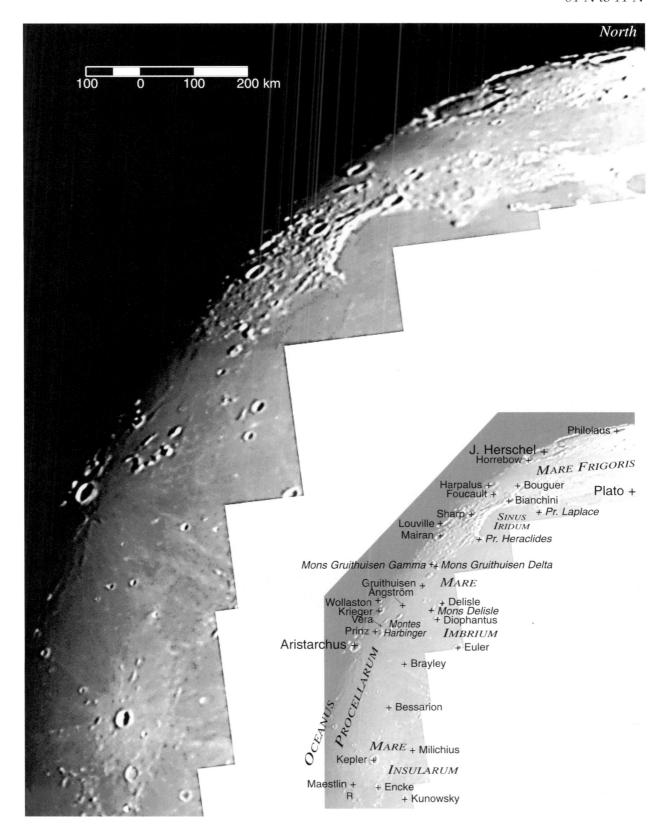

048°–C
19°N to 32°S

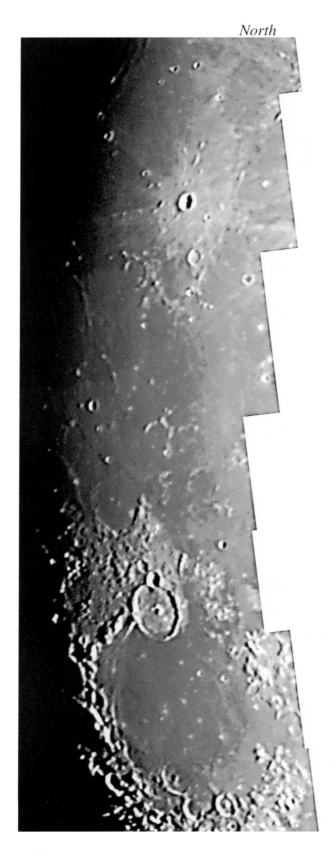

048°–S
15°S to 90°S

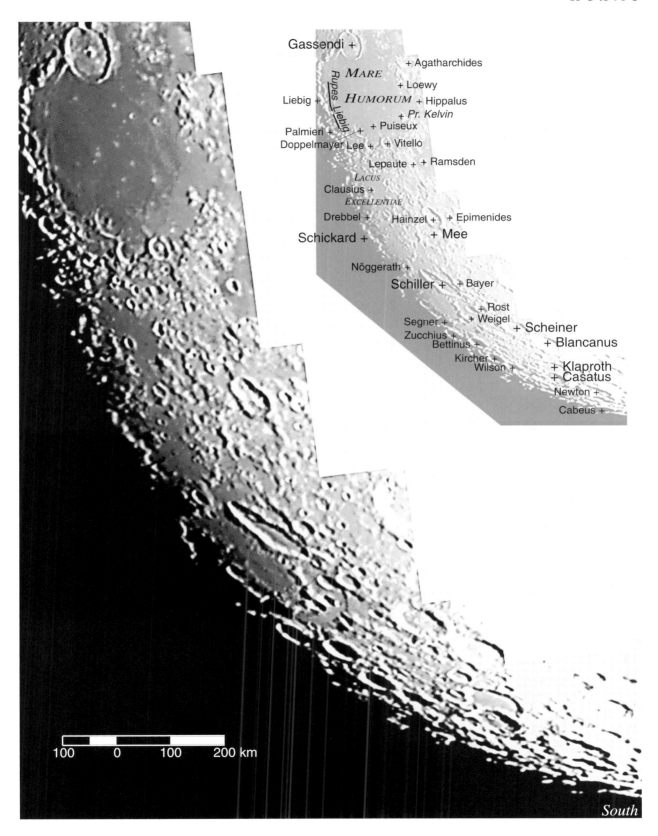

Colongitude 055°

1995 Aug 08, 05h09m UT. 28-cm Sch.-Cass., f/10, 0.08 s. W58 (green) Filter.
Colongitude = 055°.0, Solar latitude = 1°.5S, Librations = 0°.0W/5°.2S. 1.33 km/pixel.

The westward movement of the terminator has brought Sinus Roris into view; and to its south more of the western portion of Oceanus Procellarum, which contains two highly interesting volcanic regions.

The first such is the "Aristarchus Plateau," now visible in its entirety, bounded by the Montes Agricola on its northwest, and containing the Vallis Schröteri, the Moon's most prominent sinuous rille.

The second major volcanic region is about 350 km south, the "Marius Hills" west of the crater Marius and north of Reiner (Figure 1). This convoluted area contains several groups of elevations, including ridges, domes, and conical hills, along with the Rimae Marius and Galilaei.

In the highlands west of Mare Humorum the crater Mersenius is unusual, a crater with a convex floor (Figure 2). Mersenius P adjoins Mersenius on the northeast and has a dome-like central elevation. Perhaps both these features are related to the activity that created the scarps and rilles on the nearby western margin of Mare Humorum.

The Western and Southwestern Highlands are overlain by ejecta of the Orientale Basin, which will not itself be visible until near Full Phase (and even then requiring a westerly libration). The area visible now contains the crater Schickard, where the Sun is now rising on its floor (Figure 3).

Several large craters lie near the terminator south of Schickard, including the merged rings Nasmyth and Phocylides. To the west of the three similar craters Zucchius, Bettinus, and Kircher, a terminator protrusion is created by the east wall of the Earthside's largest recognized crater, Bailly; really more like a small impact basin than a conventional crater.

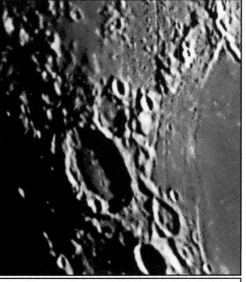

2. Mersenius. 1993 Aug 29, 04h44m UT. 28-cm Sch.-Cass., f/21, 0.05 s. Col. = 052°.0, Solar Lat. = 1°.6S, Librations = 5°.4E/4°.5S.

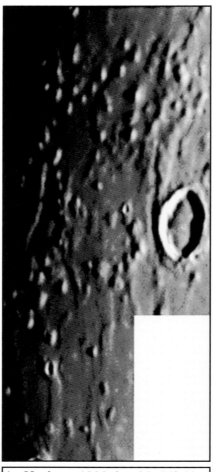

1. Marius. 1993 Jun 02, 05h18m UT. 28-cm Sch.-Cass., f/29, 0.25 s. Col. = 057°.0, Solar Lat. = 0°.0N, Librations = 3°.2E/4°.3N.

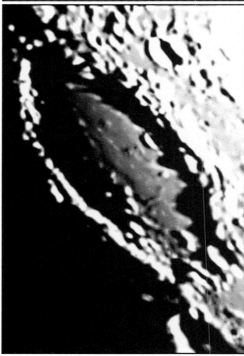

3. Schickard. 1995 Aug 08, 05h32m UT. 28-cm Sch.-Cass., f/21, 0.08 s. Col. = 055°.2, Solar Lat. = 1°.5S, Librations = 0°.1W/5°.2S.

055°–N
85°N to 15°N

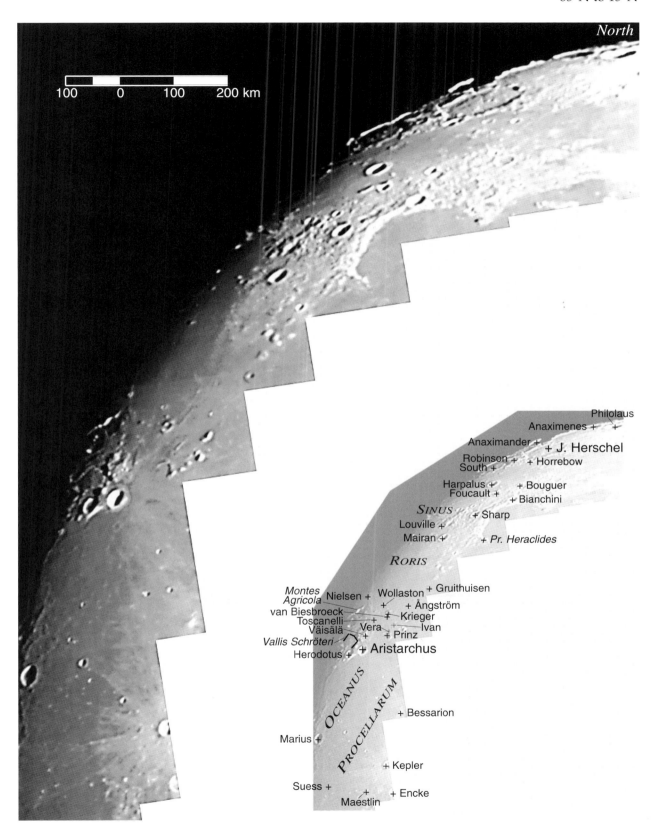

055°–C

20°N to 32°S

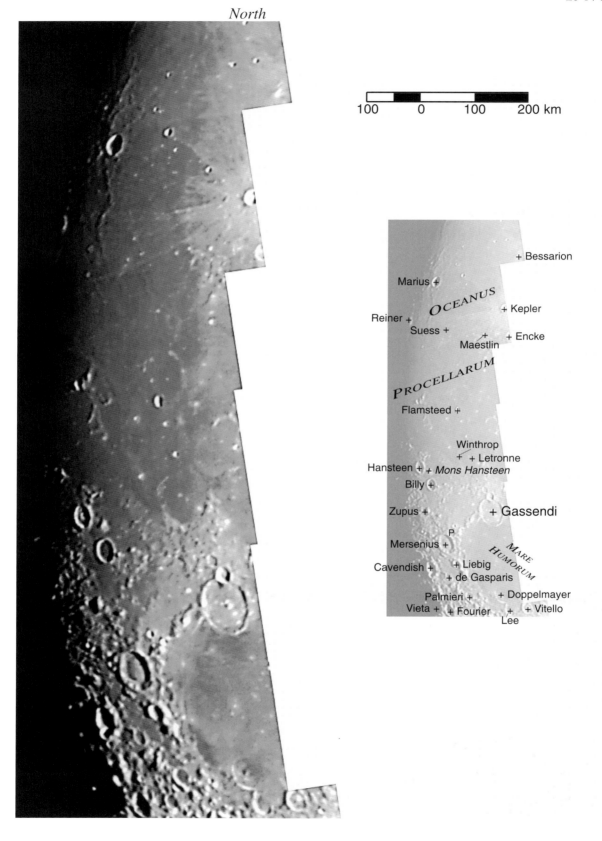

134

055°–S

26°S to 90°S

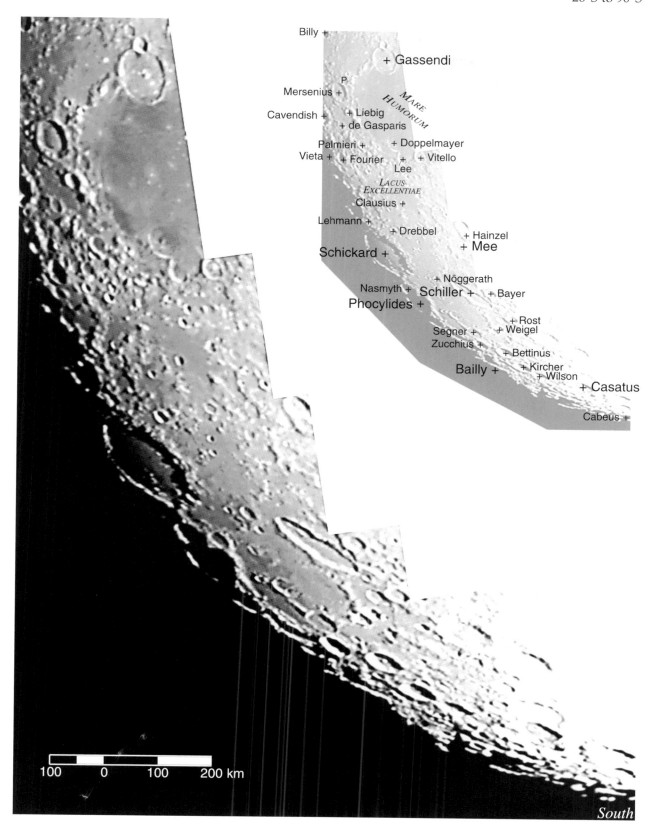

Colongitude 061°

1995 Jul 10, 06h36m UT. 28-cm Sch.-Cass., f/10, 0.07s, W58 (green) Filter.
Colongitude = 061°.3, Solar latitude = 1°.5S, Librations = 2°.2W/3°.6S. 1.32 km/pixel.

The age of the Moon is about 12 days; close enough to Full that detail along the entire length of the terminator is highly foreshortened. Because of this perspective effect, the sun angle appears to increase rapidly as one looks away from the terminator, making the zone of useful shadow detail appear to be narrow compared with earlier phases.

Two more flooded northern craters have appeared; Anaximander, lying northwest of J. Herschel, and Babbage, just west of the ring South. South of Babbage lies Sinus Roris, where the east wall of the crater Markov is catching the sunlight.

Near the vague Sinus Roris-Oceanus Procellarum boundary is a large dome complex called Mons Rümker (Figure 1), visible only under a low Sun. Mons Rümker is just one of several unusual features in the Oceanus Procellarum. South of Mons Rümker, Oceanus Procellarum has two noticeable features near the terminator: the crater Schiaparelli west of Herodotus and the "Swirl Feature" Reiner Gamma west of Reiner. Unlike almost all other lunar albedo (tonal) features, Reiner Gamma stands out even when near the terminator.

A little farther south is the southwestern "shoreline" of Oceanus Procellarum. The Western and Southwestern Highlands are carpeted and scarred with ejected material originating in the Orientale Basin, creating a chaotic jumbled landscape (Figure 2). There are a few clear-cut fresh craters like Vieta, but the majority are low, damaged rings with flat floors and sometimes even patches of mare material. Most of these irregular features have no names.

Among the more prominent features in the Southwestern Highlands is Wargentin, the Moon's best example of a crater whose floor is flooded to the rim (upper left in Figure 3). Nearby, Schickard and Phocylides are the two largest craters along the terminator. However, in the Southern Highlands, the giant crater/small basin Bailly is becoming more prominent.

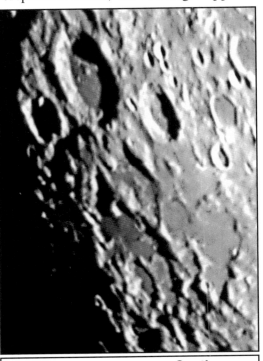

2. Vieta and Area to South.
1995 Sep 07, 04h46m UT. 28-cm Sch.-Cass., f/21, 0.30 s, W58 (green) Filter. Col. = 061°.1, Solar Lat. = 1°.1S, Librations = 3°.5E/5°.2S.

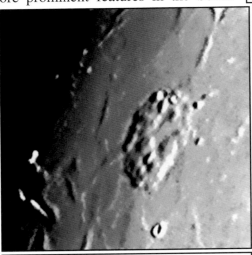

1. Mons Rümker. 1993 Mar 06, 03h43m UT. 28-cm Sch.-Cass., f/21, 0.20 s. Col. = 062°.7, Solar Lat. = 1°.5N, Librations = 3°.8W/5°.9N.

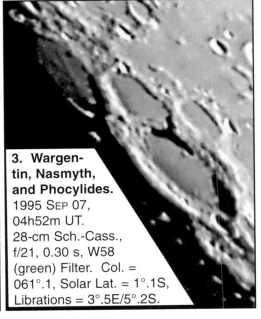

3. Wargentin, Nasmyth, and Phocylides. 1995 Sep 07, 04h52m UT. 28-cm Sch.-Cass., f/21, 0.30 s, W58 (green) Filter. Col. = 061°.1, Solar Lat. = 1°.1S, Librations = 3°.5E/5°.2S.

061°–N
86°N to 15°N

061°–C

20°N to 31°S

North

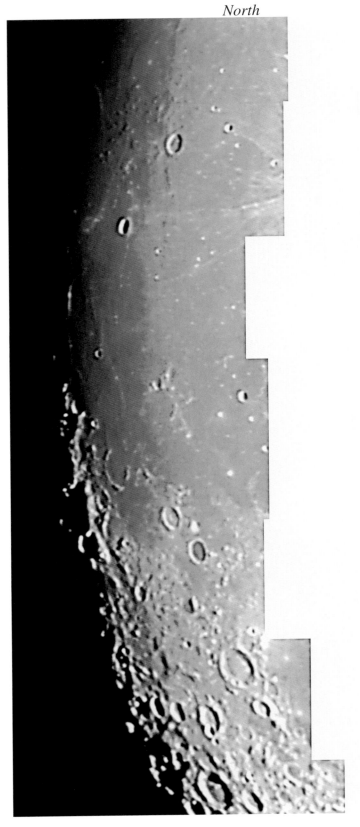

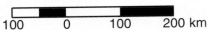

061°–S

28°S to 90°S

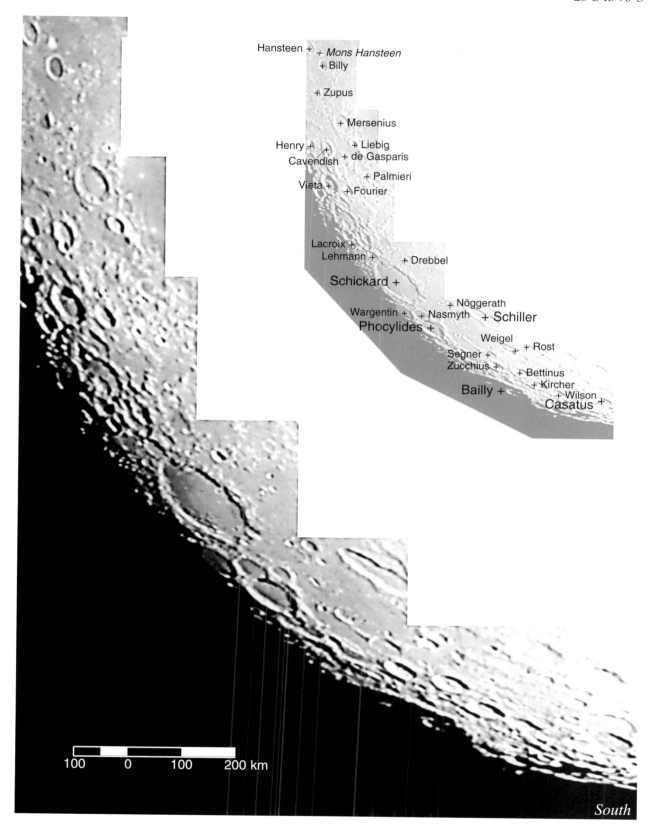

Colongitude 067°

1995 Aug 09, 05h16m UT. 28-cm Sch.-Cass., f/10, 0.20 s, W58 (green) and LPR Filters.
Colongitude = 067°.2, Solar latitude = 1°.5S, Librations = 1°.8E/5°.5S. 1.33 km/pixel.

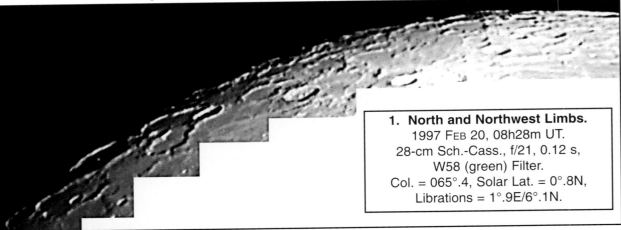

1. North and Northwest Limbs.
1997 Feb 20, 08h28m UT.
28-cm Sch.-Cass., f/21, 0.12 s,
W58 (green) Filter.
Col. = 065°.4, Solar Lat. = 0°.8N,
Librations = 1°.9E/6°.1N.

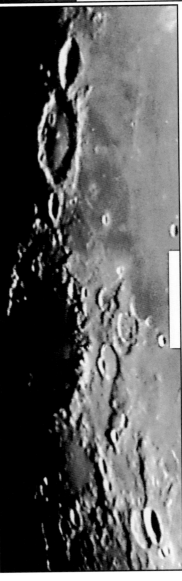

2. Hevelius - Grimaldi Region.
1994 Dec 16, 04h30m UT.
28-cm Sch.-Cass., f/21, 0.20 s,
W58 (green) Filter.
Col. = 069°.3, Solar Lat. = 1°.0N,
Librations = 0°.4E/2°.9N.

With the Moon over 95-percent illuminated, we can no longer count on seeing the terminator through a full 180° of latitude. What portion of it we do see depends on the libration, which is southerly for the Colongitude-067° mosaic. What we gain in the Southern Hemisphere, we lose in the Northern, but Figure 1 shows the view when the lunar North Pole is turned our way. In the Northern Highlands the prominent crater Pythagoras is on the terminator; the flooded rings Anaximander, Anaximander B, and Anaximander D are to its north and northeast and Babbage to its southeast.

Oceanus Procellarum has few striking features at this longitude except for the craters Lichtenberg and Seleucus on the terminator. When we reach the Western Highlands, though, there is a north-south chain of prominent craters (Figure 2). From north to south they comprise the fresh crater Cavalerius, the flat-floored crater Hevelius, the relatively small Lohrmann, and the large flooded crater Grimaldi. Grimaldi itself is the center of an impact basin; on its eastern rim is the concentric crater Damoiseau/Damoiseau M. Then, to Grimaldi's southeast is the crater pair Sirsalis-Sirsalis A, with the Rimae Sirsalis to its east and extending to its south.

The Western and Southwestern Highlands are furrowed with valleys radiating from the Oriental Basin. Among the more discernable, but usually highly eroded, features now near the terminator in this region, in north-to-south order, are: Crüger, Byrgius, Piazzi, and Inghirami; the latter just southwest of Schickard.

The rising Sun has by now illuminated the eastern portion of Bailly's floor. Bailly is a giant crater, but also a small two-ring basin, with the crater walls defining the outer ring and a circle of hills on its floor defining the inner. Poleward of Bailly in the Southern Highlands is the region of high peaks once called the "Leibnitz Mountains" (see also under Colongitude 040°). This area also contains several unusually deep craters, such as Cabeus, Newton, and Malapert.

140

067°–N

84°N to 6°N

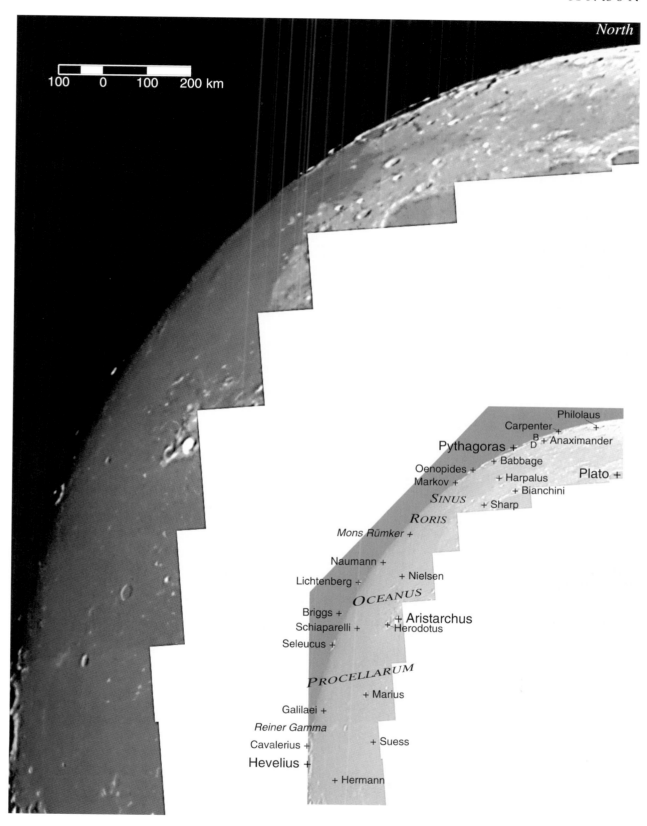

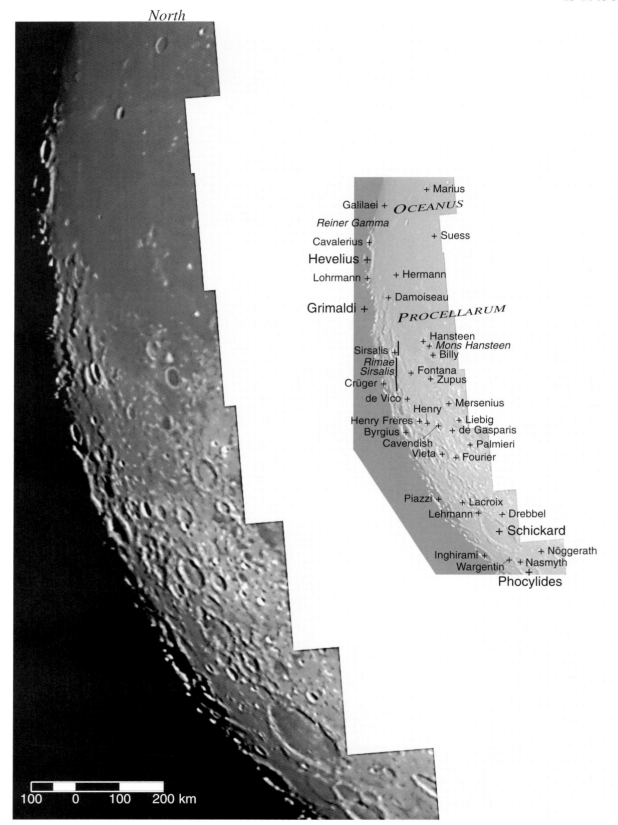

067°–S

46°S to 90°S

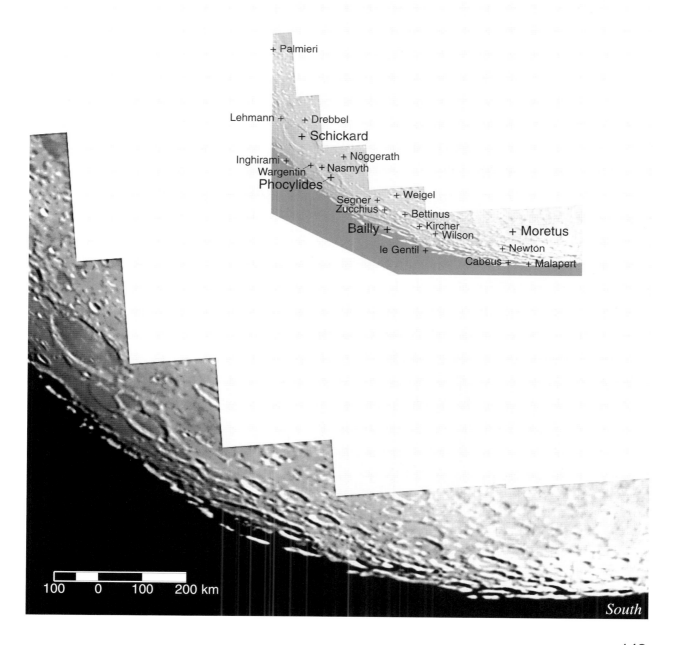

143

Colongitude 075°

1996 MAY 31, 07h28m UT. 28-cm Sch.-Cass., f/10, 0.25 s, W58 (green) and LPR Filters.
Colongitude 074°.7, Solar latitude = 1°.2S, Librations = 5°.6W/2°.6S. 1.46 km/pixel.

At this colongitude, the crater Pythagoras is the most prominent feature in the Northern Highlands, its deep interior containing a massive central peak. Just south, in or near Sinus Roris, the craters Oenopides, Markov, and Repsold stand out, the last currently upon the terminator.

Northwestern Oceanus Procellarum contains inconspicuous features, such as Dechen and Harding, but the small crater Lichtenberg is of interest because the mare to its east and northeast was flooded by lava during the Moon's Copernican Period, evidence of what perhaps was the most recent volcanism in lunar history. On the terminator south of Lichtenberg is a set of three large flooded rings with broken walls (Figure 1): Russell (northernmost, east wall sunlit), Eddington (southeast of Russell), and Struve (behind the terminator west of Eddington; merged with Russell to the north). East of this group are the smaller, fresher craters Briggs and Seleucus. South of Eddington is the crater pair Krafft and Cardanus, with Catena Krafft connecting them.

In the Western Highlands the floor of Grimaldi is now sunlit, while "behind" (west of) Grimaldi another large ring has appeared; Riccioli, containing a central peak. In the furrowed terrain farther south is a tiny mare path between Rocca and Crüger, Lacus Aestatis; a useful landmark under a high Sun. Slightly farther south, the depressions Darwin, Lamarck, and Lagrange are good examples of the ruined condition of pre-Imbrian craters in this region. Had there been anyone to see them, these would have been prominent features before the Orientale impact.

The Southwestern Highlands contain three named valleys radial to the Orientale Basin. The easternmost of them is Vallis Inghirami, extending northeast from Inghirami. In addition, on the terminator west of Inghirami is the rim of Vallis Baade. Farther south is a large ruined ring that has no name or even letter designation. It is informally called the "Pingré Basin" because it contains the crater Pingré on its southeast floor; creating confusion, some older maps call the entire basin "Pingré." Slightly larger, but in only slightly better condition than the Pingré Basin, Bailly is now fully sunlit. (Figure 2 shows the Pingré Basin and Bailly under more favorable libration and slightly higher lighting than the 075° mosaic itself.)

South of Bailly, the Southern Highlands now reveal the irregular depression le Gentil on the terminator. Between it and Cabeus are several very high peaks beyond the terminator, lying on the rims of the crater Drygalski and of the South Pole-Aitken Basin.

2. Pingré and Bailly. 1997 JUN 19, 06h52m UT. 28-cm Sch.-Cass., f/10, 0.07 s, W58 (green) & LPR Filters. Col. = 076°.2, Solar Lat. = 1°.5S, Librations = 5°.5W/5°.3S.

1. Eddington (left). 1992 DEC 08, 07h36m UT. 28-cm Sch.-Cass., f/21, 0.05 s. Col. = 074°.4, Solar Lat. = 0°.2S, Librations = 6°.0W/2°.5S.

075°–N

87°N to 15°N

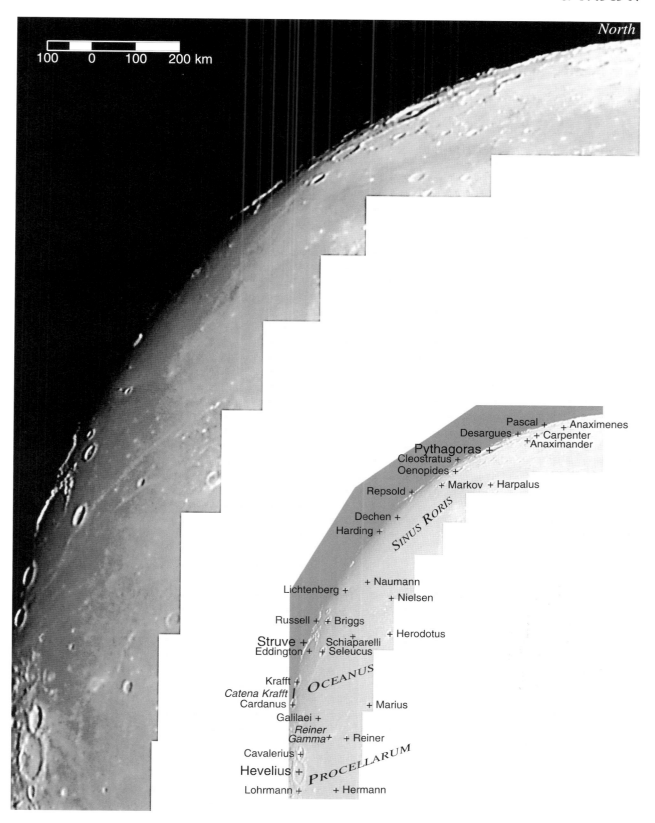

145

075°–C
24°N to 41°S

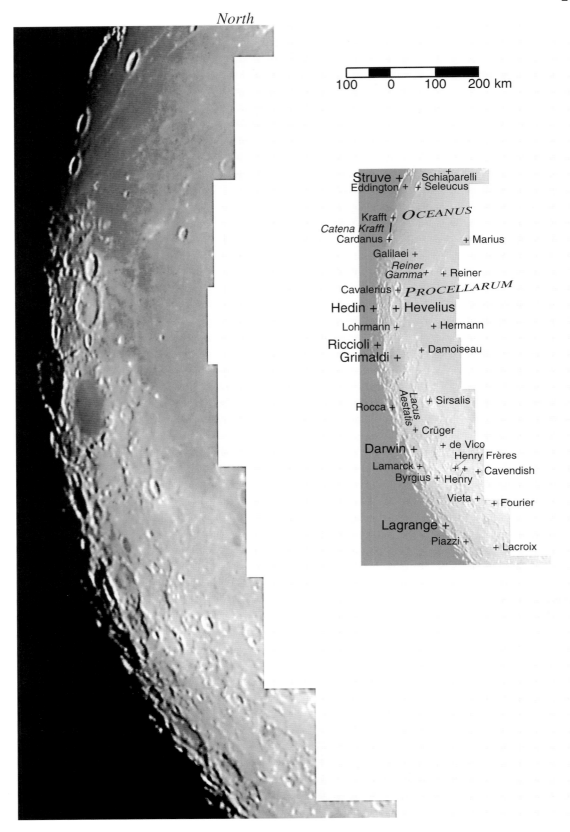

075°–S

37°S to 90°S

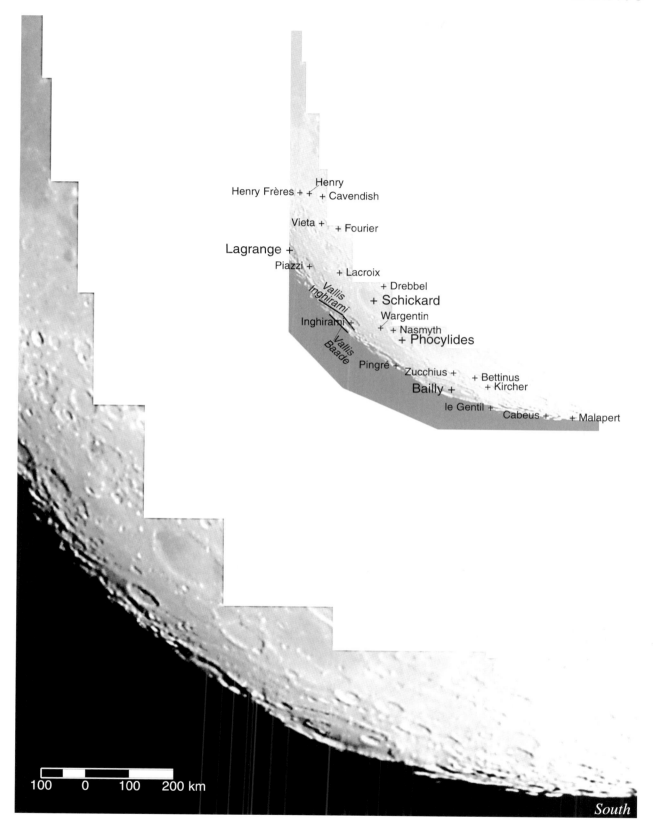

Colongitude 080°

1996 MAY 02, 06h28m UT. 28-cm Sch.-Cass., f/10, 0.15 s, W58 (green) and LPR Filters.
Colongitude = 080°.0, Solar latitude = 0°.6S, Librations = 5°.2W/0°.2S. 1.39 km/pixel.

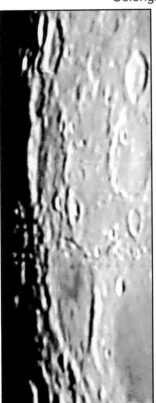

Now within a day of Full phase, the terminator has reached the *marginal zone*, the area near the limb whose visibility depends on libration.

Northwest of Pythagoras, the craters Desargues and Brianchon are on the terminator. To Pythagoras' southwest is Cleostratus, with Xenophanes' west wall southwest of it, seen as a bulge on the terminator.

The line of sunrise has reached the western shores of Sinus Roris and Oceanus Procellarum, marked by the rims of the craters Repsold, Gerard, Lavoisier, and Ulug Beigh in the Northwestern Highlands.

Within Oceanus Procellarum, the Sun has now illuminated the flat floors of Russell and Struve, showing that the two craters have merged.

A group of prominent craters is within the Western Highlands (Figure 1). Among them, Hedin, Hevelius, Olbers, Riccioli and Grimaldi are examples of ancient ruined rings. Hedin, Hevelius, and Riccioli are noteworthy for the rima systems on their floors, while Hedin contains a central peak. Besides these pre-Imbrian rings, Lohrmann and, particularly, Cavalerius are examples of more recent, sharply defined craters.

Due to the Western and Southwestern Highlands being overlain by Orientale ejecta, there are few landmark features near the terminator south of Grimaldi. A portion of the outer ring of the Oriental Basin, the Montes Cordillera, begins to the west of Rocca and extends 20 degrees of latitude (about 600 km) to the south, as far as the crater Krasnov

1. Hedin - Riccioli. 1996 JUN 30, 06h30m, 28-cm Sch.-Cass., f/21, 0.15 s, W58 (green) Filter. Col. = 080°.8, Solar Lat. = 1°.5S, Librations = 3°.3W/5°.2S.

2. Piazzi - Inghirami. (Data as for Figure 1.)

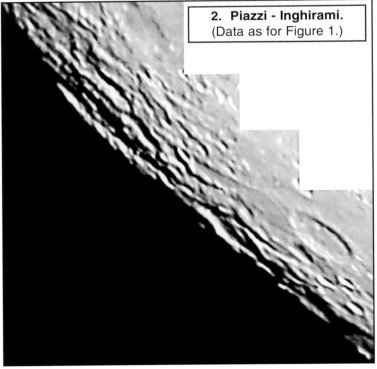

To the west and southwest of the craters Lagrange, Piazzi, and Inghirami is a particularly chaotic area of lunar terrain, where material ejected from the Orientale impact has created a pattern of alternating valleys and ridges, oriented roughly north-south (Figure 2). Three of the valleys have been named; Valles Inghirami and Baade were described in the colongitude-075° mosaic, and now Vallis Bouvard can be seen west of Piazzi.

The basin containing the crater Pingré can be seen clearly at this phase, and is comparable in size to its giant neighbor to the south, Bailly. Between these two depressions is the east rim of the crater Hausen.

The south wall of Bailly is at latitude 70° south, and in these high latitudes the terminator advances very slowly. The rising Sun now reveals more of the wall of le Gentil, showing the irregular outline of this formation. South of le Gentil, at about 80° south, the terminator is defined by the east wall of the crater Drygalski, visible only near Full Phase and then only with southwesterly libration.

080°–N

90°N to 3°N

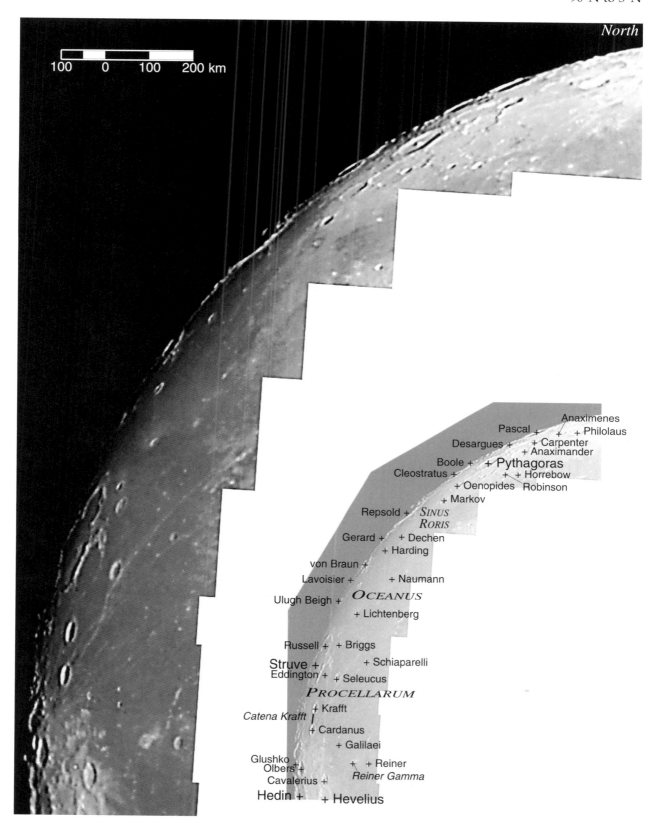

080°–C
24°N to 43°S

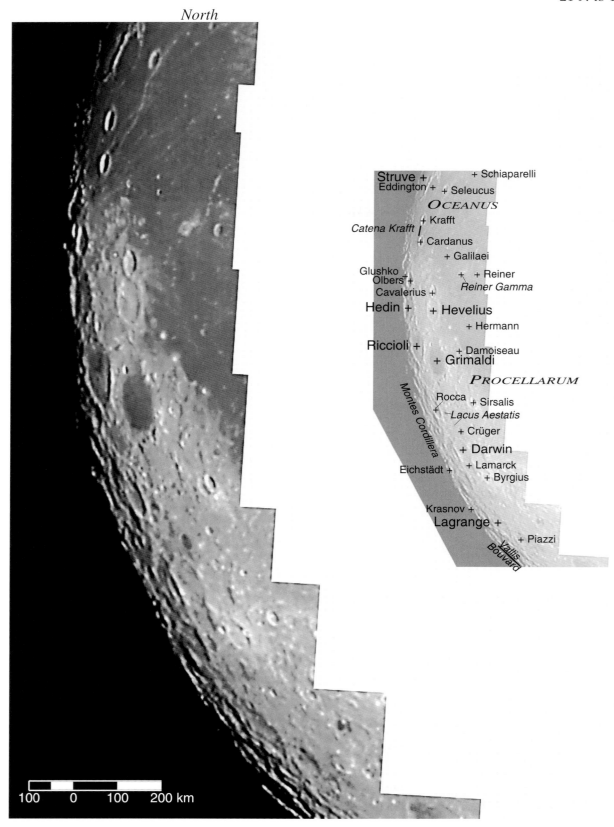

080°–S
34°S to 90°S

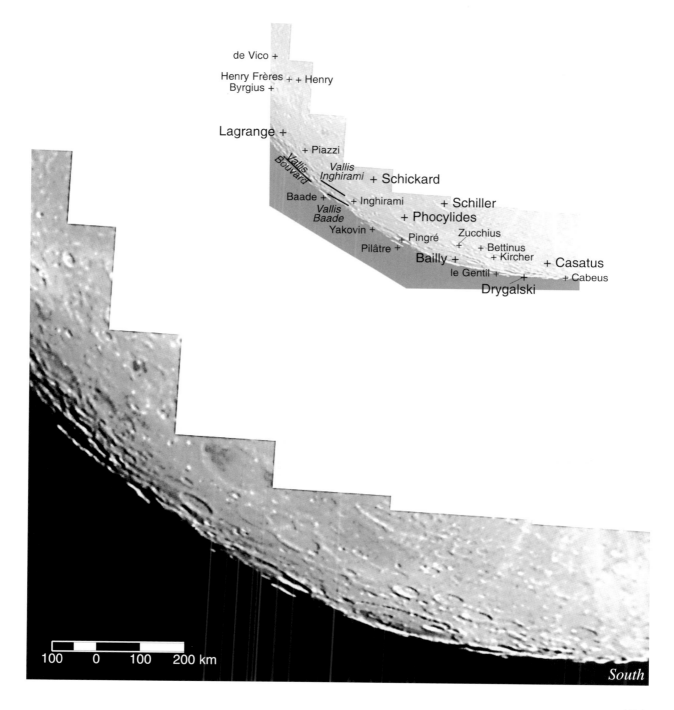

Near Full
Quadrants 1-4

Near Full Phase, the zone of shadow detail is a narrow strip near the limb, whose visibility depends on libration. Thus, the mosaics on the following nine pages are arranged differently than the other mosaics in this atlas. Here, they start at the Equator on the east limb and then proceed counterclockwise around the limb through lunar Quadrants 1, 2, 3, and 4 in order. The limb zone experiences extreme foreshortening, while the terminator zone shows extreme contrast; chiefly blacks and whites with few grey shades. Also, due to the six-year lighting/libration visibility cycle for limb areas, the northwest limb has not been favorably presented during the five years the author has used a CCD camera! Thus that area is shown with a mosaic of film photographs.

Very near Full Phase, the terminator disappears, with the Moon's appearance as shown in the figure below; lacking shadow detail, but with a wealth of tonal ("albedo") features illustrating the varying age, history, and mineral composition of the lunar surface.

1994 Nov 18, 03h44m UT. 9-cm Maksutov, f/11, 550-nm narrow-band filter, 0.07 s. Col. = 088°.3, Solar Lat. = +0°.3, Librations = 0°.8E/1°.8N.

Taken 42 minutes before first penumbral contact of penumbral lunar eclipse; phase angle = 1°.7.

Quadrant 1–E (8°S to 33°N)

1996 Dec 25, 03h35m UT. 28-cm Sch.-Cass., f/21, 0.10 s, W58 (green) Filter.
Colongitude = 089°.7, Solar latitude = 1°.6N, Librations = 4°.4E/7°.0N. 0.68 km/pixel.

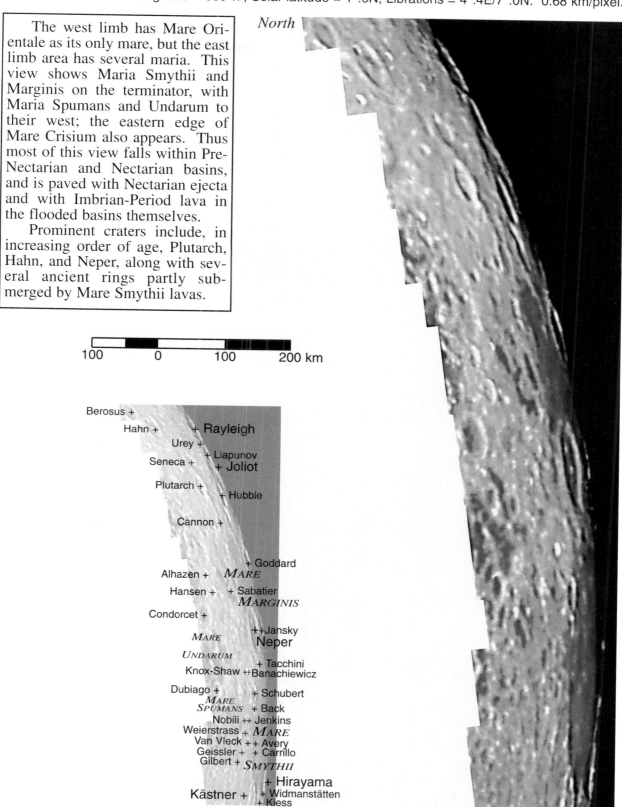

The west limb has Mare Orientale as its only mare, but the east limb area has several maria. This view shows Maria Smythii and Marginis on the terminator, with Maria Spumans and Undarum to their west; the eastern edge of Mare Crisium also appears. Thus most of this view falls within Pre-Nectarian and Nectarian basins, and is paved with Nectarian ejecta and with Imbrian-Period lava in the flooded basins themselves.

Prominent craters include, in increasing order of age, Plutarch, Hahn, and Neper, along with several ancient rings partly submerged by Mare Smythii lavas.

Quadrant 1–NE (30°N to 69°N)

(Data as for Quadrant 1-E)

Although largely highlands, this mosaic includes Mare Humboldtianum, Nectarian in age, within a double-ringed basin.

Much of this area is covered with Mare Humboldtianum ejecta, and contains several large craters, including Pre-Nectarian Riemann and Nectarian-Period Gauss and Bel'kovich, the last on the northeast rim of Mare Humboldtianum. One prominent fresh crater in this zone is Hayn, superimposed on the northwest rim of Bel'kovich.

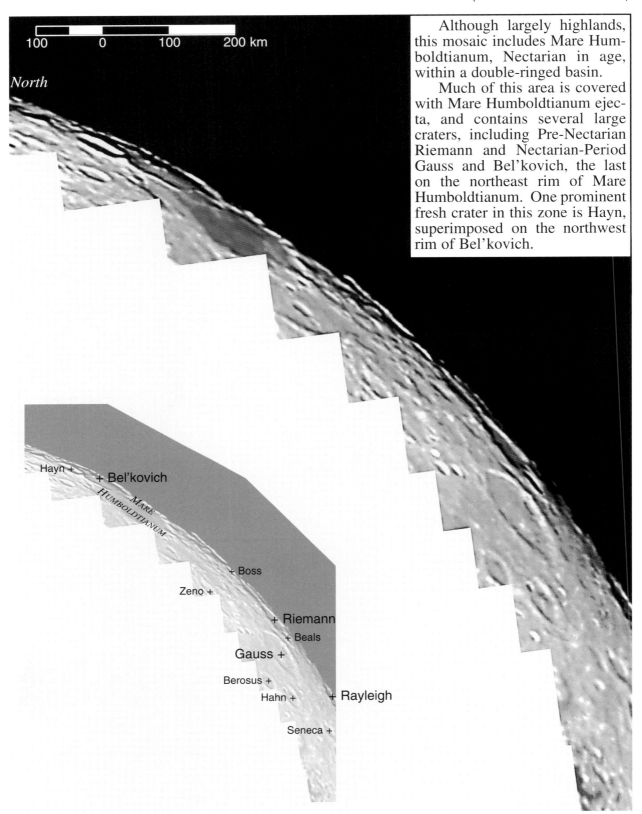

Quadrant 1–N (68°N [E. of pole] to 80°N [W. of pole])

(Data as for Quadrant 1-E; Lunar north at top.)

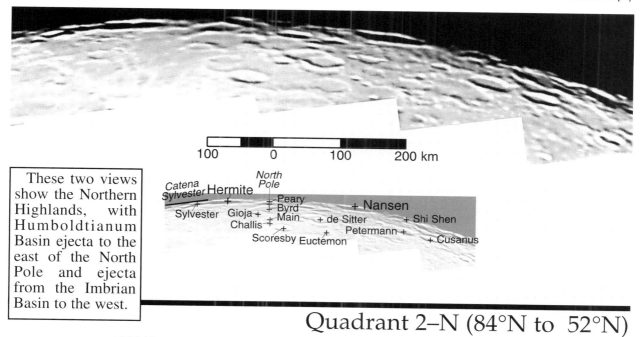

These two views show the Northern Highlands, with Humboldtianum Basin ejecta to the east of the North Pole and ejecta from the Imbrian Basin to the west.

Quadrant 2–N (84°N to 52°N)

1982 Nov 01, 06h45m UT. 25-cm Cass., f/73, 1/8 s, 3M 640-T Film. Lunar north at top.
Colongitude = 090°.9, Solar latitude = 1°.3N, Librations = 4°.2W/6°.8N. Scanned at 0.87 km/pixel.

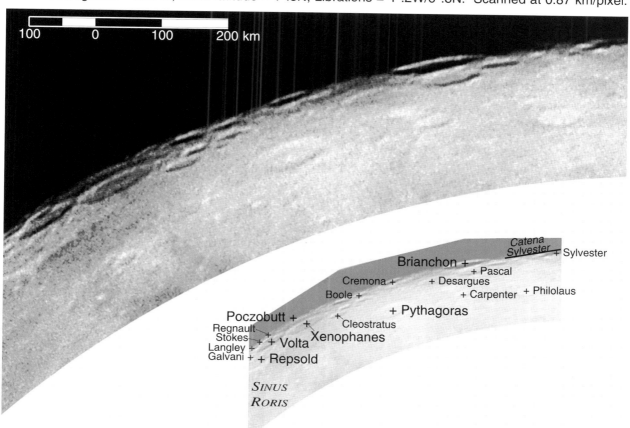

Quadrant 2–NW (54°N to 7°N)

(Data as for Quadrant 2–N)

The Northwestern Highlands contain Imbrian materials and several large Nectarian and pre-Nectarian craters including a "cluster" involving Xenophanes, Volta, Repsold, Galvani, and several smaller craters.

South of Bartels are the Western Highlands, dominated by Orientale ejecta. Farther south, Einstein can be identified by its central crater. The crater Bohr lies to Einstein's southeast, while Vallis Bohr extends to the south of Einstein.

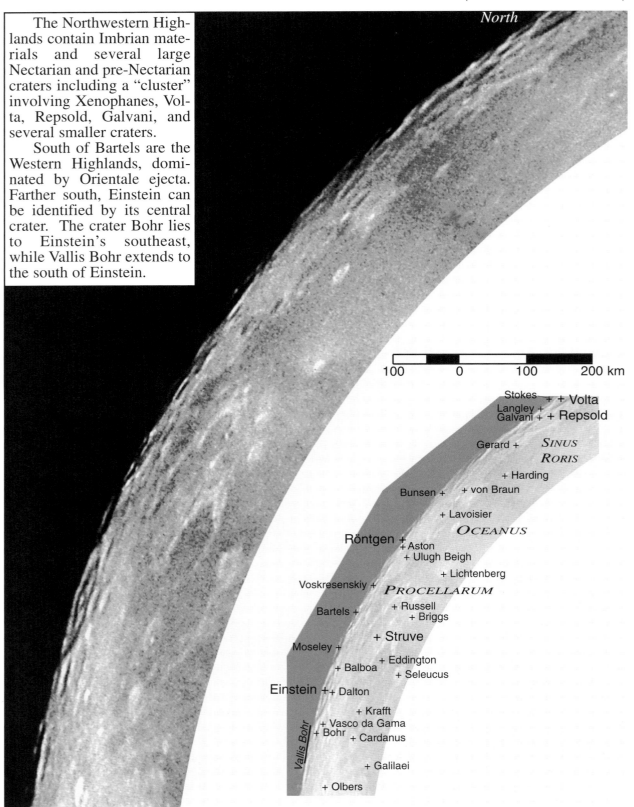

Quadrant 3–W (5°N to 34°S)

1997 May 22, 05h39m UT. 28-cm Sch.-Cass., f/21, 0.20 s, W58 (green) Filter.
Colongitude = 093°.4, Solar latitude = 1°.4S, Librations = 4°.6W/5°.2S. 0.60 km/pixel.

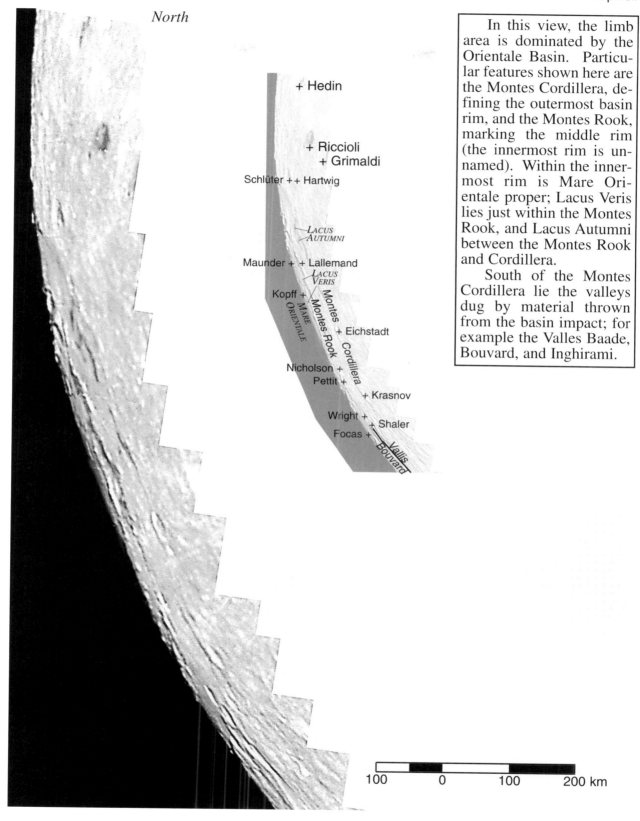

In this view, the limb area is dominated by the Orientale Basin. Particular features shown here are the Montes Cordillera, defining the outermost basin rim, and the Montes Rook, marking the middle rim (the innermost rim is unnamed). Within the innermost rim is Mare Orientale proper; Lacus Veris lies just within the Montes Rook, and Lacus Autumni between the Montes Rook and Cordillera.

South of the Montes Cordillera lie the valleys dug by material thrown from the basin impact; for example the Valles Baade, Bouvard, and Inghirami.

Quadrant 3–SW (28°S to 73°S)

(Data as for Quadrant 3–W)

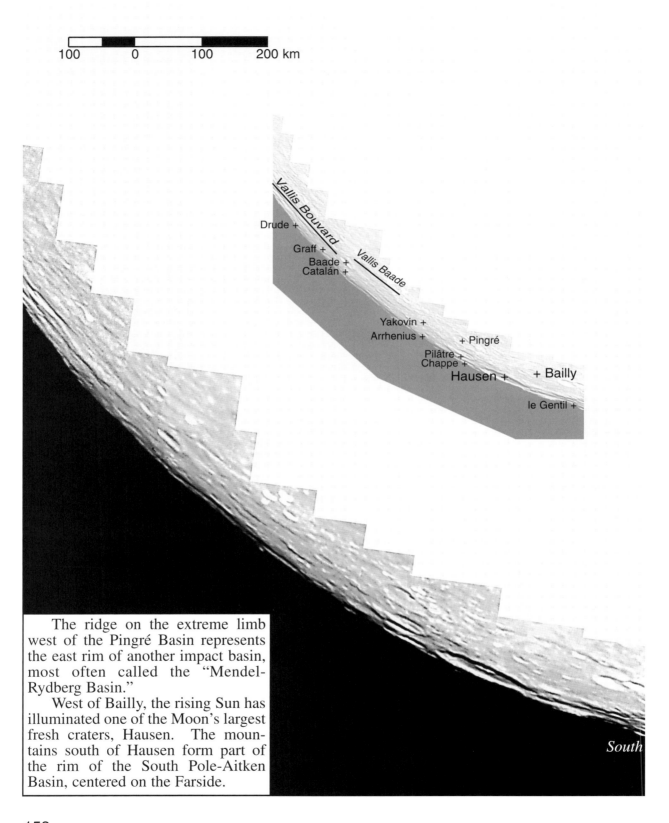

The ridge on the extreme limb west of the Pingré Basin represents the east rim of another impact basin, most often called the "Mendel-Rydberg Basin."

West of Bailly, the rising Sun has illuminated one of the Moon's largest fresh craters, Hausen. The mountains south of Hausen form part of the rim of the South Pole-Aitken Basin, centered on the Farside.

Quadrant 3–S (72°S [W. of pole] to 76°S [E. of pole])

(Data as for Quadrant 3–W; lunar south at bottom)

The rugged terrain in these two views is due to the presence of the rim of the South Pole-Aitken Basin and to several large, deep craters, such as Drygalski, Scott, Amundsen, and Demonax.

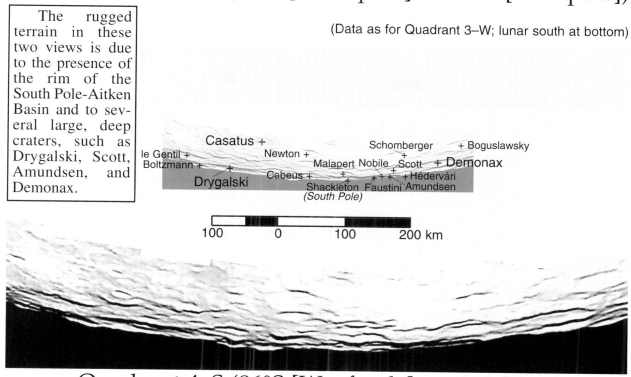

Quadrant 4–S (86°S [W. of pole] to 62°S [E. of pole])

1994 Jul 23, 07h11m UT. 36-cm Sch.-Cass., f/11, 0.05 s, V Filter.
Colongitude = 090°.7, Solar latitude = 1°.5S, Librations = 5°.1E/5°.4S. 1.30 km/pixel.

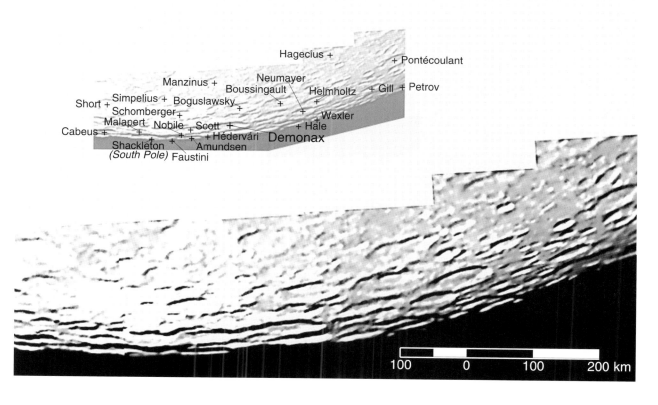

Quadrant 4–SE (65°S to 13°S)

(Data as for Quadrant 4–S)

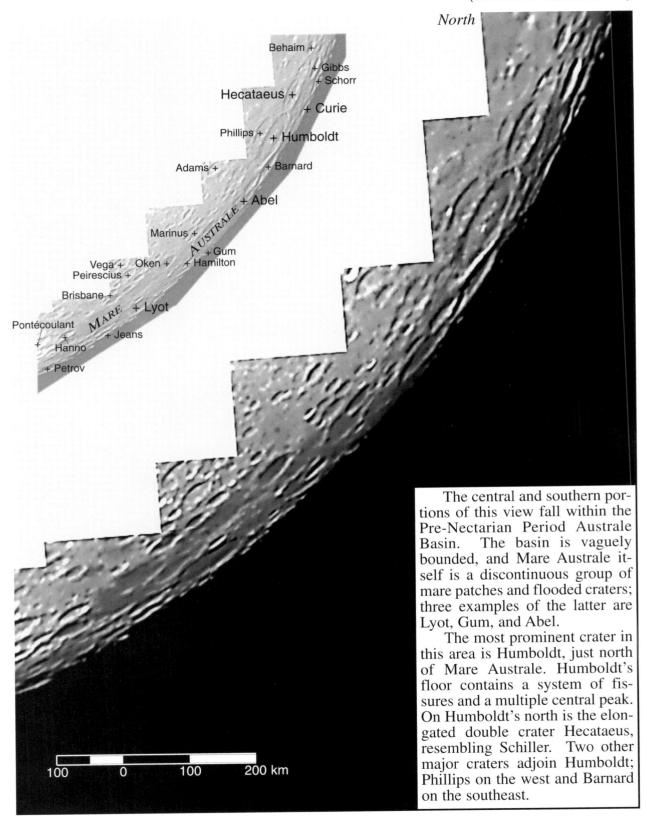

The central and southern portions of this view fall within the Pre-Nectarian Period Australe Basin. The basin is vaguely bounded, and Mare Australe itself is a discontinuous group of mare patches and flooded craters; three examples of the latter are Lyot, Gum, and Abel.

The most prominent crater in this area is Humboldt, just north of Mare Australe. Humboldt's floor contains a system of fissures and a multiple central peak. On Humboldt's north is the elongated double crater Hecataeus, resembling Schiller. Two other major craters adjoin Humboldt; Phillips on the west and Barnard on the southeast.

Quadrant 4–E (16°S to 27°N)

(Data as for Quadrant 4–S)

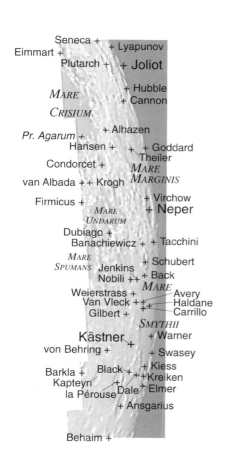

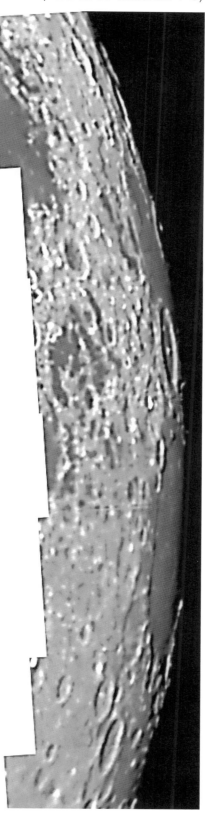

This view completes the counter-clockwise circuit of the limb areas near Full Phase, and most of this region was described in the view of Quadrant 1–E. Note here the two craters la Pérouse and Ansgarius southwest of Mare Smythii, which approximately mark the subtle outer rim of the Mare Smythii Basin.

161

Colongitude 099°

1995 Jul 13, 08h16m UT. 36-cm Sch.-Cass., f/11, 0.50 s, 905-nm narrow-band Filter.
Colongitude 098°.7, Solar latitude = 1°.5S, Librations = 3°.8E/5°.5S. 1.00 km/pixel.

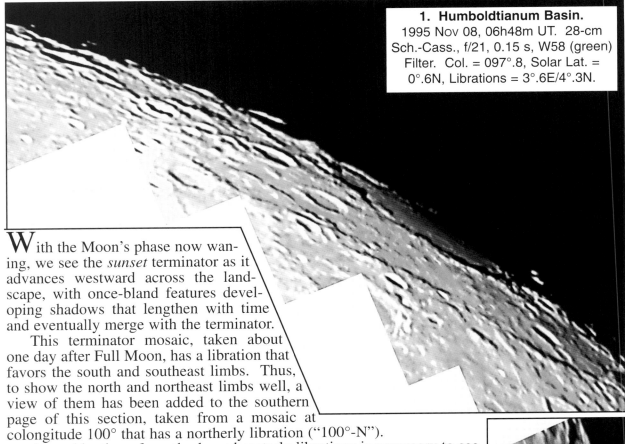

1. Humboldtianum Basin.
1995 Nov 08, 06h48m UT. 28-cm Sch.-Cass., f/21, 0.15 s, W58 (green) Filter. Col. = 097°.8, Solar Lat. = 0°.6N, Librations = 3°.6E/4°.3N.

With the Moon's phase now waning, we see the *sunset* terminator as it advances westward across the landscape, with once-bland features developing shadows that lengthen with time and eventually merge with the terminator.

This terminator mosaic, taken about one day after Full Moon, has a libration that favors the south and southeast limbs. Thus, to show the north and northeast limbs well, a view of them has been added to the southern page of this section, taken from a mosaic at colongitude 100° that has a northerly libration ("100°-N").

A combination of northerly and easterly librations is necessary to see the Humboldtianum Basin well (Figure 1). With this lighting both the inner and outer rim of the basin are clearly visible. The crater Hayn straddles the inner rim, which encloses Mare Humboldtianum itself.

The large crater on the terminator south of the basin is Gauss, whose floor contains craters, hills, and a rille system. The craters Berosus and Hahn are west of Gauss; Hahn contains a prominent central peak.

The terrain east and southeast of Mare Crisium is scarred by ejecta from that basin and the older limb basins, particularly Mare Smythii. For example, a prominent, if unnamed, valley lies just west of Banachiewicz, radiating from Mare Crisium (upper portion of Figure 2).

The southeast portion of the terminator passes near several large craters: Gilbert and Kästner (Figure 2), la Pérouse, Ansgarius, the elongated depression Hecataeus, Phillips, and the "walled plain" Humboldt with its central peak, a rille system, and dark patches on its floor.

This mosaic's lighting-libration combination shows well the discontinuous nature of Mare Australe, its "shoreline" marked by flooded craters.

Several large craters near the South Pole can also be seen, the libration allowing us to trace the terminator as far west of the pole as Drygalski.

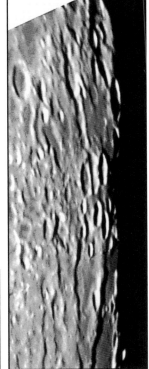

2. Banachiewicz - Kästner Area.
1995 Nov 08, 06h31m UT. 28-cm Sch.-Cass., f/21, 0.15 s, W58 (green) Filter. Col. = 097°.6, Solar Lat. = 0°.6N, Librations = 3°.6E/4°.3N.

099°–N

52°N to 4°S

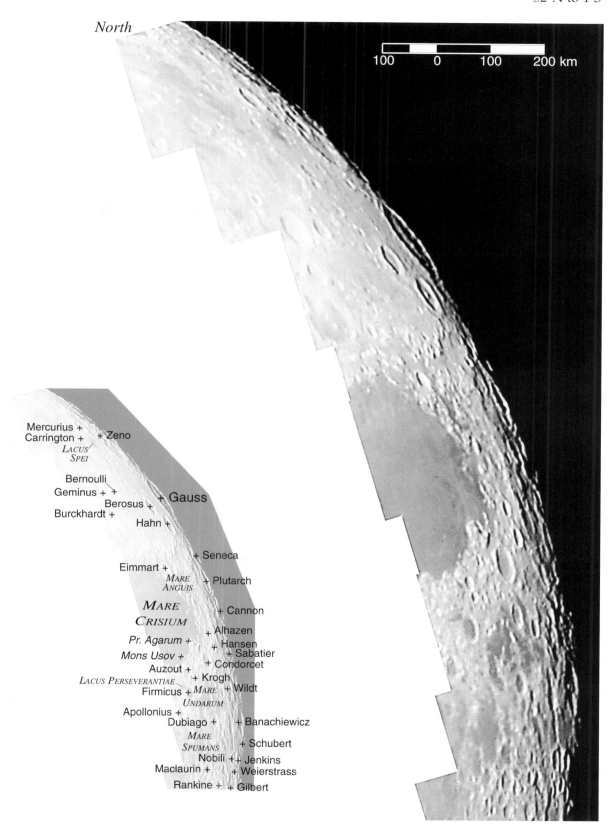

163

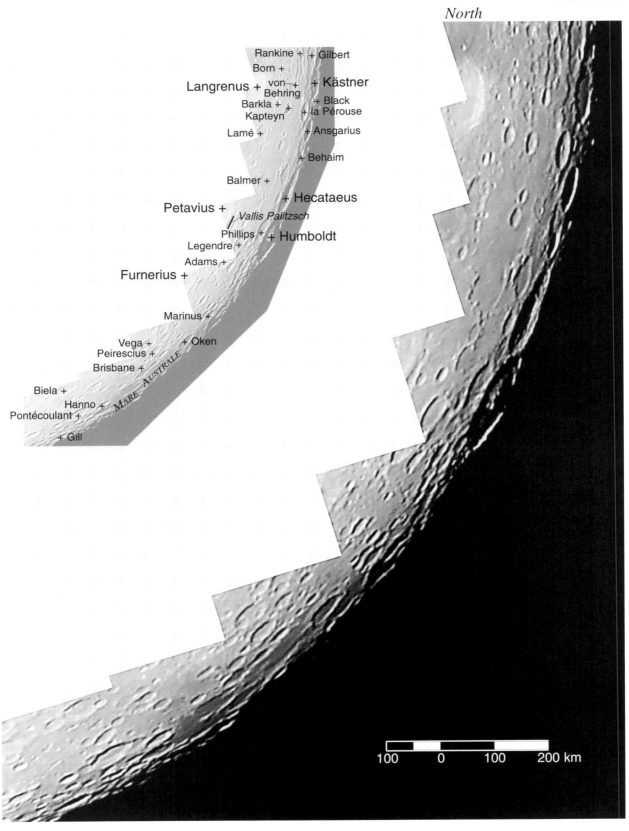
099°–C
2°S to 65°S

099°–S/100°–N

62°S to 90°S/50°N to 90°N

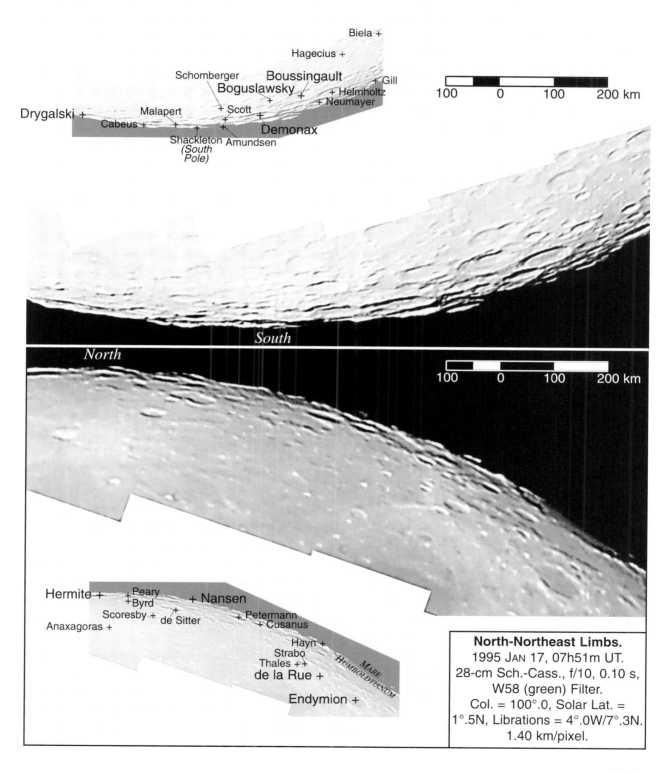

North-Northeast Limbs.
1995 Jan 17, 07h51m UT.
28-cm Sch.-Cass., f/10, 0.10 s,
W58 (green) Filter.
Col. = 100°.0, Solar Lat. =
1°.5N, Librations = 4°.0W/7°.3N.
1.40 km/pixel.

Colongitude 105°

1995 Aug 12, 08h11m UT. 36-cm Sch.-Cass., f/11, 0.06 s, B Filter.
Colongitude = 105°.2, Solar latitude =1°.4S, Librations = 5°.8E/3°.9S. 0.98 km/pixel.

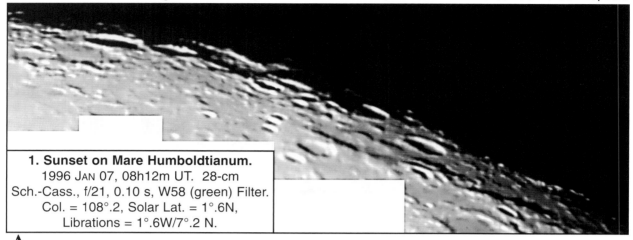

1. Sunset on Mare Humboldtianum.
1996 Jan 07, 08h12m UT. 28-cm Sch.-Cass., f/21, 0.10 s, W58 (green) Filter.
Col. = 108°.2, Solar Lat. = 1°.6N,
Librations = 1°.6W/7°.2 N.

As with the colongitude-099° mosaic, this mosaic's libration favors the south and southeast limbs, so a view of the north and northeast limbs at colongitude 106° with a northerly libration has been added to the southern page of this section ("106°-N"). That insert shows Cusanus and Petermann on the terminator, as well as the outer ring of the Mare Humboldtianum Basin under low lighting; this area is shown at a larger scale in Figure 1. To the basin's southwest lies the dark-floored crater Endymion,, while northwest is the distinctive crater pair Strabo and Thales.

In the highlands southeast of Endymion is the crater Zeno. The large crater on the terminator just south of Zeno, and now more prominent, has neither a name or even a letter designation; this is not unusual for low, ruined formations that stand out only under low lighting. In this area, note two mare patches: Lacus Spei and an unnamed similar unit north of Berosus.

The lighting is now suitable for viewing the rugged eastern edge of Mare Crisium, including its irregular northeast extension, Mare Anguis, as well as Promontorium Agarum and the craters Alhazen, Hansen, and Condorcet. Hansen has a central peak, while Condorcet's floor is flooded.

In the Eastern Highlands, Mare Undarum is now under a low Sun, while Mare Spumans lies to its southwest. The area to the south of the first is marked by several valleys. some radial to the Mare Crisium Basin, such as the one extending north of Kapteyn. The parallel sinuous valleys north of von Behring, however, appear unrelated to any basin. Slightly farther south, extending north of Balmer, is a good example of "light colored plains material;" smooth and mare-like in appearance, but light in albedo and thus not easily identified under a higher sun angle. This area lies within a possible two-rimmed pre-Nectarian impact basin, named the "Balmer-Kapteyn Basin."

In the zone east and southeast of Petavius are several interesting features. One is the Vallis Palitzsch, at the base of the east outer wall of Petavius, a chain of secondary craters originating from the Crisium impact. The Vallis Snellius, radial to the Nectaris Basin, runs north of Furnerius, while a second, unnamed Nectarian valley runs south of Adams. Finally, a chain of four craters, beginning with Legendre as its westernmost, apparently protrudes beyond the terminator.

In the Southeastern Highlands, the Sun is setting on Marinus, Oken, and the Mare Australe, bringing briefly into prominence the low walls of the flooded rings that fringe the mare basin. Somewhat better preserved are the craters Hanno and Pontécoulant, just southwest of the mare.

This mosaic's libration and lighting show favorably the southernmost portion of the Southern Highlands. One of its most distinctive features is the large double-walled crater Boussingault, whose inner ring is called Boussingault A. Northwest of this landmark is the crater Helmholtz, with Neumayer near the terminator to the southeast. Oddly, the distinctive crater northwest of Boussingault is simply named "Boussingault E." However, immediately west of Boussingault is Boguslawsky, with a flooded floor, and to that crater's south is Demonax, now on the terminator. Southwest of Demonax is Scott, adjoining Amundsen, which lies on the limb at latitude 85° south.

105°–N

68°N to 1°N

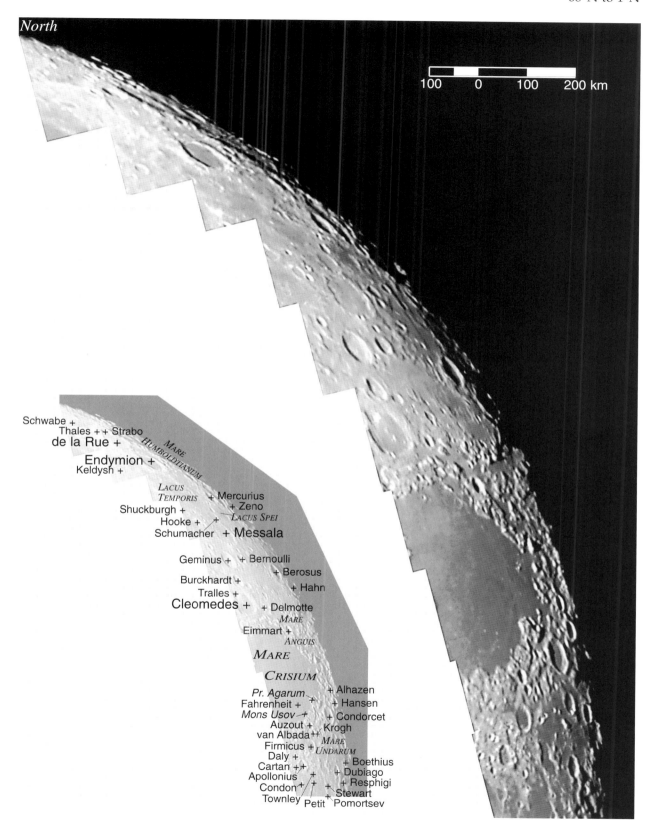

167

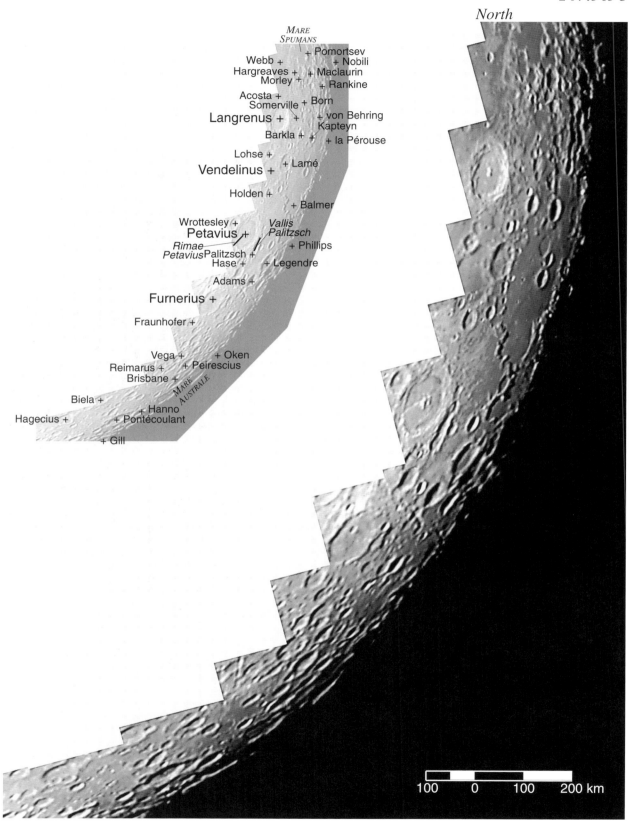

105°–S/106°–N
50°S to 90°S/54°N to 90°N

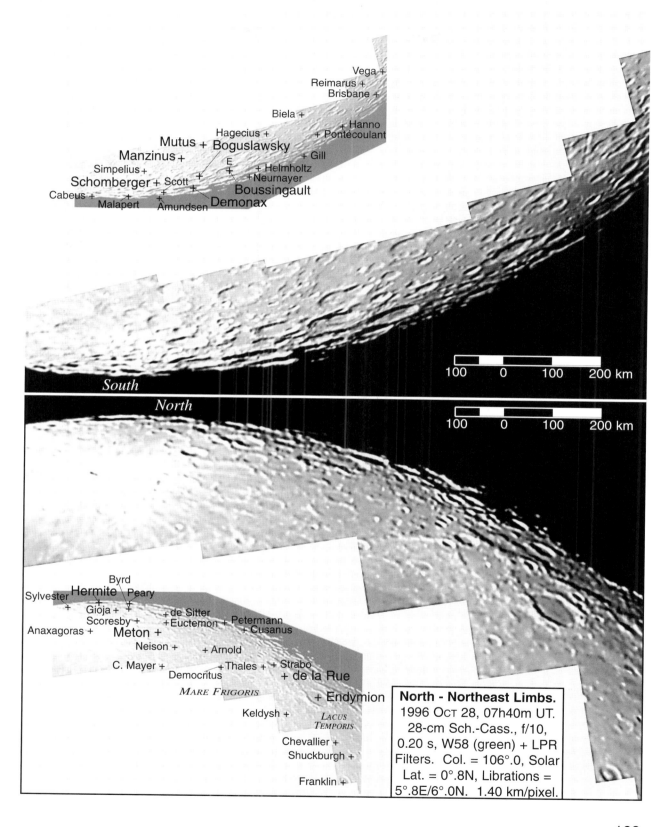

North - Northeast Limbs. 1996 Oct 28, 07h40m UT. 28-cm Sch.-Cass., f/10, 0.20 s, W58 (green) + LPR Filters. Col. = 106°.0, Solar Lat. = 0°.8N, Librations = 5°.8E/6°.0N. 1.40 km/pixel.

Colongitude 112°

1995 Jul 14, 09h30m UT. 36-cm Sch.-Cass., f/11, 0.50 s, 905-nm narrow-band Filter.
Colongitude = 111°.5, Solar latitude = 1°.5S, Librations = 5°.3E/5°.2S. 1.02 km/pixel.

This is the best phase for viewing Endymion before its dark floor is covered by shadow, highlighting its resemblance to the slightly smaller but better-known crater Plato. As the sun angle decreases, Endymion's northern and larger neighbor, the ruined ring de la Rue, is becoming more evident. South of Endymion, two other flat-floored craters, similar in size, are also shown well at this phase: Messala and Cleomedes.

Mare Crisium is still fully sunlit, but the low Sun on its eastern floor reveals the Dorsa Tetyaev and Dorsa Harker, the eastern components of the circumferential ridges that characterize this mare. South of Mare Crisium, the surfaces of Maria Undarum and Spumans retain their smooth appearance even under a setting Sun.

The most prominent "feature" at this phase is a multiple one; the four-crater chain comprising Langrenus on the north, Vendelinus and Petavius, and Furnerius on the south (Figure 1). All at a similar longitude, and thus observable at the same phase, and of similar sizes, they are of considerably different ages: Langrenus is the youngest (Copernican Period), with a prominent central peak, terraced walls, and secondary craters in nearby Mare Fecunditatis. Vendelinus and Furnerius are the oldest of the quartet (pre-Nectarian). Petavius is intermediate in age (Imbrian), and is noteworthy for its multiple central peak, the radial Rimae Petavius on its southwestern floor, an inner ring at the base of its inner wall, and the Vallis Palitzsch adjoining its outer wall on the east. In this region, note a possible large ancient ring on the terminator east of Vendelinus. Even in the lunar highlands, it is not unusual to find hints of ancient ruined craters, or even basins, that are visible only when straddling the terminator.

Finally, in the Southeastern Highlands, the valley system radial to the Nectarian Basin is clearly visible now, particularly near the craters Reimarus and Mallet.

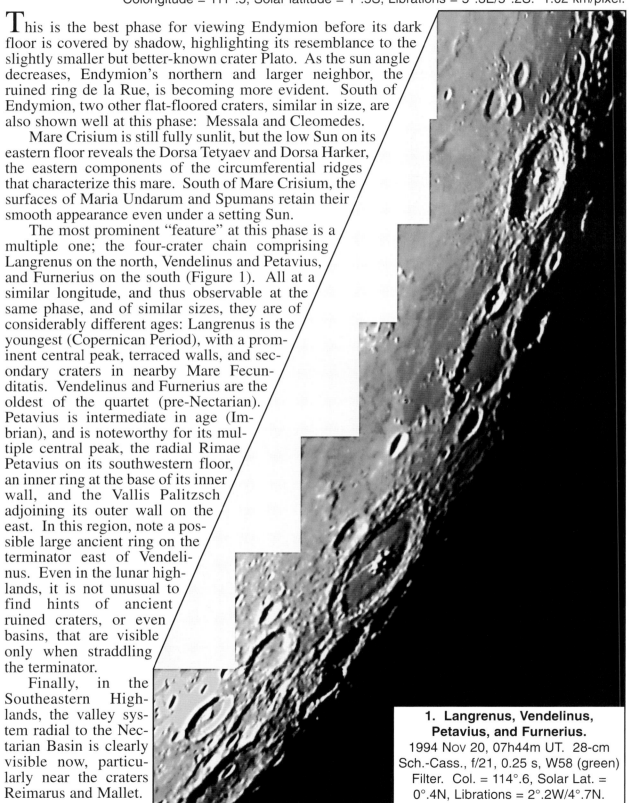

1. **Langrenus, Vendelinus, Petavius, and Furnerius.**
1994 Nov 20, 07h44m UT. 28-cm Sch.-Cass., f/21, 0.25 s, W58 (green) Filter. Col. = 114°.6, Solar Lat. = 0°.4N, Librations = 2°.2W/4°.7N.

112°–N
65°N to 5°N

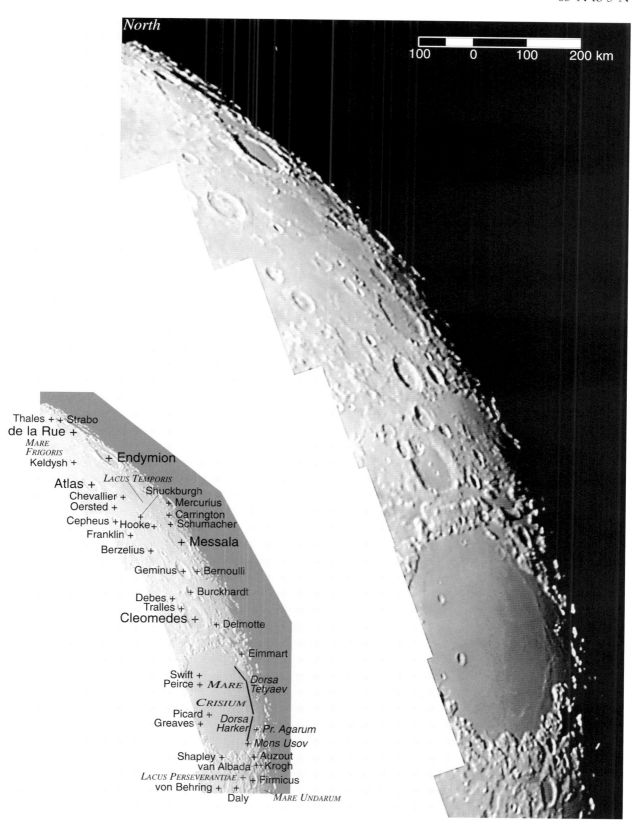

112°–C

6°N to 50°S

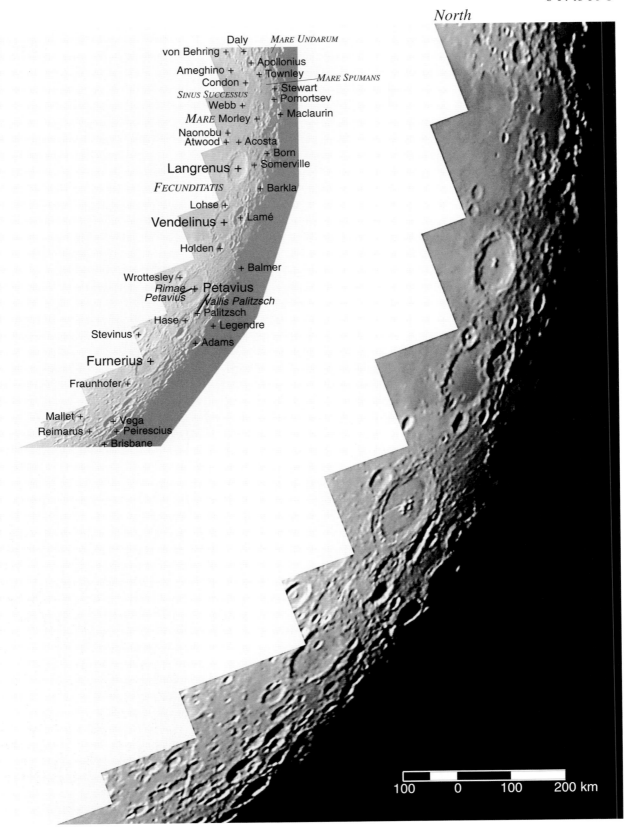

112°–S
49°S to 90°S

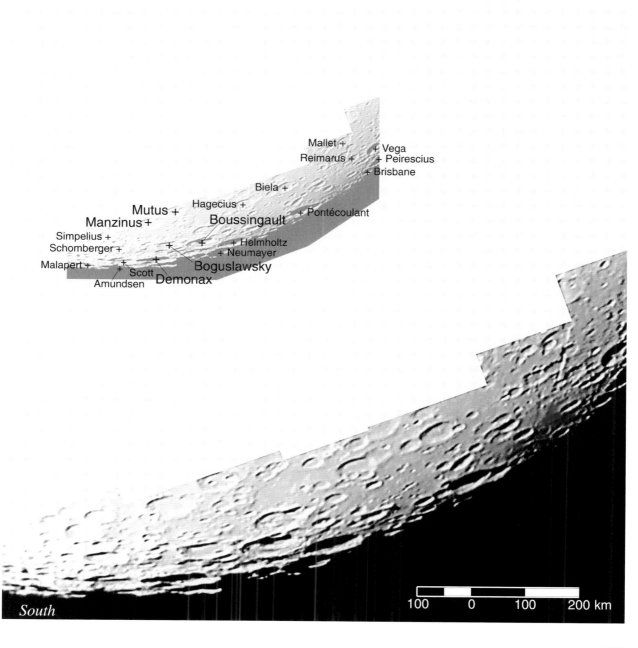

Colongitude 119°

1995 Aug 13, 10h26m UT. 36-cm Sch.-Cass., f/11, 0.08 s, B Filter.
Colongitude = 118°.6, Solar latitude = 1°.4S, Librations = 6°.1E/2°.7 S. 0.98 km/pixel.

The shallow ruined crater de la Rue is seen best at this phase, when its floor is still sunlit but the low sun angle brings out its interior hills and craters. By contrast, the floor of de la Rue's southern neighbor, Endymion, is filled with shadow. East of Endymion, the distinctive serrated feature on the terminator is part of the outer ring of the Humboldtianum Basin.

Messala's floor is also becoming shadow-filled. Between it and Endymion a small zone of mare material, Mare Temporis, appears smooth even when next to the terminator. Of this area's major craters, only Cleomedes' floor remains sunlit, showing its off-center "central" peak (Figure 1).

The terminator is now moving across eastern Mare Crisium, highlighting Dorsum Termier, the north-south ridge that bisects the mare. The unnamed radial valley lying northwest of Apollonius, actually a crater chain radial to the Crisium Basin, is also very distinct under the current lighting.

To the south of the Moon's equator, the present sun angle effectively brings out the ridges and craters on the surface of Mare Fecunditatis. In the northwestern portion of the mare is the famous twin feature Messier-Messier A (Figure 2; Messier is the one on the east). There have been several explanations proposed for this pair (actually, triplet, because Messier A is itself superimposed on another crater), some of them quite odd. One theory is that the pair was formed by a simultaneous double impact, with the impacting bodies approaching from the east at a low angle, which would also explain the "comet ray" extending west of Messier A

The floors of Langrenus, Vendelinus, and Furnerius are in darkness, but the present lighting effectively brings out the features of Petavius' interior, including its inner ring and detail on the crater's floor (Figure 3). The ejecta of the two newest craters of the chain of four, Langrenus and Petavius, is also highlighted.

As the terminator moves westward across the Southeastern Highlands, the Vallis Rheita, radial to the Nectaris Basin, is beginning to show shadow. The crater Young marks the southeast end of the valley; but note at least two more, unnamed, valleys lying between Young and the terminator. Indeed several more unnamed shallow valleys radial to Nectaris can be seen in the Southern Highlands to the east and northeast of the crater Biela.

In the high southern latitudes the terminator moves only slightly between successive mosaics in this atlas, but by now has enveloped the craters Helmholtz, Neumayer, and Demonax. However, the late-afternoon Sun shows well the interior double-ringed structure of Boussingault. Rather than being an under-sized double-ringed basin, Boussingault simply represents a fairly large crater that happened to have a large subsequent impact occur in its interior, creating the feature called "Boussingault A."

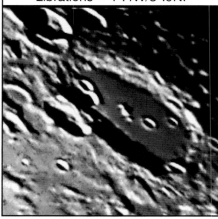

1. Cleomedes.
1993 Dec 01, 07h04m UT.
28-cm Sch.-Cass., f/21, 0.07 s.
Col. = 118°.5, Solar Lat. = 0°.2N,
Librations = 4°.4W/3°.9N.

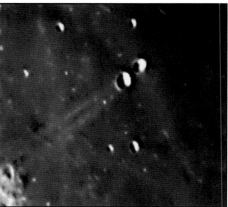

2. Messier. 1993 Dec 01, 06h30m UT. 28-cm Sch.-Cass., f/21, 0.07 s. Col. = 118°.2, Solar Lat. = 0°.2N, Librations = 4°.4W/3°.8N.

3. Sunset on Petavius. 1993 Dec 01, 06h40m UT. 28-cm Sch.-Cass., f/21, 0.10 s. Col. = 118°.3, Solar Lat. = 0°.2N, Librations = 4°.4W/3°.9N.

119°–N

87°N to 24°N

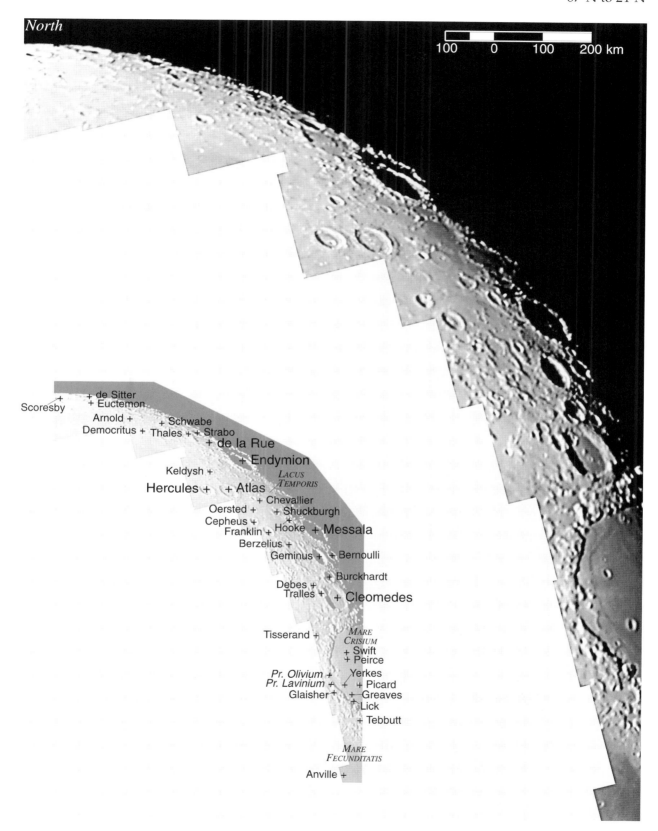

119°–C

28°N to 28°S

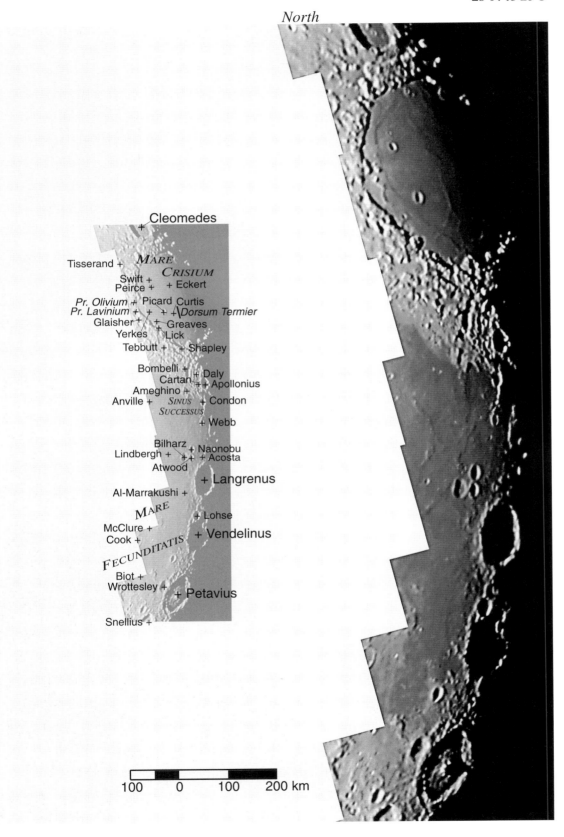

119°–S

27°S to 90°S

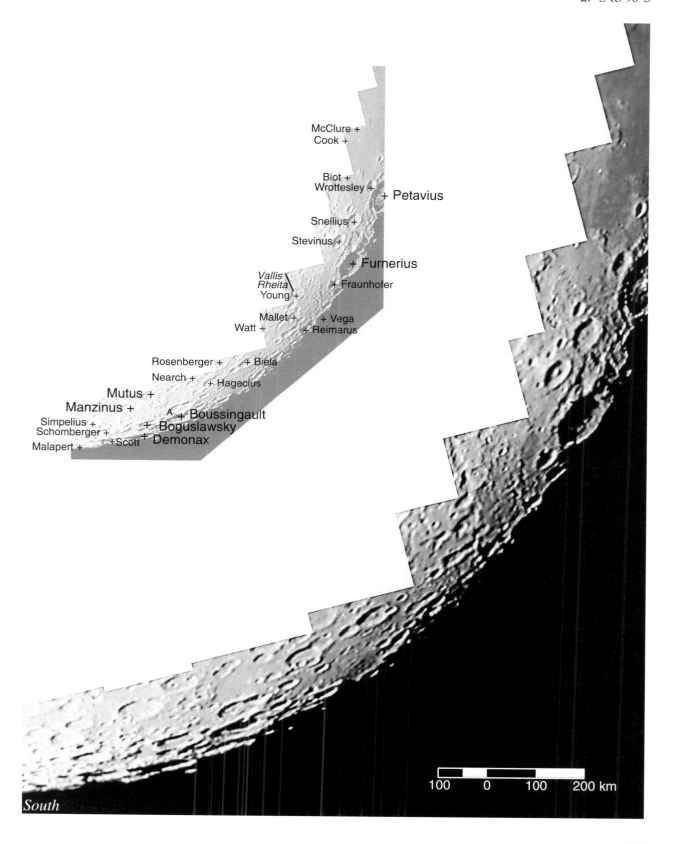

Colongitude 124°

1995 JUL 15, 10h22m UT. 36-cm Sch.-Cass., f/11, 0.08 s, W58 (green) Filter.
Colongitude = 124°.1, Solar latitude = 1°.5S, Librations = 6°.4E/4°.6S. 1.04 km/pixel.

Several of the Northern and Northeastern Highlands' ancient flooded craters can now be seen again, including Bauillaud, Neison, Arnold, and Gärtner. Farther east, de la Rue's floor detail is brought out by the setting Sun, while Endymion has become only a protrusion from the terminator.

Southeast of the Atlas-Hercules crater pair, the Northeastern Highlands show a variety of terrain types. The first is Lacus Temporis, southeast of Atlas, consisting of several scattered mare units. The second is the mountainous area between Franklin and Newcomb, probably shaped by ejecta from the Humboldtianum Basin. Finally, between Newcomb and the northern rim of Mare Crisium is an area of mixed plains and mountains, with several valleys radial to the Crisium Basin.

In western Mare Crisium, the low Sun now highlights the Dorsum Oppel ridge, the flooded crater Yerkes, and the post-mare craters Swift, Peirce, Picard, and Greaves. The flooded crater Lick, however, is in shadow. Sunlight still shines through the "strait" between Promontoria Olivium and Lavinium, showing how the "O'Neill's Bridge" illusion occurred (see Col. 309°); the shaded face of a mound east of the gap mimics the shadow of an arch connecting the shadows of the two promontories (Figure 1).

Bisected by the terminator, Mare Fecunditatis shows a wealth of detail (Figure 2). The Dorsa Cayeux, Geikie and Mawson may be concentric to the ancient Fecunditatis Basin, although the pattern of ridges is complex in this region. Northwest of Dorsa Mawson, Bilharz is on the terminator, and the crater appears to be merely the southern portion of an elongated depression. Southeast of Bilharz, the Sun is setting on Langrenus' west wall, as well as bringing out the chaotic pattern of ejecta southwest of Langrenus and the low north-south ridges in the southern mare.

At this phase the Southeastern Highlands are interesting in that they display two intersecting patterns of radial valleys. For example, north-south valleys and crater chains occur north of Rheita and southeast of Stevinus, which are possibly radial to the Crisium Basin, while Vallis Rheita and the unnamed valley north of the Steinheil-Watt pair are radial to the Nectaris Basin.

The Southern Highlands region south of the craters Rosenberger and Biela consists largely of pre-Nectarian "intercrater terrane," not associated with the deposits of any recognized basin, and thus constitutes one of the most ancient landscapes on the lunar Nearside. This area illustrates effectively the concept of "crater saturation," where almost always the more recent craters like Schomberger (Imbrian), or Biela or Demonax (Nectarian) overlap their pre-Nectarian predecessors.

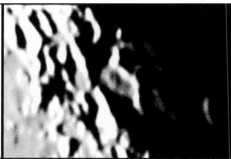

1. Promontoria Olivium and Lavinium. 1997 APR 25, 10h27m UT. 28-cm Sch.-Cass., f/21, 0.06 s. Col. = 126°.1, Solar Lat. = 0°.9S, Librations = 5°.5W/5°.4S.

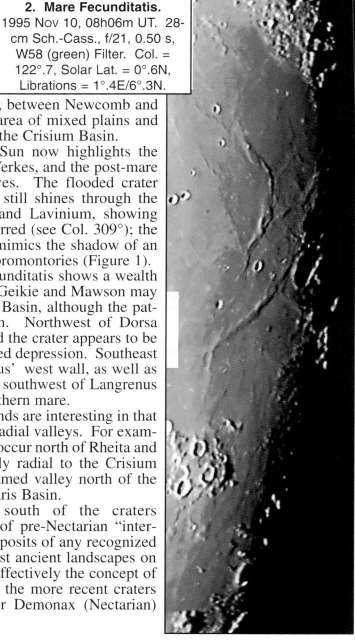

2. Mare Fecunditatis. 1995 Nov 10, 08h06m UT. 28-cm Sch.-Cass., f/21, 0.50 s, W58 (green) Filter. Col. = 122°.7, Solar Lat. = 0°.6N, Librations = 1°.4E/6°.3N.

124°–N
85°N to 8°N

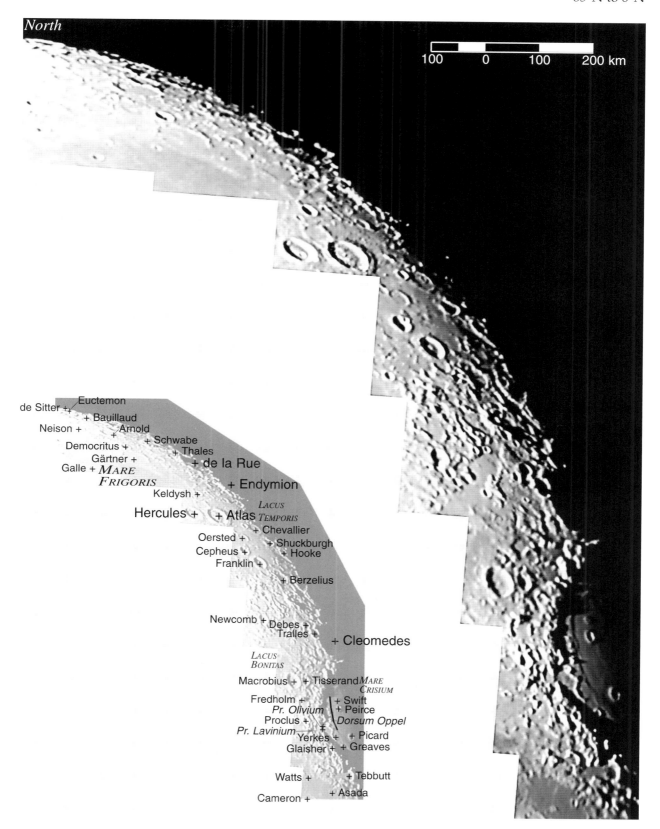

124°–C
10°N to 43°S

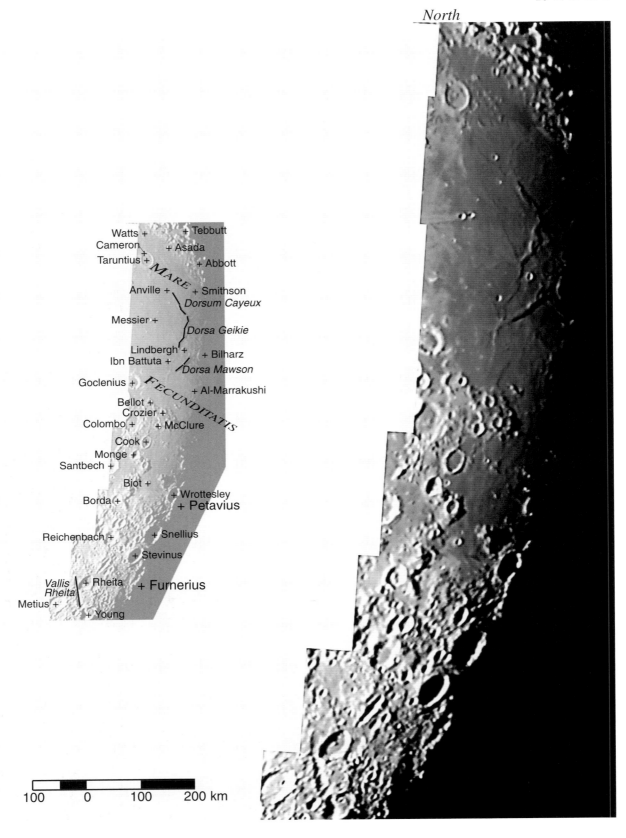

124°–S
38°S to 90°S

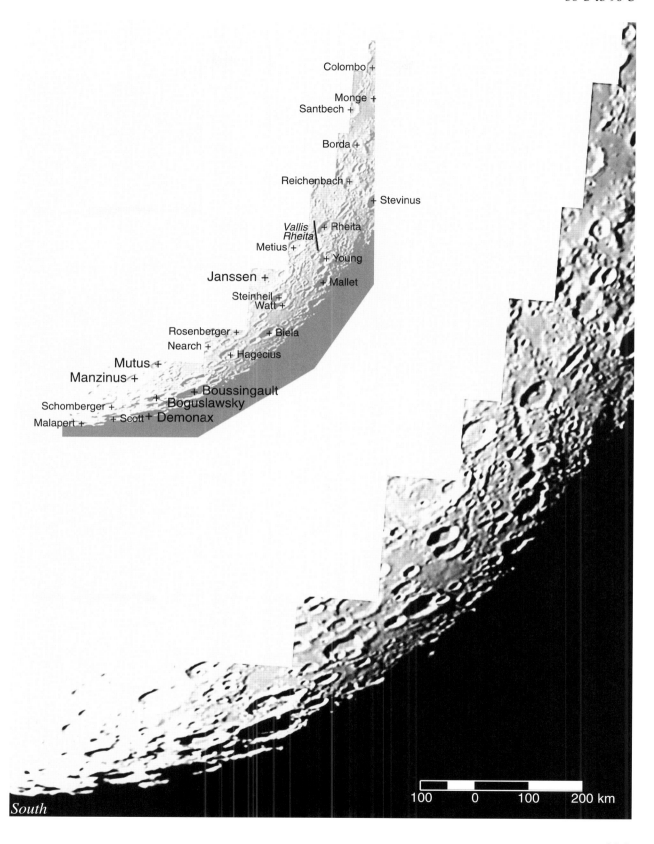

Colongitude 131°

1995 Aug 14, 10h38m UT. 28-cm Sch.-Cass., f/10, 0.10 s, W58 (green) Filter.
Colongitude = 130°.9, Solar latitude = 1°.4S, Librations = 6°.5E/1°.3S. 1.40 km/pixel.

Several mare units appear in our mosaic for the first time since the waxing-crescent phase. From north to south, they are Lacus Mortis, Lacus Somniorum, Palus Somnii, Sinus Amoris, Lacus Bonitatis, Sinus Concordiae, Mare Tranquillitatis, and Mare Nectaris. Also, the terminator has reached the east "shore" of Mare Frigoris and the northwestern portion of Mare Fecunditatis

The crater pair Atlas-Hercules is now quite prominent. Hercules (Eratosthenian) is more recent than Atlas (Imbrian), but both have ejecta patterns that are shown well by the low Sun.

Between Lacus Somniorum and Sinus Amoris are the ill-defined Montes Taurus, containing a mixture of highlands geologic types. There is little obvious structure here except for portions of the rims of the Serenitatis and Crisium Basins, and the region appears to contains highlands material from both basins.

The area between Newcomb and Macrobius contains several Imbrian Plains units; volcanic, but lighter in tone than mare materials (Figure 1). The area also contains a striking ejecta valley radial to Mare Crisium, best seen at this phase.

The concentric-walled central-peaked crater Taruntius lies in the borderland between Mare Tranquillitatis and Mare Fecunditatis (Figure 2). The Montes Secchi are scattered patches of highlands extending southwest of this crater and may mark the rim of the Fecunditatis Basin, the Tranquillitatis Basin, or both.

Between Maria Nectaris and Fecunditatis, centered on the crater Santbech, is a region containing both mare plains material (east of Santbech) and Imbrian Plains highlands material (west of Santbech); under a low Sun it is difficult to distinguish between the two types of material. The Montes Pyrenaeus lie northwest of this area and represent part of the inner rim of the Nectaris Basin.

Southeast of Rheita, the Vallis Rheita is now on the terminator. A ridge extends north-northwest from, and including, the northeast wall of Brenner, parallel to the Vallis Rheita as is a ridge from the northwest wall of Janssen; all radial to Mare Nectaris. There also appears to be a linear structure extending south-southeast from Neander, perhaps an eroded basin rim, and evident only under the current lighting.

Although the Moon does not experience plate tectonics like the Earth, tectonic activity is evident in some areas, represented by fault scarps, by parallel structures, and possibly by ridges. The hexagonal crater Janssen is a case in point (Figure 3). Two ridges and a rille are parallel to the northwest wall, itself parallel to the southeast wall. Indeed, these all parallel two ridges inside Fabricius, indicating crustal movement after the formation of both craters (Janssen is pre-Nectarian, but Fabricius is Eratosthenian).

The terrain between craters is obviously smoother south of Janssen than to its north, indicating that Janssen approximately marks the southern limit of basin ejecta.

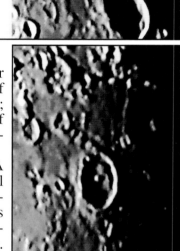

1. Newcomb - Macrobius Area. 1993 Dec 02, 08h00m UT. 28-cm Sch.-Cass., f/21, 0.10 s. Col. = 131°.1, Solar Lat. = 0°.2N, Librations = 4°.6W/5°.2N.

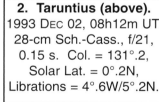

2. Taruntius (above). 1993 Dec 02, 08h12m UT. 28-cm Sch.-Cass., f/21, 0.15 s. Col. = 131°.2, Solar Lat. = 0°.2N, Librations = 4°.6W/5°.2N.

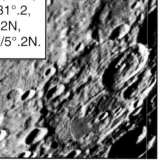

3. Janssen (to right). 1993 Jan 11, 09h 44m UT. 28-cm Sch.-Cass., f/21, 0.05 s. Col. = 128°.7, Solar Lat. = 0°.8N, Librations = 1°.5E/7°.0N.

131°–N

89°N to 20°N

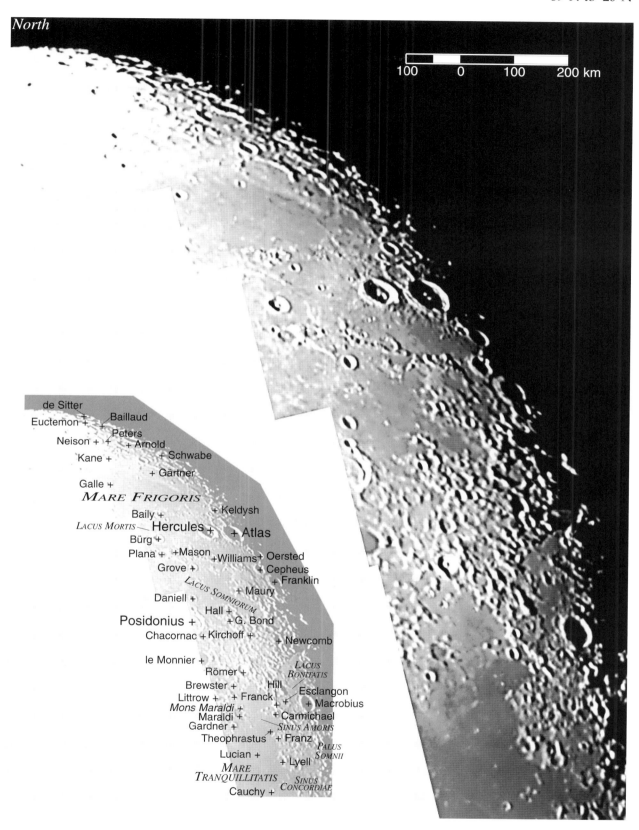

131°–C
28°N to 26°S

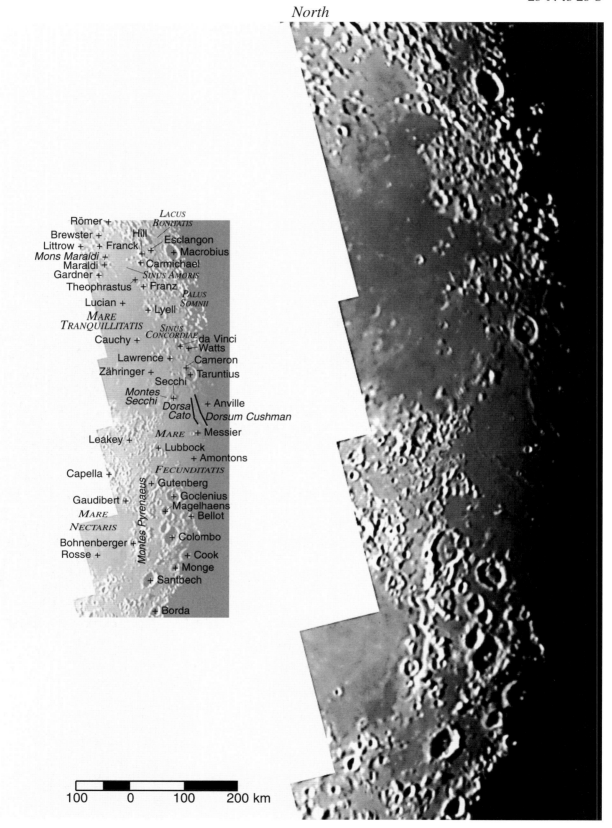

131°–S
10°S to 90°S

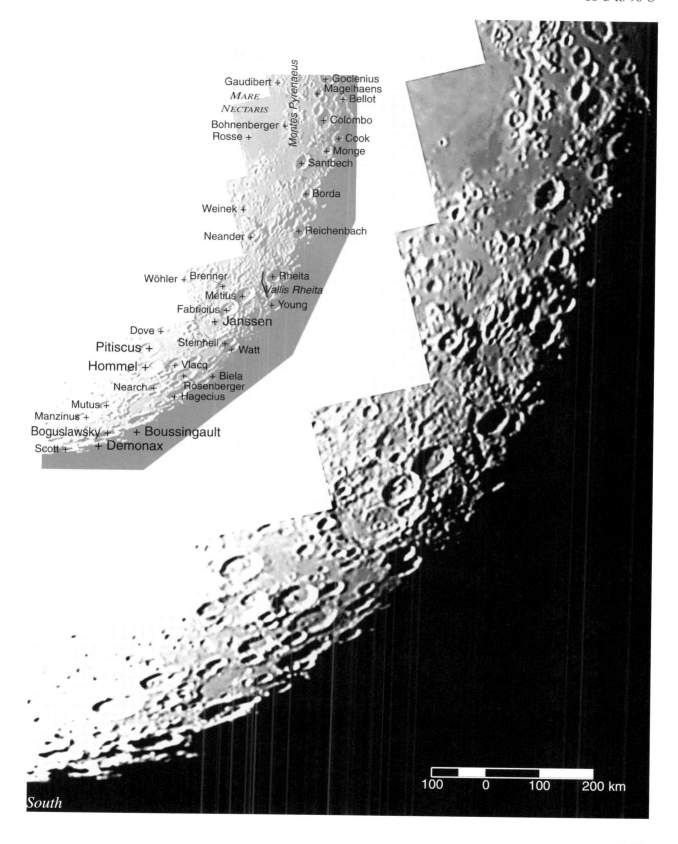

Colongitude 135°

1995 Nov 11, 08h44m UT. 28-cm Sch.-Cass., f/10, 0.05 s, W58 (green) Filter.
Colongitude = 135°.2, Solar latitude = 0°.6N, Librations = 0°.2E/6°.9N. 1.48 km/pixel.

A very favorable northerly libration at this phase brings out many large flooded craters in the Northern Highlands. These include pre-Nectarian Hermite, Peary, and Byrd; Nectarian Gioja, Main, Challis, de Sitter, and Euctemon; and finally Meton, Baillaud. Neison, Arnold, Moigno. Kane. and Gärtner, which are dated approximately to anywhere between the pre-Nectarian and Imbrian. All these craters have level floors composed of Imbrian-age materials.

South of the prominent Atlas-Hercules pair, the rugged aspect of the Montes Taurus region is evident at this phase. The topography of this area is confused because three impact basins overlap here; Crisium, Serenitatis, and Tranquillitatis. One characteristic of this region is large numbers of heavily damaged craters.

South of the Palus Somni highlands region, little Sinus Concordiae shows well, along with the Montes Secchi on the eastern border of Mare Tranquillitatis.

Another region that is currently prominent is the Pyrenaeus Highlands, between Mare Nectaris and Mare Fecunditatis (Figure 1). The "backbone" of this region, the Montes Pyrenaeus themselves, parallel the terminator and form part of the inner ring of the Nectarian Basin. Gutenberg and several unnamed ruined rings adjoin the mountain ridge on its east.

Although of highlands material, much of the region west and southwest of Santbech remains mare-like in texture even under grazing illumination. Part of the outer Nectarian ring can be traced between Santbech and Piccolomini. Beyond this outer ring are rugged highlands punctuated by valleys radial to the Nectaris Basin.

Janssen remains the most prominent crater in the Southeastern Highlands. Northwest of the crater, the low lighting hints at two rings outside the crater rim itself. Janssen also is on the northeast rim of a probable basin, the Mutus-Vlacq Basin (Vlacq is on its east rim, and Mutus on its southeast rim). The suspected feature is centered near Baco and will become more evident as the terminator moves westward.

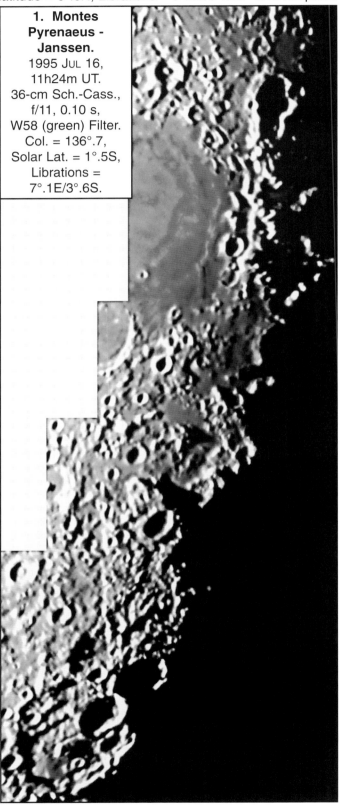

1. Montes Pyrenaeus - Janssen. 1995 JUL 16, 11h24m UT. 36-cm Sch.-Cass., f/11, 0.10 s, W58 (green) Filter. Col. = 136°.7, Solar Lat. = 1°.5S, Librations = 7°.1E/3°.6S.

135°–N
90°N to 28°N

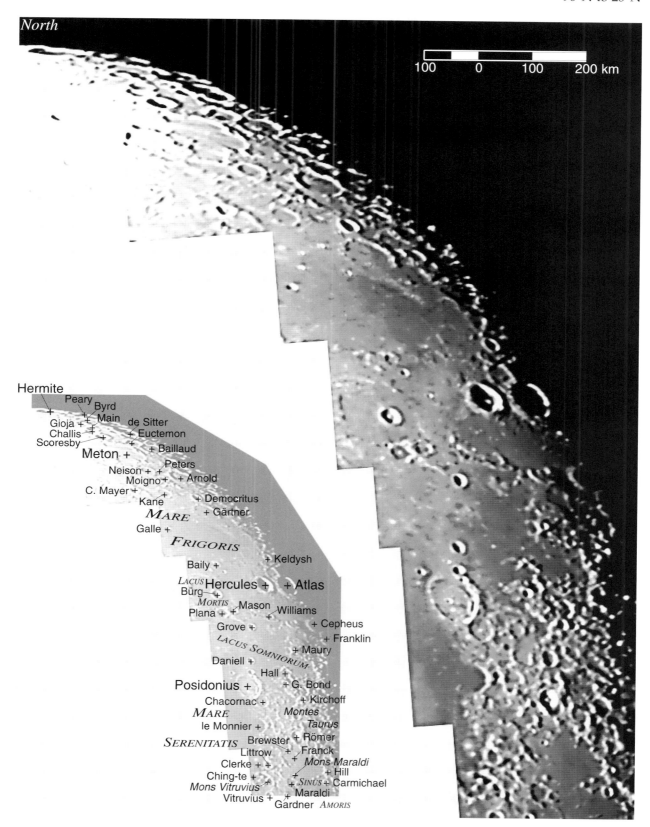

135°–C
34°N to 17°S

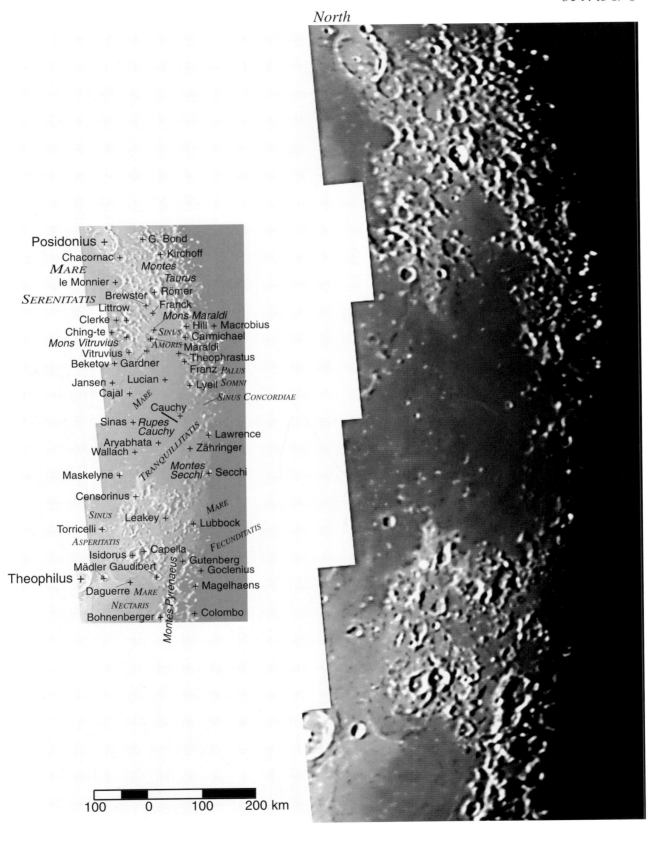

135°–S

14°S to 83°S

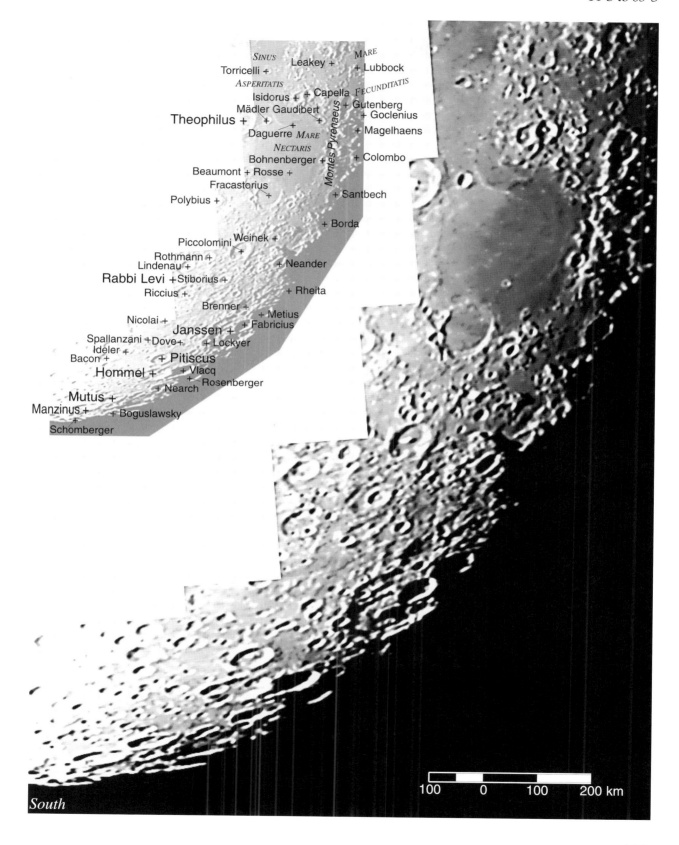

Colongitude 143°

1995 Oct 13, 10h58m UT. 28-cm Sch.-Cass., f/10, 0.05 s, W58 (green) Filter.
Colongitude = 143°.2, Solar latitude = 0°.2S, Librations = 1°.7E/6°.0N. 1.49 km/pixel.

As with the previous mosaic, a northerly libration favors the Northern Highlands. Within them, the merged rings that make up Meton appear smooth-floored even under grazing sunlight. The low walls of the similarly flooded craters Goldschmidt, Barrow, and W. Bond are beginning to cast shadows. Gärtner, on the northern edge of Mare Frigoris, has ridges and a crater on its floor; low ridges and hills are also found in the mare to its south.

East of Bürg in the eastern Lacus Mortis is a shallow southeast-northwest furrow, visible only under low lighting. A deeper valley also appears to extend southeast from Mason to Williams.

South of Lacus Somniorum is Posidonius with its concentric wall; to its southeast the Sun is setting on the Montes Taurus.

The terminator is now crossing eastern Mare Tranquillitatis (Figure 1), and the Rupes Cauchy fault scarp is particularly prominent at sunset, as is now the case. The apparent peak south of the scarp, just within the terminator, is actually the dome Cauchy τ. In this area are several very low ridges that are visible only under a solar elevation of a few degrees. Considerably more prominent in this area are the flooded, partial-walled crater Aryabhata, a similar ring to its south (Maskelyne F), and several isolated highlands-material hills. In contrast, an extremely shallow "ghost crater" straddles the terminator just southeast of Maskelyne F.

The Pyrenaeus Highlands are thrown into relief by the low Sun, and show hints of two rims of the Nectaris Basin; the innermost is the northern shore of the mare itself, while another rim is suspected slightly south of Censorinus; the southern rim of the Tranquillitatis Basin is also traced in this area.

Mare Nectaris has no circumferential system of ridges, although a north-south ridge can be found north of Beaumont, with another north of Rosse; both are unnamed (Figure 2). Several pre-mare craters on the mare's edge show varying signs of age. Daguerre and the unnamed crater touching it on the northwest have been completely covered by mare lava, but still protrude slightly above the general surface. The larger craters Beaumont and Fracastorius are preserved better but have lost their northern walls to the mare basalt that has covered their floors.

A southwestern rim of the Nectaris Basin, the Montes Altai, is now clearly seen, extending northwest of Piccolomini. Another rim, between the Montes Altai and the mare edge, is also visible, passing near Polybius.

Janssen has been enveloped by the terminator, but the two outer rings west of the crater are clearly shown at this phase. With Janssen no longer visible, Hommel is currently the most prominent crater near the terminator in the Southern Highlands, the late-afternoon Sun highlighting the hills and ridges on its floor and the craters on its floor and walls.

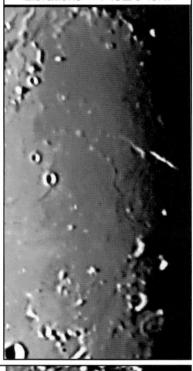

1. Eastern Mare Tranquillitatis. 1995 Oct 13, 11h38m UT. 28-cm Sch.-Cass., f/21, 0.10 s. Col. = 143°.5, Solar Lat. = 0°.2S, Librations = 1°.3E/6°.0N.

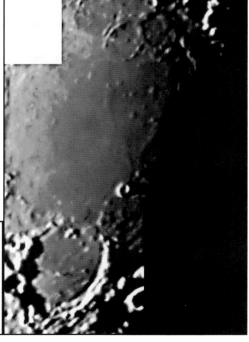

2. Mare Nectaris. 1995 Oct 13, 11h30m UT. 28-cm Sch.-Cass., f/21, 0.05 s. Col. = 143°.4, Solar Lat. = 0°.2S, Librations = 1°.4E/6°.1N.

143°–N
90°N to 28°N

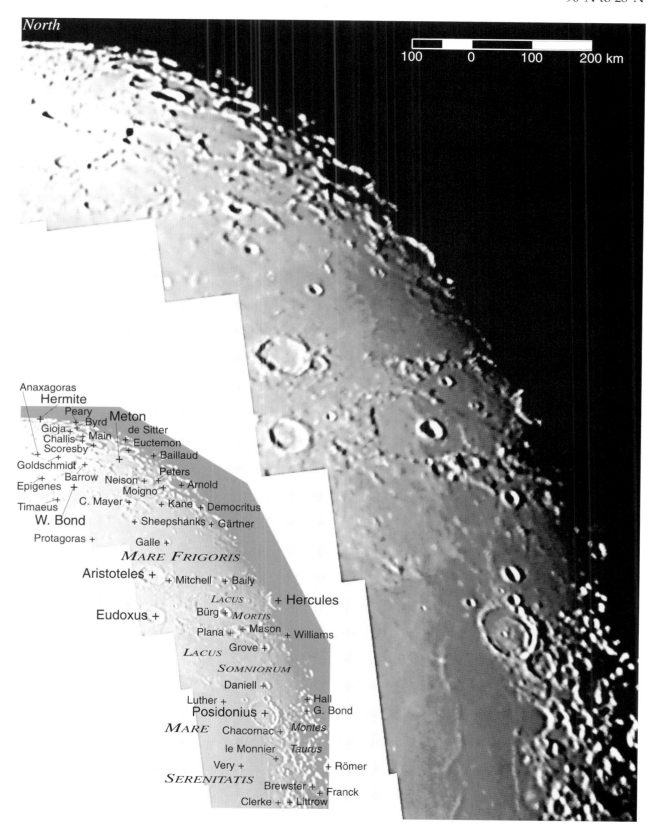

143°–C
29°N to 19°S

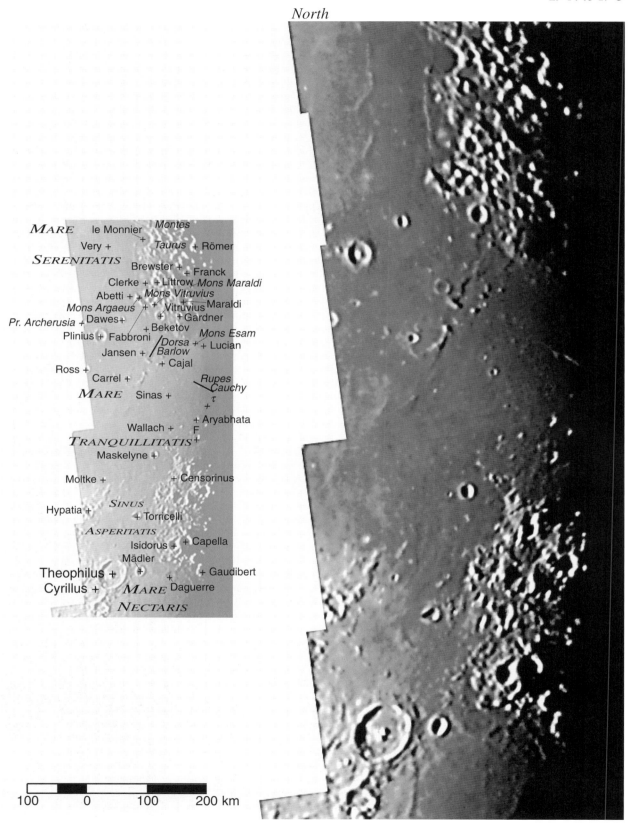

143°–S
16°S to 84°S

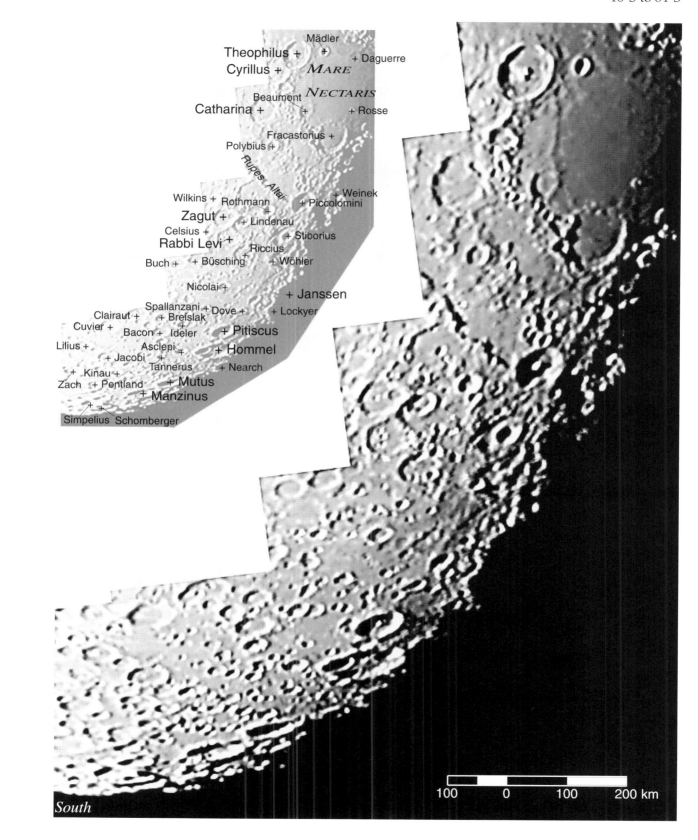

Colongitude 148°

1996 Nov 30, 08h30m UT. 28-cm Sch.-Cass., f/10, 0.10 s, W58 (green) Filter.
Colongitude = 148°.0, Solar latitude = 1°.4N, Librations = 1°.8E/5°.9N. 1.48 km/pixel.

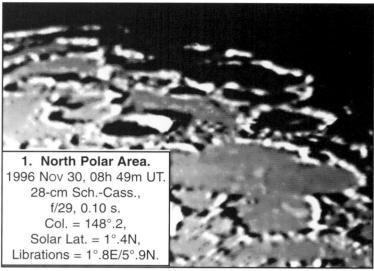

1. North Polar Area.
1996 Nov 30, 08h 49m UT.
28-cm Sch.-Cass.,
f/29, 0.10 s.
Col. = 148°.2,
Solar Lat. = 1°.4N,
Librations = 1°.8E/5°.9N.

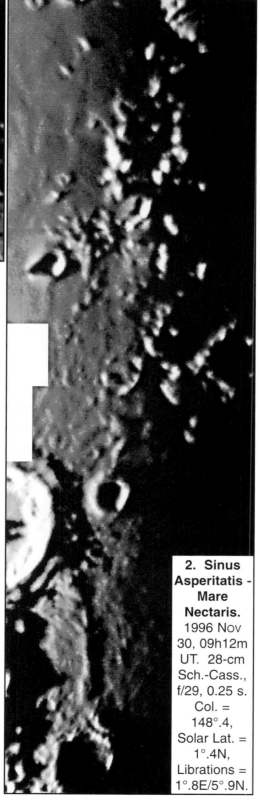

2. Sinus Asperitatis - Mare Nectaris. 1996 Nov 30, 09h12m UT. 28-cm Sch.-Cass., f/29, 0.25 s. Col. = 148°.4, Solar Lat. = 1°.4N, Librations = 1°.8E/5°.9N.

Once again, the libration lets us look beyond the lunar North Pole (Figure 1). Meton is the largest of this area's low, flooded rings. Meton's walls show a "cloverleaf" pattern and the remnants of the west wall of its eastern component, Meton D, are now faintly visible.

South of Mare Frigoris, Lacus Mortis consists of the flooded remains of an ancient crater, with the post-mare crater Bürg superimposed on its floor. Plana, on the ring's southern rim, appears to also be on the rim of an ancient ring that was engulfed by Lacus Somniorum.

The Sun is setting on Posidonius and the eastern edge of Mare Serenitatis. In the mare, three wrinkle-ridge systems are visible; the Dorsa Smirnov, Lister, and Aldrovandi; note a possible dome northwest of Littrow.

In central Mare Tranquillitatis, the Dorsa Barlow and the hills near Cajal are prominent in the northern part, and the hills and ridges northeast of Maskelyne are well presented in the south-central portion of the mare.

Lying between Mare Tranquillitatis and Mare Nectaris, Sinus Asperitatis' hills and ridges are clearly shown at this phase (Figure 2). Within Sinus Asperitatis, the double crater Torricelli appears to lie in the eastern portion of a largely buried pre-mare crater.

In Mare Nectaris, the unnamed ridge north of Beaumont is prominent at this sun angle, along with the ejecta pattern around Theophilus. The three-crater group, Theophilus, Cyrillus, and Catharina is best seen about now. Unfortunately, the current sun angle is about the lowest suitable for viewing the floor and central elevation of a "fresh" (i.e., deep) crater like Theophilus; a few hours later and its floor will be entirely in shadow.

The Rupes Altai are also best viewed now, as is the far less obvious Mutus-Vlacq Basin, centered near Bacon.

148°–N
90°N to 23°N

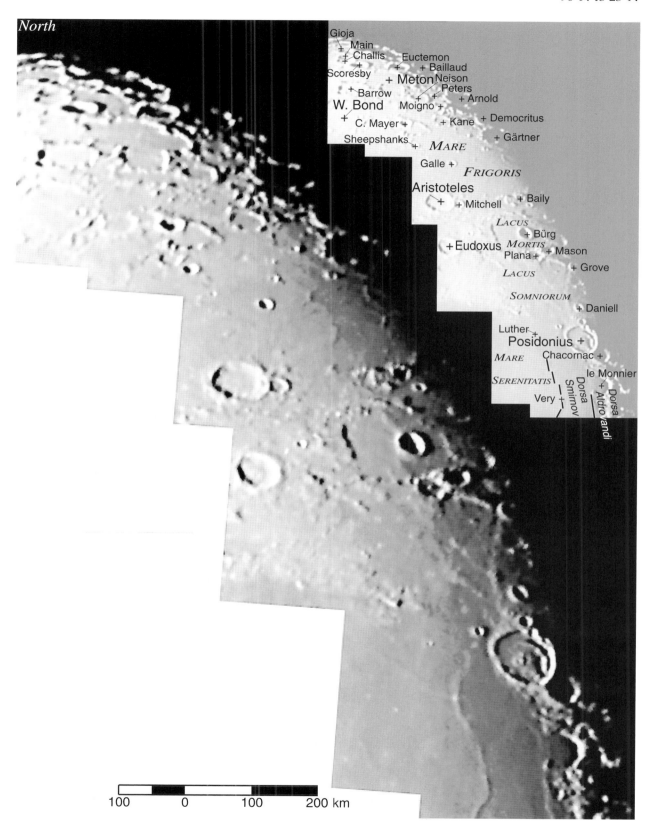

148°–C
24°N to 16°S

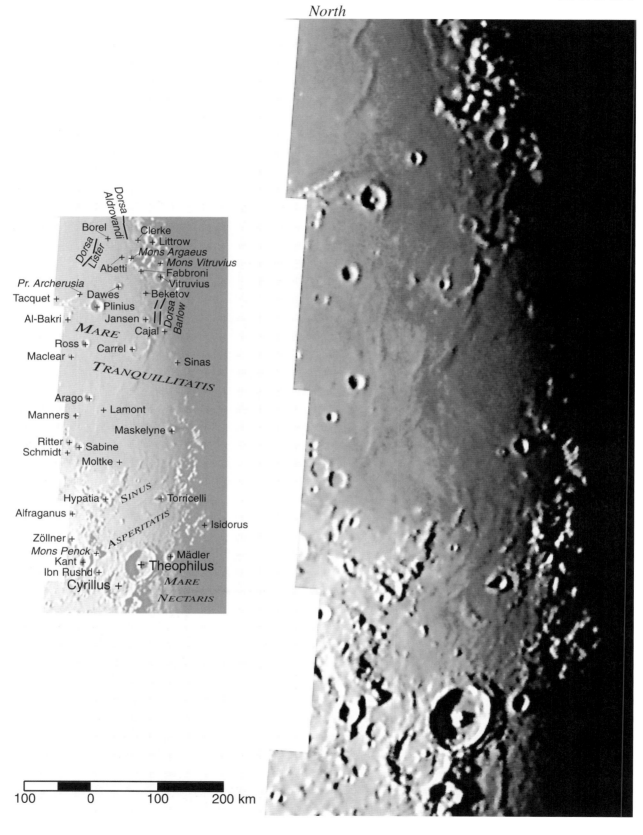

148°–S

15°S to 84°S

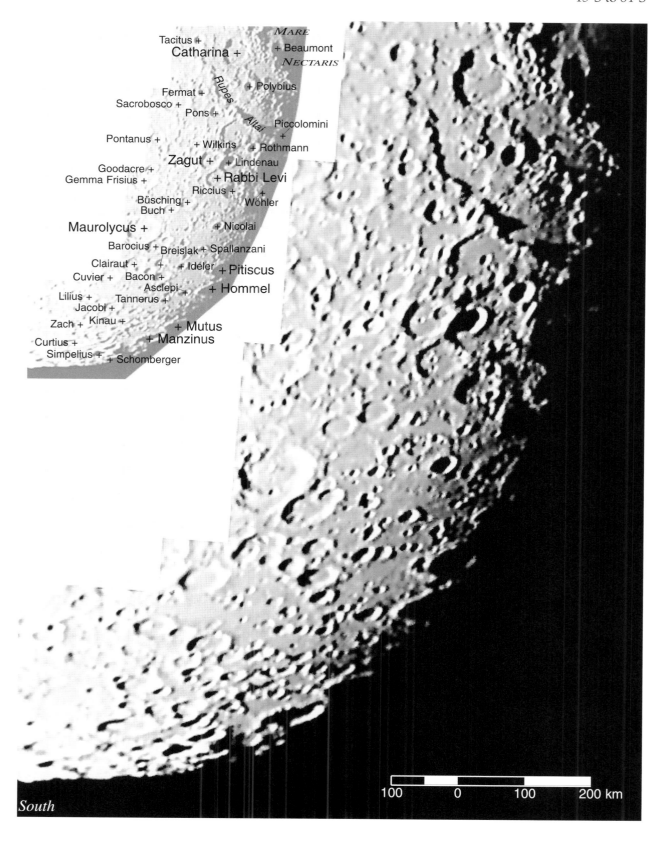

197

Colongitude 154°

1995 Oct 14, 08h54m UT. 28-cm Sch.-Cass., f/10, 0.15 s, W58 (green) Filter.
Colongitude = 154°.3, Solar latitude = 0°.1S, Librations = 1°.0E/6°.4N. 1.48 km/pixel.

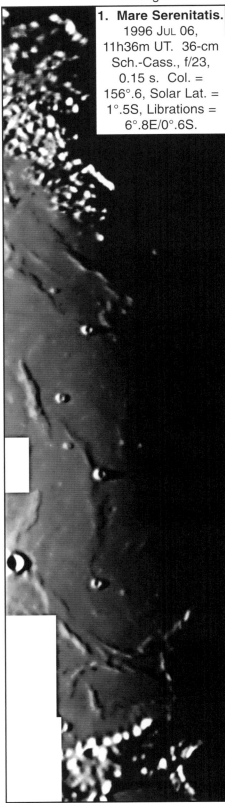

1. Mare Serenitatis. 1996 Jul 06, 11h36m UT. 36-cm Sch.-Cass., f/23, 0.15 s. Col. = 156°.6, Solar Lat. = 1°.5S, Librations = 6°.8E/0°.6S.

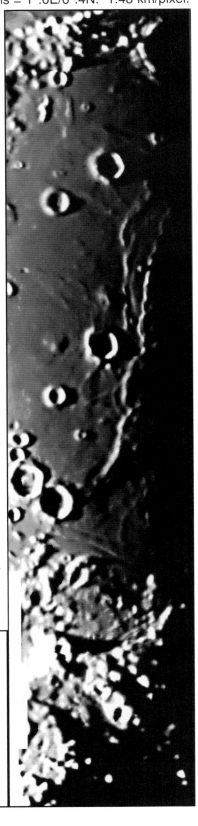

2. Mare Tranqullitatis. 1996 Jul 06, 11h14m UT. 36-cm Sch.-Cass., f/23, 0.15 s. Col. = 156°.4, Solar Lat. = 1°.5S, Librations = 6°.9E/0°.7S.

The frequent isolated hills near Sheepshanks, Galle, and Protagoras in Mare Frigoris suggest that the mare's lava layer is thin in this area.

South of the prominent crater pair Aristoteles-Eudoxus, the ridge systems of two maria can presently be seen well. Those in central Mare Serenitatis (Figure 1) tend to run north-south, but have bends and branches. Most of the ridges in western Mare Tranquillitatis (Figure 2) also are approximately north-south, and several appear associated with the Lamont formation. Also of interest in western Mare Tranquillitatis are the Rimae Hypatia, the Arago α and β domes, and a chain of three small domes between Arago α and Maclear.

Western Sinus Asperitatis contains furrows and ridges formed by ejecta from Theophilus, although the interior of Theophilus itself, together with those of the neighboring craters Cyrillus and Catharina, are now filled with shadow.

The west rim of the Nectaris Basin can be seen by tracing a line southwest through the ridge northeast of Hypatia to Mons Penck and the mountain ridge to its north, then on along the Rupes Altai.

This illumination also reveals the west rim of an ancient crater, curving from the wall of Sacrobosco to Pontanus and possibly on to Wilkins.

154°–N

90°N to 25°N

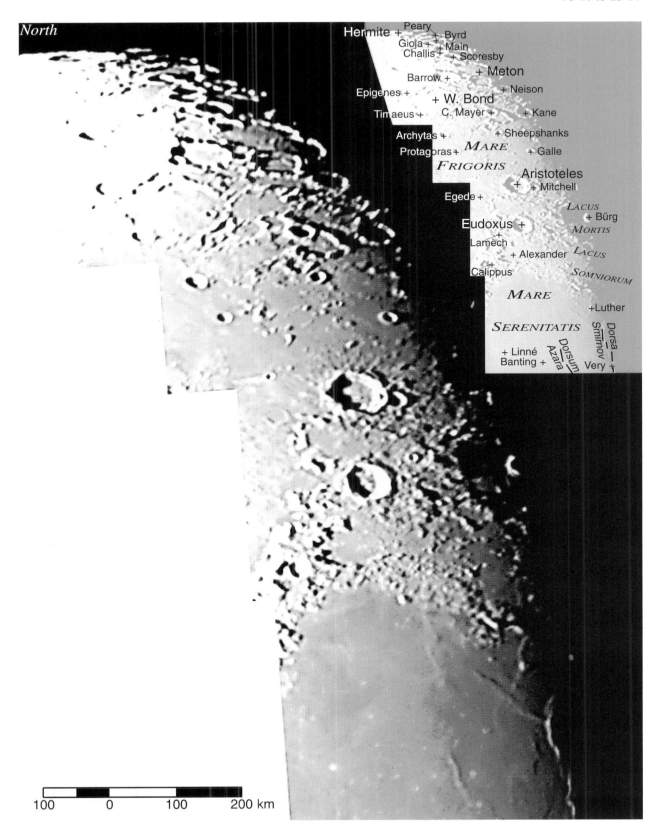

154°–C

27°N to 14°S

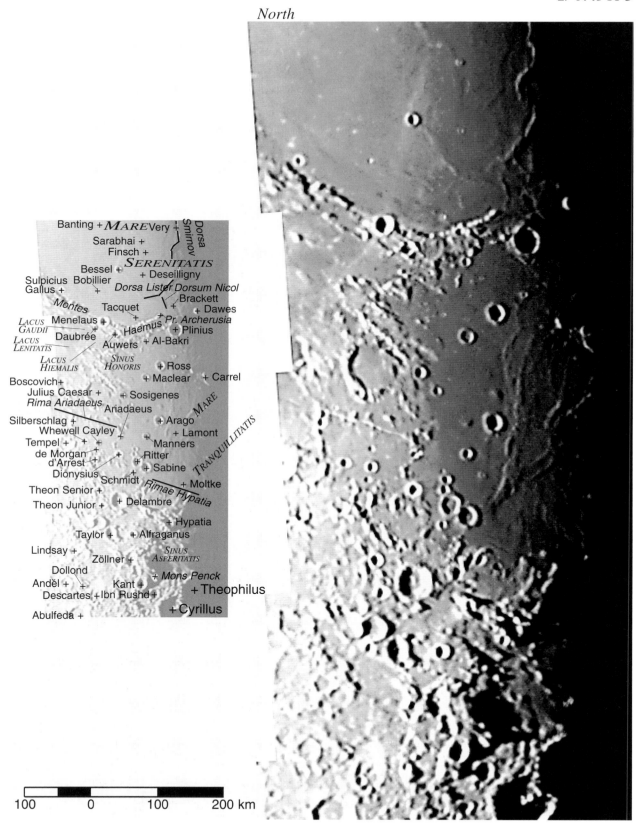

154°–S

12°S to 84°S

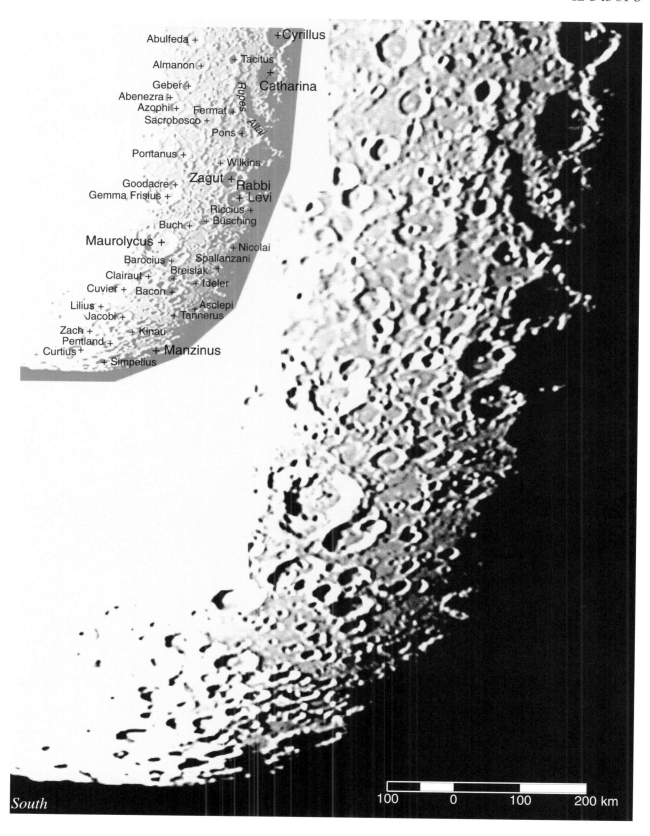

Colongitude 161°

1995 Nov 13, 11h03m UT. 28-cm Sch.-Cass., f/10, 0.07 s, W58 (green) Filter.
Colongitude = 160°.7, Solar latitude = 0°.7N, Librations = 2°.5W/7°.4N. 1.48 km/pixel.

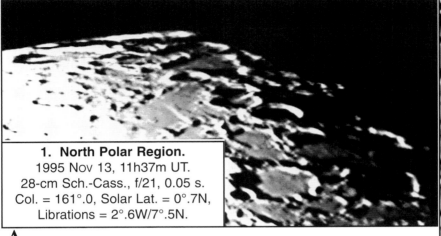

1. North Polar Region.
1995 Nov 13, 11h37m UT.
28-cm Sch.-Cass., f/21, 0.05 s.
Col. = 161°.0, Solar Lat. = 0°.7N,
Librations = 2°.6W/7°.5N.

Another look past the lunar North Pole this time features the large farside crater Rozhdestvenskiy on the limb, with Hermite adjoining it (Figure 1). The present lighting brings out possible tectonic features such as the straight, parallel walls of Peary and Byrd. Also in the Northern Highlands, the rays of the setting Sun are grazing the floor of Meton, bringing out its low relief.

At the present phase, the only mare areas on the terminator are Mare Frigoris and Mare Serenitatis. As in the previous mosaic, numerous low hills and mounds can be seen in Mare Frigoris, north of Aristoteles, along with the ejecta halo of that crater.

Southwest of Eudoxus is the irregular depression Alexander; pre-Imbrian in age with Imbrian-Period mare flooding on its floor.

Mare Serenitatis' north-south medial ridge is on the terminator now. Its oddly unnamed segments run the entire diameter of the mare basin from north to south, passing through the crater Bessel.

Highlands areas near the terminator may at first glance appear as confusing masses of shadow. However, as the Central Highlands now show, they contain numerous features of interest (Figure 2). One example is "light plains" areas such as that between Taylor and Lindsay. These areas are mare-like in texture but consist of light-toned highlands materials. They may represent basin impact-induced melting, debris flows, or both. The floors of some craters, such as Abulfeda, may contain similar deposits. Also, a hummocky and pitted deposit lies immediately southwest of Almanon, filling the floors of a chain of ancient craters.

In the Southern Highlands, Sacrobosco lies on the northern floor of an older disintegrated depression, evident only under low lighting. Farther south, the crater Barocius is shadow-filled, but the floor of its western neighbor Maurolycus is still sunlit, with its complex central elevation, together with the partly demolished southern component of Maurolycus. A bit more south, the western rim of the possible Mutus-Vlacq Basin is outlined by the craters Bacon, Cuvier, Lilius and Jacobi.

2. Central Highlands.
1996 Jun 07, 11h17m UT.
28-cm Sch.-Cass., f/21, 0.15 s.
Col. = 161°.9, Solar Lat. = 1°.4S,
Librations = 5°.2E/3°.4S.

161°–N

90°N to 26°N

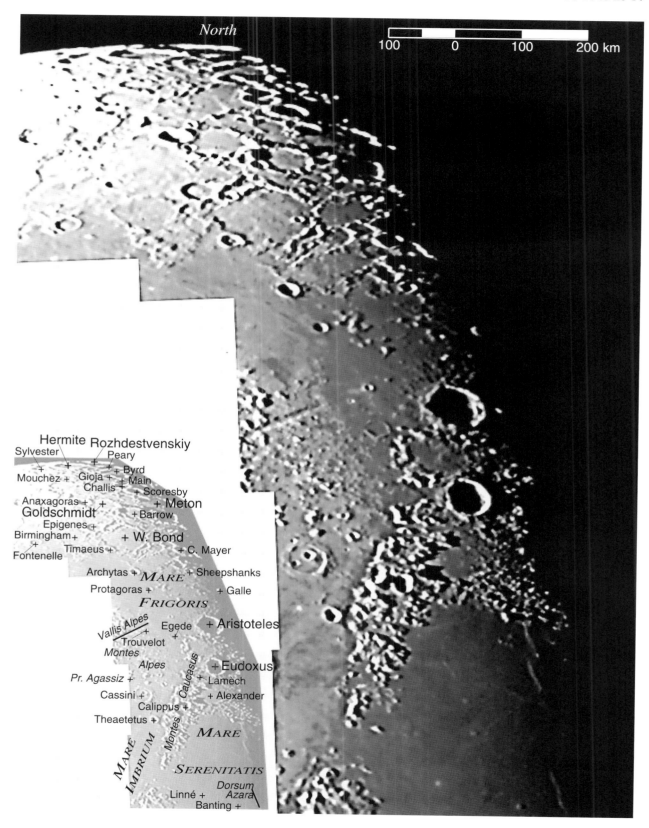

203

161°–C
28°N to 15°S

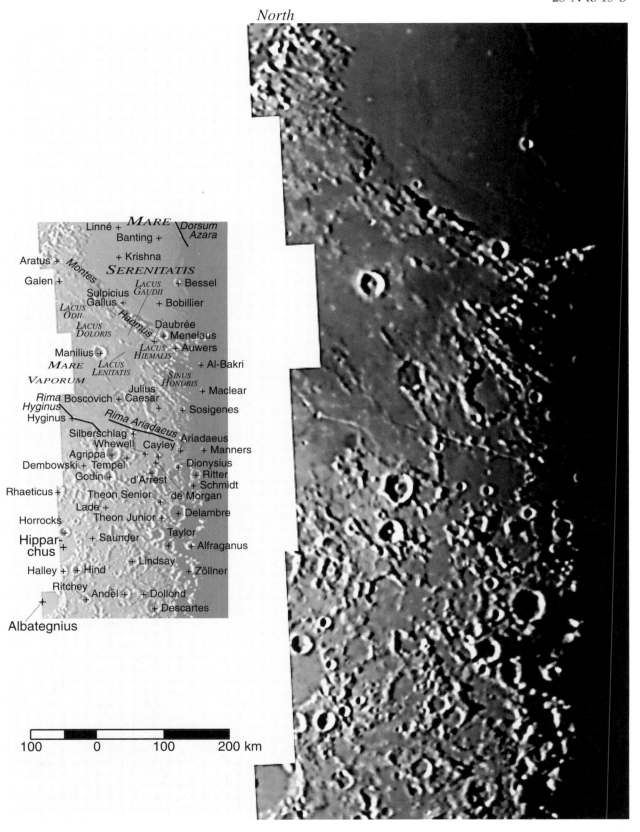

161°–S
11°S to 83°S

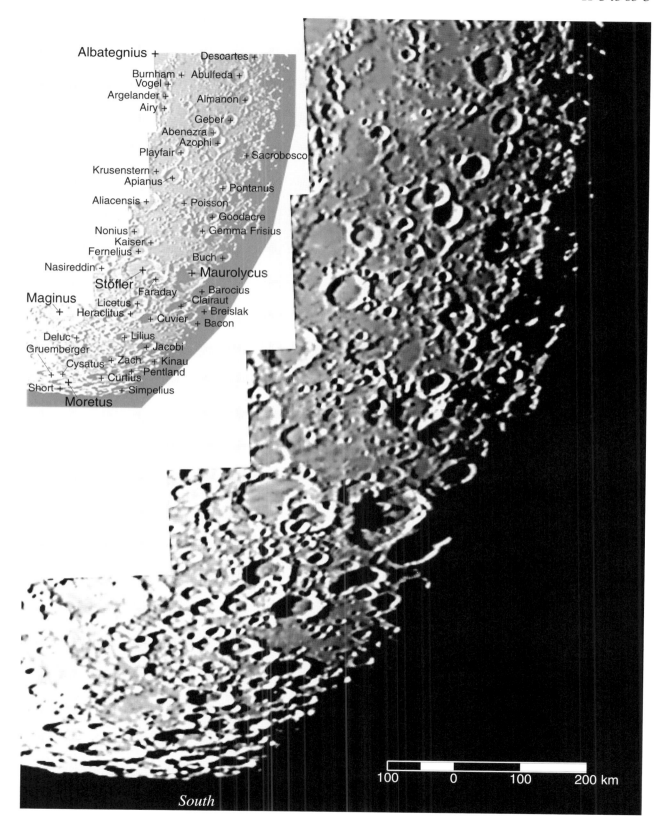

205

Colongitude 167°

1995 Jun 19, 11h22m UT. 28-cm Sch.-Cass., f/10. 0.20 s, W58 (green) Filter.
Colongitude = 166°.8, Solar latitude = 1°.3S, Librations = 8°.0E/3°.5S. 1.41 km/pixel.

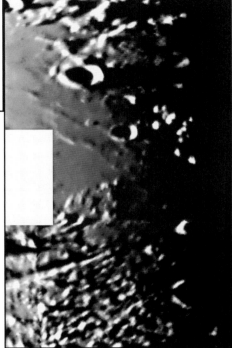

1. Mare Frigoris - Montes Alpes. 1996 Jul 07, 11h56m UT. 36-cm Sch.-Cass., f/23, 0.10 s. Col. = 168°.9, Solar Lat. = 1°.5S, Librations = 7°.4E/0°.9N.

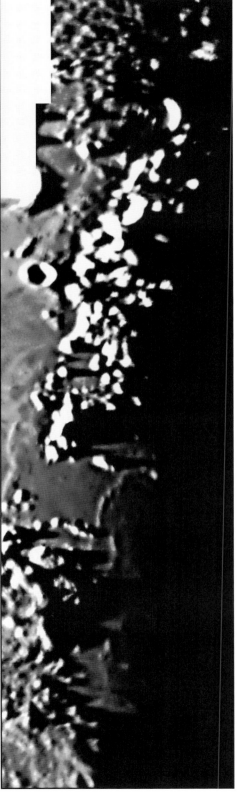

2. Montes Caucasus - Apenninus (to right). 1996 Jul 07, 11h28m UT. 36-cm Sch.-Cass., f/23, 0.10 s. Col. = 168°.7, Solar Lat. = 1°.5S, Librations = 7°.5E/0°.9N.

In the Northern Highlands the Sun is setting on W. Bond, a polygonal crater with a segment of a possible outer ring to its southeast.

The ridges south of Archytas in Mare Frigoris, trending southeast to northwest, may represent traces of a rim of the Imbrian Basin (Figure 1).

In the Montes Alpes, southeast of the Vallis Alpes, the somewhat chaotic terrain shows alignments radial to the Imbrian Basin, parallel to the valley (Figure 1).

Just northeast of the strait between Mare Imbrium and Mare Serenitatis is the "Valentine Dome" (Figure 2). On the Serenitatis side of the strait is a wrinkle ridge; the "ridge" on the Imbrium side appears to be the lobe of a lava flow from Mare Serenitatis.

The ejecta valleys and crater chains southeast of Mare Vaporum are quite evident at this phase. To their south the Sun is setting on the Rima Ariadaeus.

In the Central Highlands, a series of patches of light-plains material begins at the lunar equator near the breached crater Lade, and extends in patches south to Krusenstern, a distance of about 800 kilometers. The flat floors of the older craters in this area, such as Hipparchus, Albategnius, and Apianus, appear covered with similar deposits. The crater density in these patches is lower than in the surrounding highlands, meaning the plains are more recent and are probably filled with Imbrian-Basin ejecta.

Maurolycus is on the terminator. To its west is Stöfler, whose floor shows similar light-plains deposits. However, the plain area immediately northeast of Stöfler appears rougher and thus older than the Imbrian ejecta deposits, probably Nectarian or even pre-Nectarian in age.

167°–N

86°N to 15°N

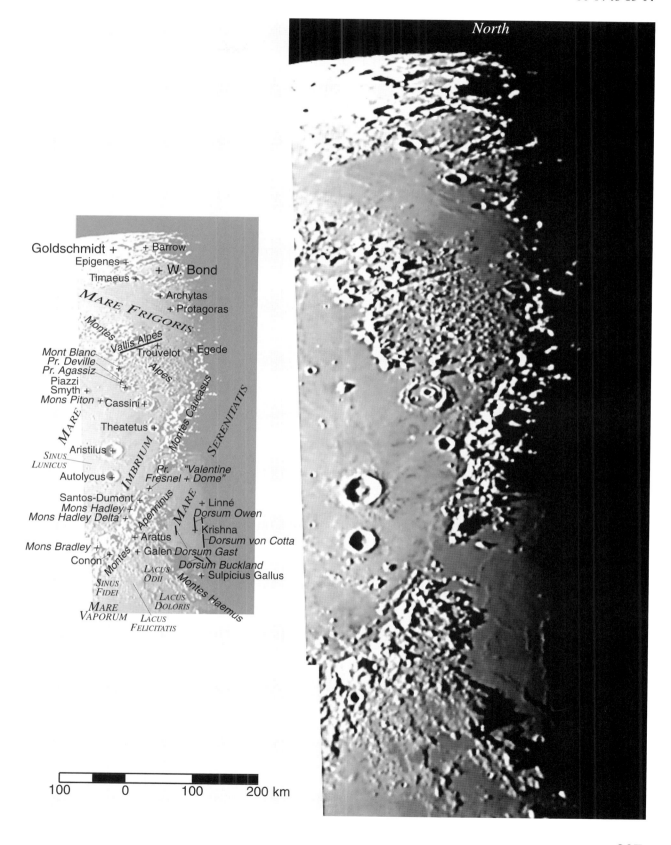

167°–C

17°N to 24°S

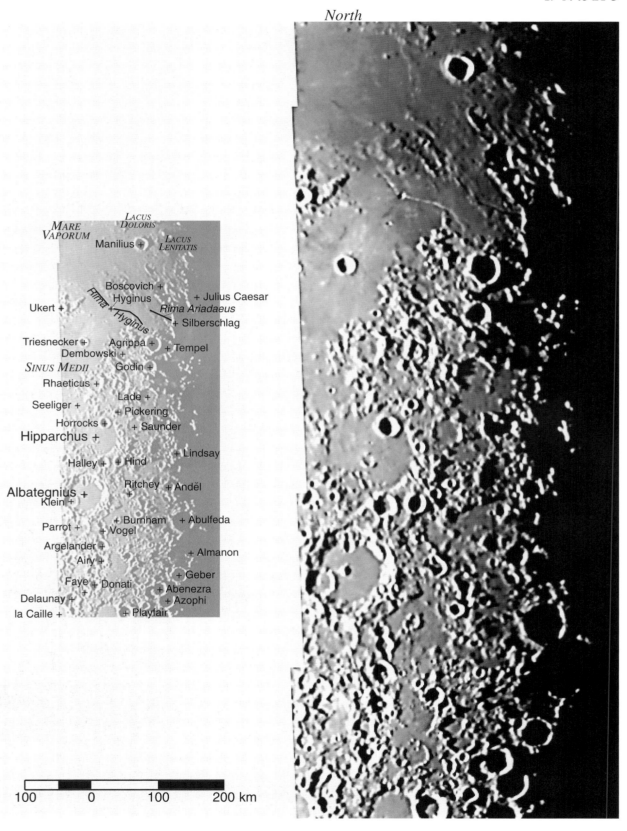

167°–S
22°S to 90°S

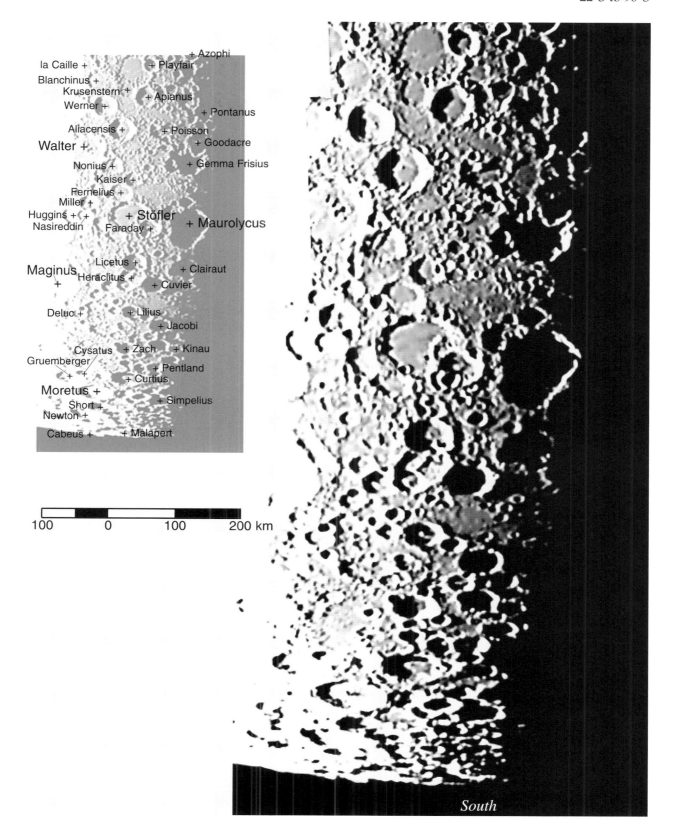

Colongitude 174°

1995 Sep 16, 11h44m UT. 36-cm Sch.-Cass., f/11, 0.15 s, W58 (green) Filter.
Colongitude = 174°.3, Solar Latitude = 0°.9S, Librations = 1°.8E/6°.0N. 1.13 km/pixel.

This is an ideal phase for studying the flat floors of several ancient flooded craters. Within the Northern Highlands, Goldschmidt and W. Bond both have a variety of low relief on their floors (Figure 1). W. Bond contains several "saucers" similar to those in Ptolemaeus. In the Central Highlands, the floors of the craters Hipparchus, Albategnius, la Caille, Blanchinus, and Walter show features ranging from low mounds to massive central peaks.

The portion of Mare Frigoris near the terminator shows the ridges concentric to the Imbrian Basin, mentioned earlier, but now also reveals a system of ridges radial to the Basin in the area south of Timaeus.

The region from the Montes Apenninus south to Sinus Medii and beyond contains many features formed by ejecta from the Imbrian Basin (Figure 2; in which, incidentally, we are poised exactly above the terminator). The Montes Apenninus themselves, of course, are one of the clearest examples of a basin rim, characterized by a steep scarp facing the interior of the basin with a more gentle slope facing the exterior.

The floor of Mare Vaporum apparently was formed in Eratosthenian times, and thus shows little detail beyond peripheral ridges. In contrast, north of Hyginus, the so-called "Schneckenberg" ("Snail Mountain") formation also appears furrowed by material from the Imbrian Basin, but the furrows are superimposed on an older volcanic formation. The famous Rima Hyginus, and the crater Hyginus itself, are usually interpreted as volcanic collapse features. Similarly, and not far away, layered lava flows can be seen north of Triesnecker.

In the Central Highlands south of Sinus Medii (and thus the lunar equator), the area near the major craters Ptolemaeus, Hipparchus, and Albategnius contains numerous valleys and crater chains radial to the Imbrian Basin; for example on the west side of Gyldén and southeast of Réaumur. In this area, though, is one non-conforming crater chain, that extending northwest of Müller. The chain is not obviously aligned with any recognized impact basin and its sharply defined components appear to be primary, rather than secondary, craters. It is possible that this unnamed feature resulted from the impact of a fragmented comet, similar to the Comet Shoemaker-Levy 9 impact with Jupiter in 1994.

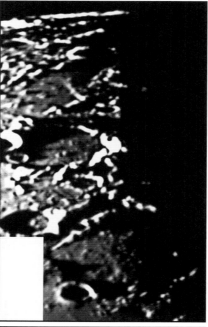

1. Northern Highlands (above).
1995 Sep 16, 12h20m UT.
36-cm Sch.-Cass., f/23, 0.10 s.
Col. = 174°.6, Solar Lat. = 0°.9S,
Librations = 1°.6E/6°.1N.

2. Montes Caucasus - Sinus Medii (to right).
1997 Sep 23, 12h49m UT.
28-cm Sch.-Cass., f/21, 0.15 s.
Col. = 171°.6, Solar Lat. = 0°.3N,
Librations = 8°.4E/7°.5N.

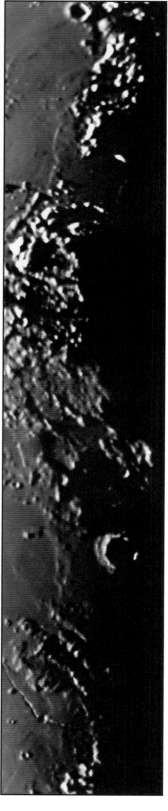

174°–N

90°N to 24°N

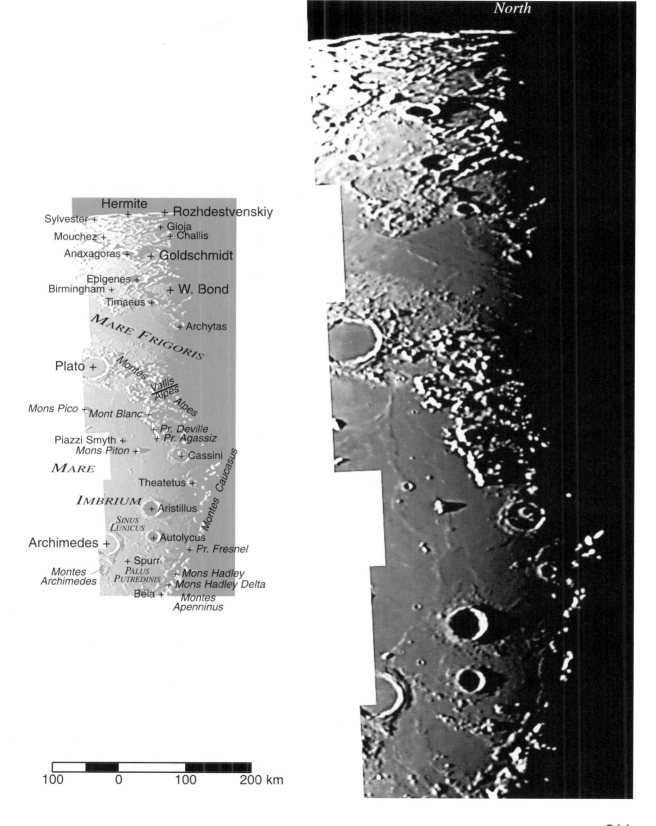

174°–C
26°N to 15°S

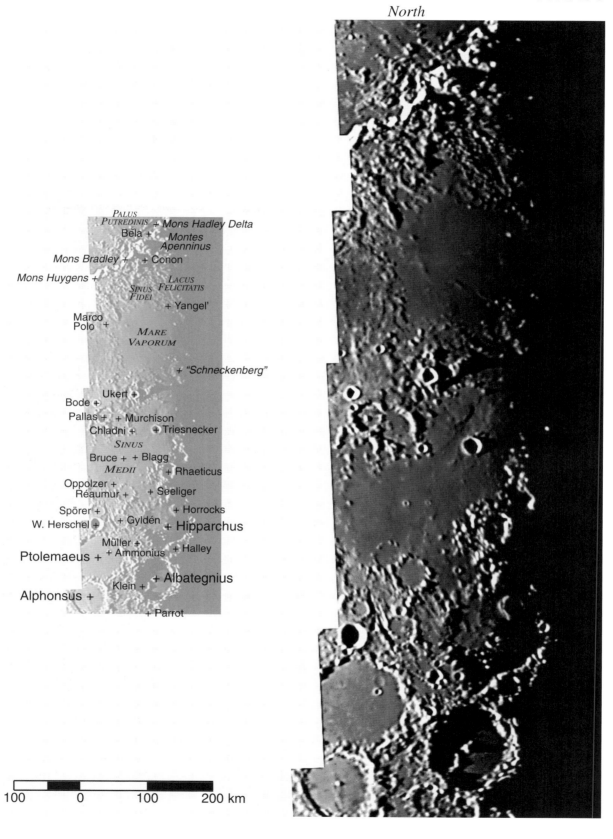

174°–S
13°S to 84°S

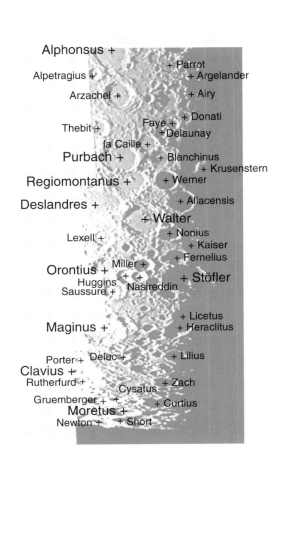
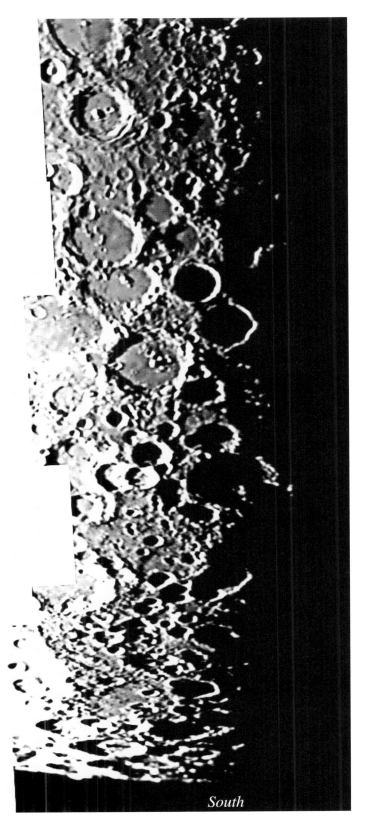

South

213

Colongitude 179°

1995 Jun 20, 11h22m UT. 28-cm Sch.-Cass., f/10, 0.25 s, W58 (green) Filter.
Colongitude = 179°.0, Solar latitude = 1°.3S, Librations = 7°.2E/2°.3S. 1.44 km/pixel.

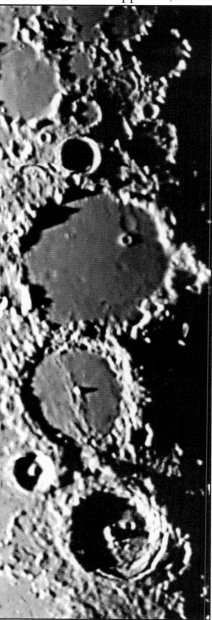

Three-quarters of the way through the lunation, our line of sight is perpendicular to the direction of illumination. This geometry brings out well the relief within eastern Mare Imbrium (Figure 1): a system of concentric wrinkle ridges, isolated highlands peaks marking perhaps two inner rims, the ejecta apron around the crater Aristillus, and the highlands-volcanism feature called the "Apennine Bench."

The lighting is also favorable in the Central Highlands (Figure 2). Several domes are now apparent on the floors of Flammarion and Oppolzer, while the "saucers" on the floor of Ptolemaeus and the central elevation and medial ridge of Alphonsus are also evident. The crater Arzachel also has a central peak and medial ridge, along with a concentric valley on the southeast and southwest rims of its terraced wall. An additional distinctive feature is the semicircular valley just southwest of Herschel.

1. Eastern Mare Imbrium (to left).
1995 Aug 18, 11h12m UT. 28-cm Sch.-Cass., f/21, 0.15 s. Col. = 180°.0, Solar Lat. = 1°.4S, Librations = 4°.1E/4°.0N.

2. Flammarion to Arzachel (to right).
1995 Jun 20, 11h44m UT. 28-cm Sch.-Cass., f/21, 0.15 s. Col. = 179°.2, Solar Lat. = 1°.3S, Librations = 7°.2E/2°.3S.

179°–N
88°N to 16°N

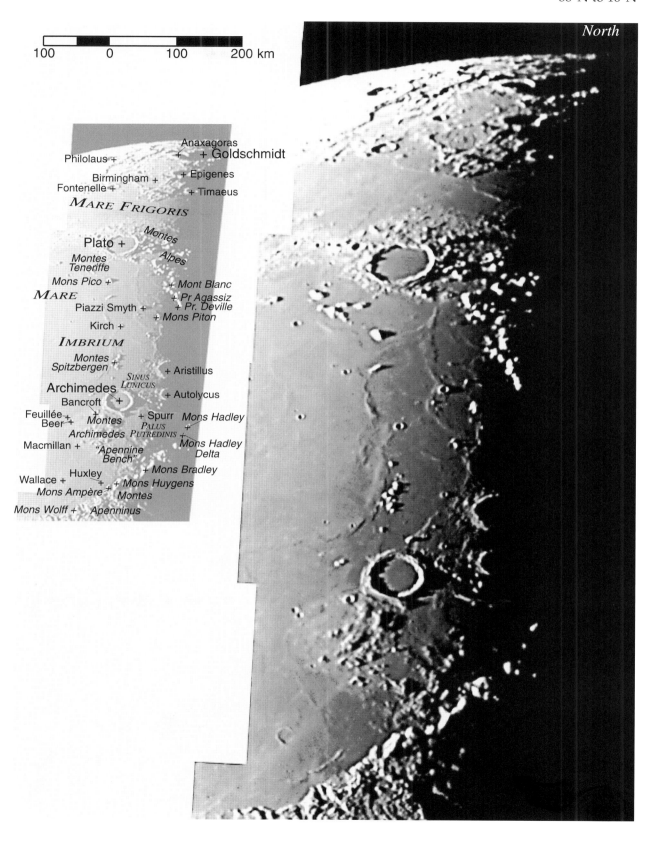

179°–C

17°N to 23°S

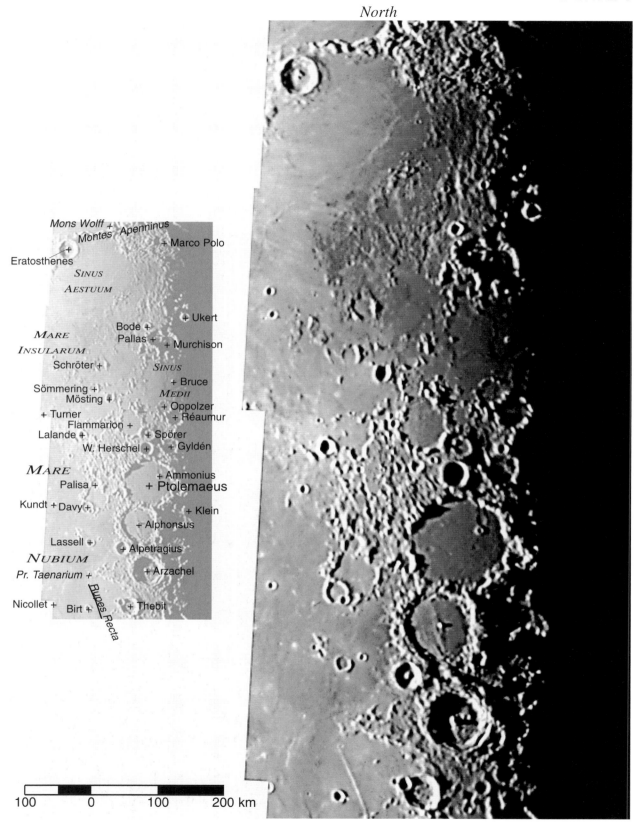

216

179°–S

22°S to 90°S

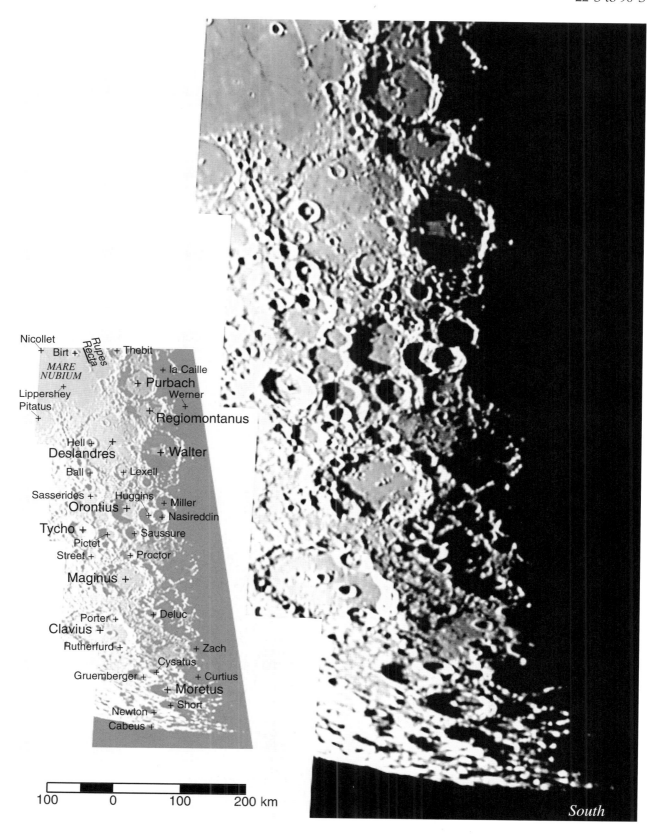

South

217

Colongitude 183°

1994 Dec 25, 12h34m UT. 28-cm Sch.-Cass., f/10, 0.10 s, W58 (green) Filter.
Colongitude = 182°.6, Solar latitude = 1°.2N, Librations = 5°.5W/5°.2N. 1.38 km/pixel.

The mere four degrees of terminator advance since the last mosaic has altered the appearance of Mare Imbrium (Figure 1), bringing into prominence ridges, hills, and mountains that mark an inner Imbrian rim running from Pico south through the crater Kirch and the Montes Spitzbergen and Archimedes. Note also the ghost crater between Plato and Pico, once called "Newton," and a dome on the western face of the ridge just west of the Montes Spitzbergen. This is also a good phase for seeing the small craters on the floors of the craters Plato and Archimedes.

For about one day near this phase, the peaks of the Apennine scarp protrude into the night hemisphere. South of the Montes Apenninus, the smooth surface of Sinus Aestuum contrasts with the rugged hills and ridges of Imbrian ejecta to its southeast, north of the crater Schröter.

In the Southern Highlands the floor detail of the craters Deslandres and Maginus is brought into prominence by the setting Sun (Figure 2). Two southeast-northwest chains of hills are found in Deslandres, parallel to its northeast wall. One of these ridges forms the northeast wall of a quadrangular enclosure inside Deslandres, just northwest of Lexell. Finally, note the traces of a concentric scarp inside Deslandres' northwest wall.

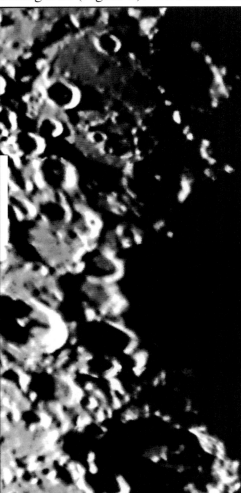

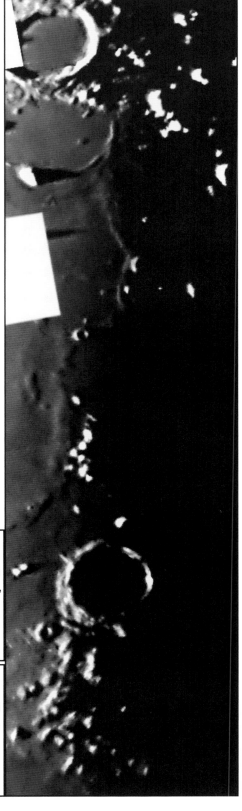

1. Eastern Mare Imbrium (to right).
1997 Sep 24, 13h12m UT. 28-cm Sch.-Cass., f/21, 0.20 s. Col. = 184°.0, Solar Lat. = 0°.4N, Librations = 8°.1E/6°.9N.

2. Deslandres - Maginus (to left).
1997 Sep 24, 12h38m UT. 28-cm Sch.-Cass., f/21, 0.15 s. Col. = 183°.8, Solar Lat. = 0°.4N, Librations = 8°.2E/6°.8N.

183°–N
90°N to 24°N

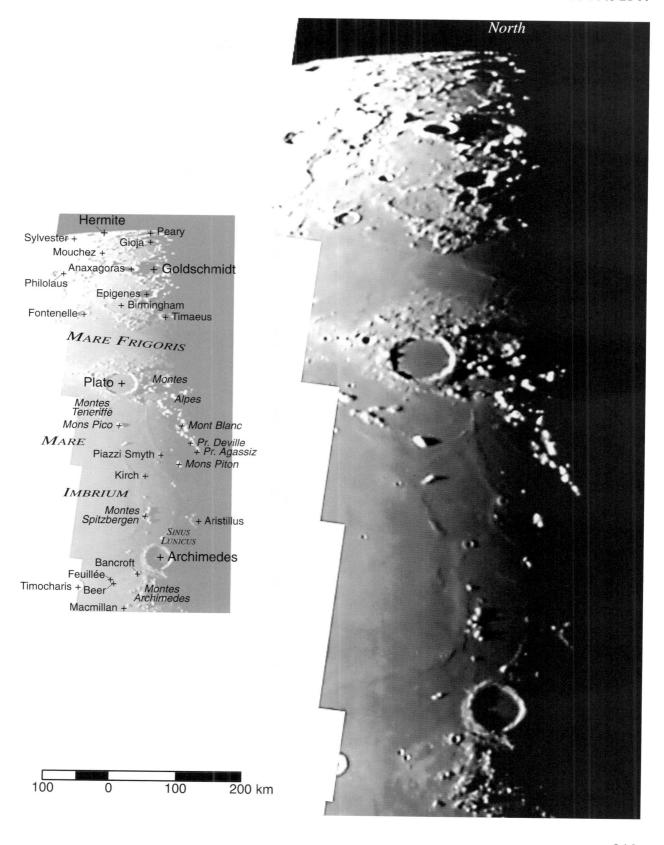

183°–C
26°N to 16°S

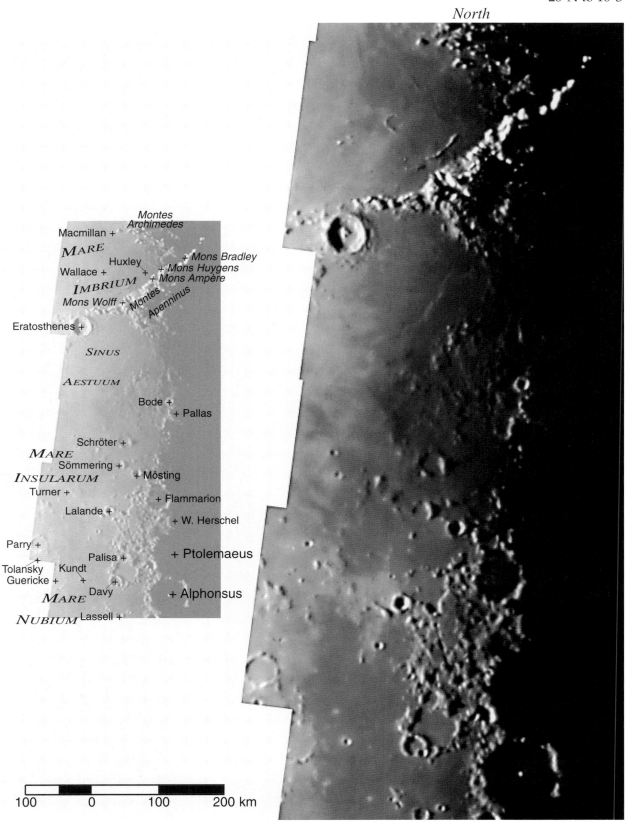

183°–S
14°S to 85°S

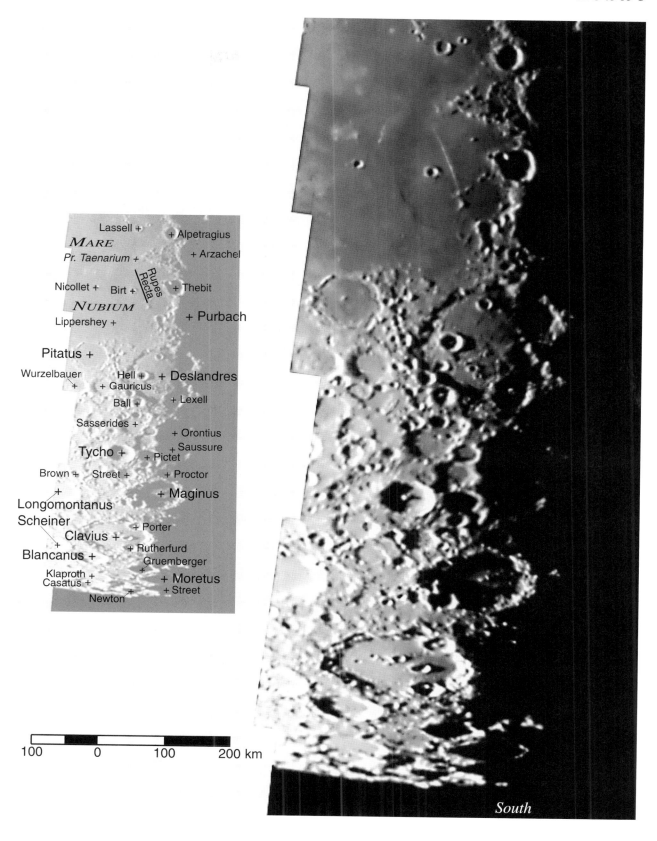

South

221

Colongitude 191°

1995 JUN 21, 11h34m UT. 28-cm Sch.-Cass., f/10, 0.30s, W58 (green) Filter.
Colongitude = 191°.3, Solar latitude = 1°.3S, Librations = 7°.1E/1°.0S. 1.46 km/pixel.

Although the libration is slightly to the south, an interesting feature is currently visible in the Northern Highlands; a shallow valley or chain of depressions running north-south from just northwest of Anaxagoras to slightly north of Fontenelle. Roughly parallel to that feature is a scarp or chain of hills running from Mouchez to Philolaus and then south from Philolaus to a point west of Fontenelle.

In northern Mare Imbrium the Montes Recti and Montes Teneriffe define a portion of the inner rim of the Imbrian Basin. The southern boundary of the Imbrian Basin and Mare is marked by the Montes Carpatus. In the mare itself there are few craters but several wrinkle ridges, such as the Dorsum Grabau, Dorsum Higazy, and several unnamed ridges.

West of Eratosthenes, near the ghost crater Stadius, the ejecta from Copernicus have formed groups of ridges near Copernicus but have created several chains of secondary craters farther to the east (Figure 1).

The lighting is also good to view the highlands "island" within Mare Insularum, which includes the ruined craters Fra Mauro, Parry, Bonpland, and Guericke. The highlands consist of Imbrian ejecta (the *Fra Mauro Formation*) on the west and north and volcanic plains on its eastern and southern portions.

Mare Nubium also contains some notable features. Near colongitude 186°-187° the west face of the Rupes Recta is at its most prominent. The ridges on the western rim of the mini-basin that encloses the scarp and the crater Birt are also easily seen at this time (Figure 2).

When the Sun advances a few more degrees west an intricate pattern of ridges appears in Mare Nubium south of Nicollet (Figure 3), perhaps marking additional pre-Mare flooded craters. Nearby is the unusual feature Wolf, a possible volcanic caldera.

On the southern rim of Mare Nubium, Pitatus is a well-known example of a crater with multiple, concentric walls. To its southwest is Wurzelbauer, another concentric-walled crater.

In the Southern Highlands, the terminator has reached Clavius, now the area's "show-piece" feature.

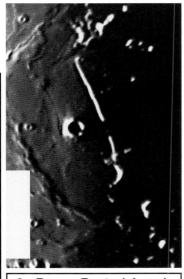

1. Stadius and Vicinity (above). 1995 AUG 19, 11h07m UT. 28-cm Sch.-Cass., f/21, 0.15 s. Col. = 192°.2, Solar Lat. = 1°.4S, Librations = 2°.9E/5°.0N.

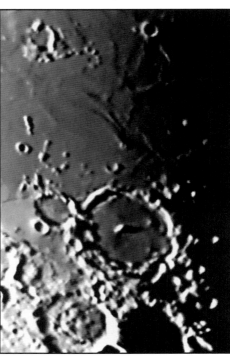

2. Rupes Recta (above). 1996 DEC 03, 13h52m UT 28-cm Sch.-Cass., f/29, 0.75 s. Col. = 187°.2, Solar Lat. = 1°.4 N, Librations = 2°.8W/3°.0N.

3. Mare Nubium (left). (Same data as for Figure 1.)

191°–N

89°N to 17°N

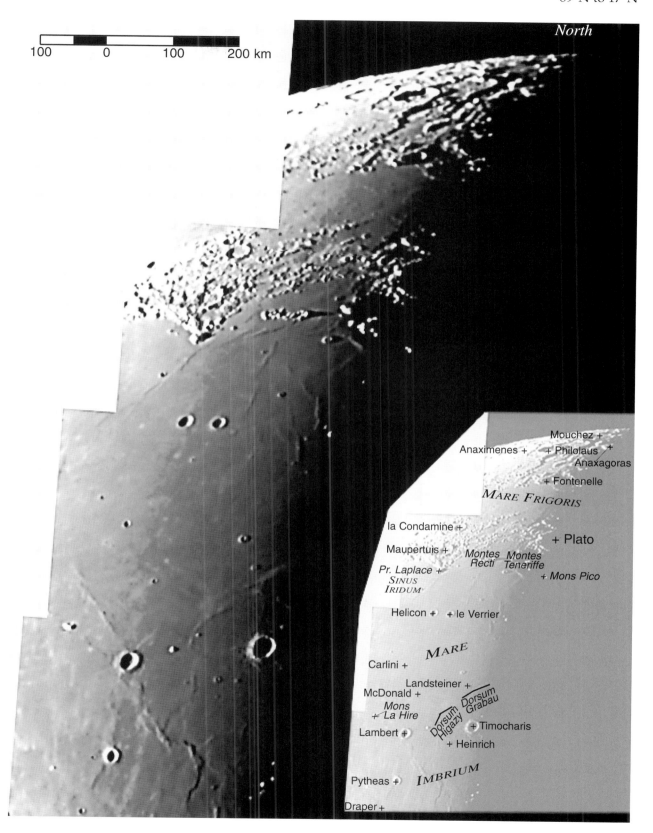

223

191°–C
19°N to 22°S

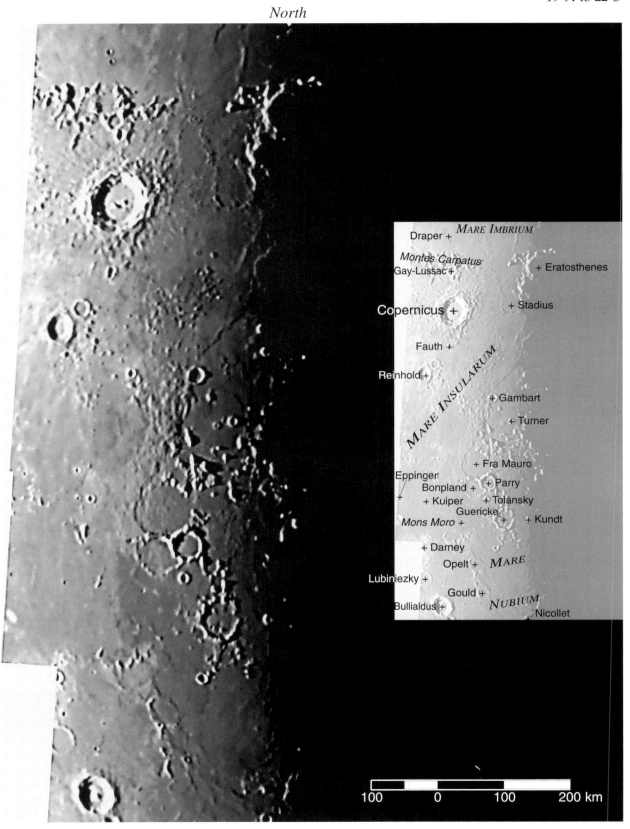

191°–S

20°S to 90°S

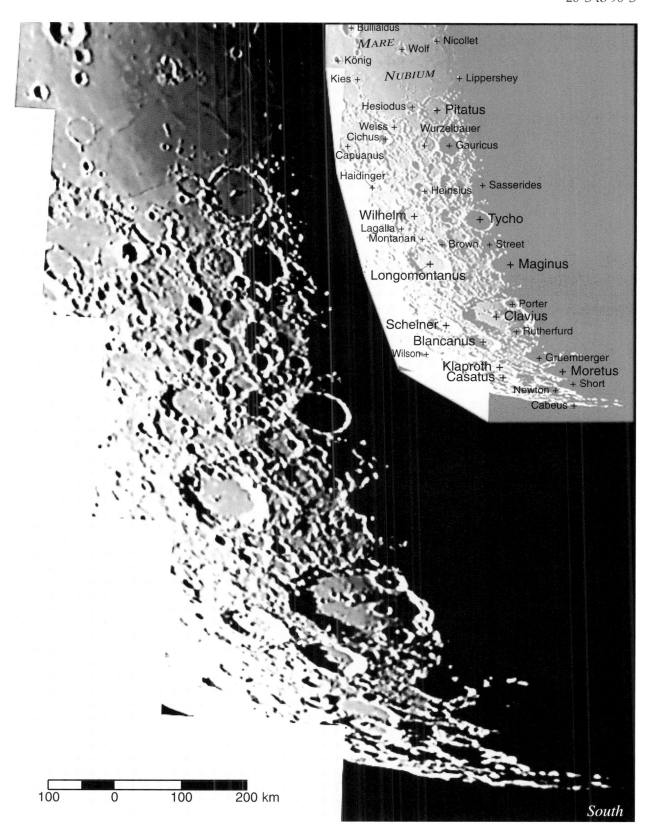

South

Colongitude 201°

1996 Feb 13, 13h22m UT. 28-cm Sch.-Cass., f/10, 0.20s, W58 (green) Filter.
Colongitude = 200°.7, Solar latitude = 1°.2N, Librations = 4°.9W/4°.3S.   1.36 km/pixel.

The north-south valley in the Northern Highlands described for Colongitude 191° is now filled with shadow, but still evident, as is the parallel chain of hills to its west.

The Montes Jura enclose Sinus Iridum on the west, north, and east, while the ridge that defines its southern boundary with Mare Imbrium is now visible as well. The latter appears to connect with ridges extending east to the Montes Teneriffe, as well as with ridges running south to C. Herschel and beyond to Lambert, joining the Dorsum Helm-Dorsum Zirkel system.

Only the western half of the Montes Carpatus remains sunlit, but forms a striking pattern of light and shadow (Figure 1). Copernicus lies directly on the terminator, the low Sun dramatically highlighting the complex ejecta apron on its western flank. South of Copernicus, the crater Reinhold and the patch of hilly highlands material surrounding it are also evident.

The scattered hills east of the Montes Riphaeus mark the boundary between Mare Insularum and Mare Cognitum and the southern rim of the suspected Mare Insularum Basin. The Montes Riphaeus form the northwest boundary of Mare Cognitum, itself possibly a large ghost crater (or small basin) whose southern rim is marked by the hills near the crater Darney.

The fresh crater Bullialdus is the most prominent feature in western Mare Nubium (Figure 2). Two flooded craters lie nearby; Lubiniezky to the northwest of Bullialdus and Kies to its south; note also the dome Kies π just west of Kies and the shallow valley extending northwest of Bullialdus.

South of the Campanus-Mercator crater pair, Rima Hesiodus crosses northern Palus Epidemiarum. On the southern margin of the Palus, its lava has covered the floor of the crater Capuanus.

In the Southern Highlands the great craters Wilhelm, Montanari and Longomontanus remain visible only as protrusions of shadow from the terminator; Clavius is entirely gone. To their west, however, the descending Sun now brings out the relief of the multiple craters Hainzel and Schiller. Scheiner is near the terminator and the low relief on its floor is visible. This is also the case with the crater pair to its south, Klaproth and Casatus, but whose interiors contain floors with little relief beyond a few craters.

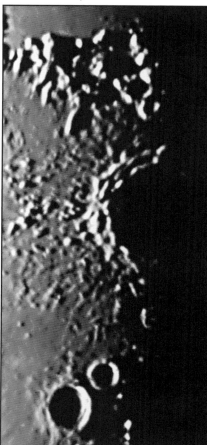

1. Sunset on Copernicus.
1996 Feb 13, 13h56m UT.
28-cm Sch.-Cass., f/21, 0.15s.
Col. = 201°.0, Solar Lat. = 1°.2N,
Librations = 5°.0W/4°.3S.

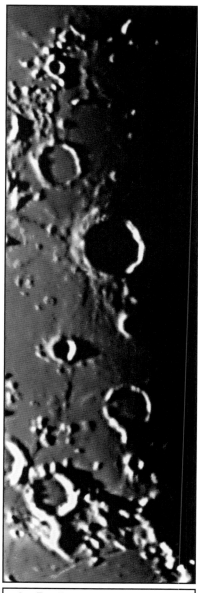

2. Bullialdus and Vicinity.
(Same data as for Figure 1.)

201°–N
86°N to 15°N

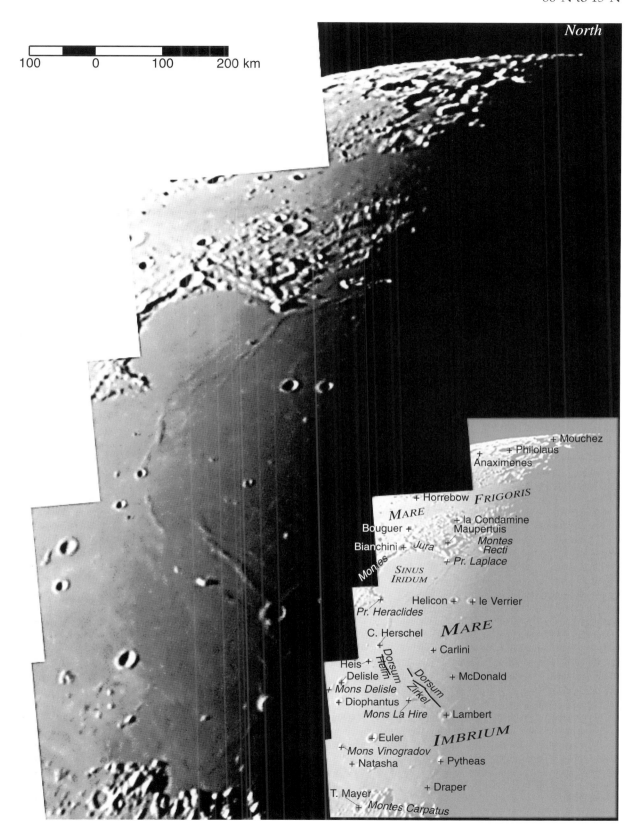

201°–C

16°N to 24°S

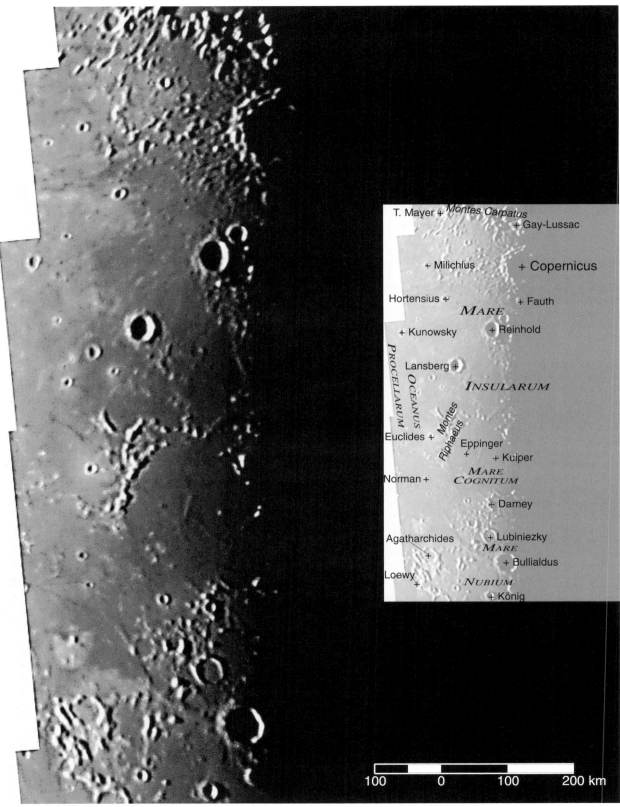

201°–S

23°S to 90°S

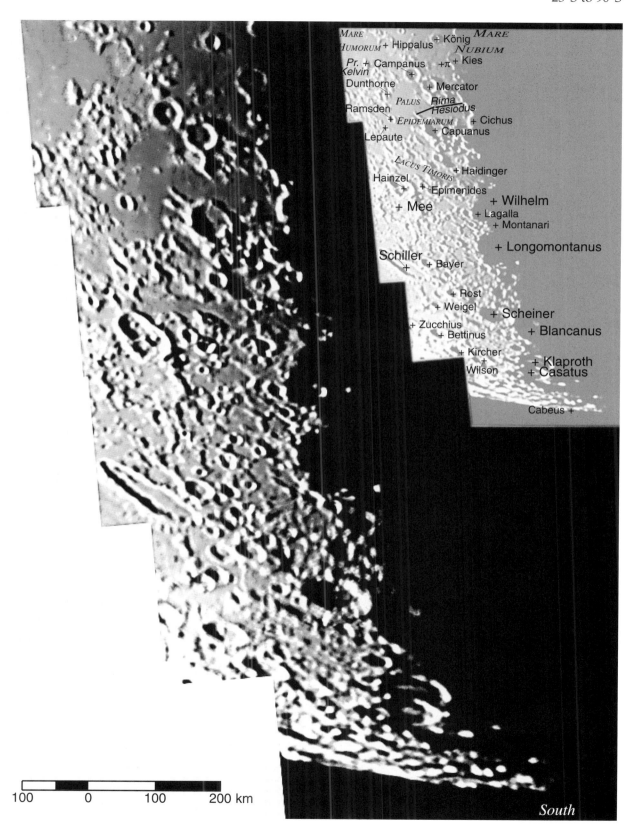

Colongitude 205°

1995 Oct 18, 12h02m UT. 28-cm Sch.-Cass., f/10, 0.30s, W47 (blue) Filter.
Colongitude = 204°.6, Solar latitude = 0°.0S, Librations = 3°.6W/6°.8N. 1.46 km/pixel.

The libration for this mosaic is particularly favorable for viewing the northwestern quadrant of the Moon; an area where hexagonal structures abound. The most evident is Sinus Iridum, whose outline is completed on the south by ridges demarcating the Sinus from Mare Imbrium. Other six-sided formations include J. Herschel, Anaximenes, and Philolaus C (the feature enclosing Philolaus). Also, there is a possible ancient hexagonal crater, large, ruined and unnamed, bounded in part by J. Herschel, Anaximander, Carpenter, Anaximenes, and Philolaus.

In the area west and southwest of what remains visible of the Montes Carpatus the low sun angle now brings out a well-known "dome field" (Figure 1). For example, a large number of these low-relief features occur southwest of T. Mayer, west of Milichius, and southeast, north, and northeast of Hortensius. Two other eye-catching features in this area are a square enclosure east of Milichius and a semicircle of peaks catching the Sun just north of the crater T. Mayer C (right of center of Figure 1).

The northern components of the Montes Riphaeus outline a partial ring and an irregular enclosure, both to the northeast of the crater Euclides (Figure 2). The shadows cast by the extremely low Sun make these low-lying mountains appear as spires.

The region east of Mare Humorum includes a mixture of features (Figure 3). First are several partial, flooded rings of which Agatharchides is the largest example. Northeast of that crater is the broad northwest-southeast trending valley Bullialdus W; extended northwest, this valley becomes a ridge that can be traced at least as far as the crater Herigonius. Finally, southeast of Agatharchides is Palus Epidemiarum and the Hippalus Rimae system, concentric to the rim of the Humorum Basin; the crater Dunthorne is superimposed on two of these rilles.

In the Southern Highlands, this phase is favorable for viewing the Hainzel complex, the ruined crater Mee, and the elongated depression Schiller. Schiller itself lies on the east rim of the Schiller-Zucchius Basin, which is slowly becoming visible again as the terminator gradually approaches it.

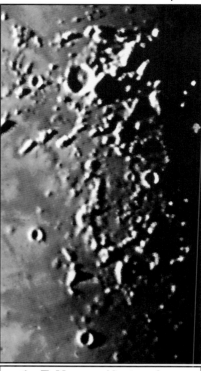

1. T. Mayer - Hortensius.
1995 Oct 18, 12h30m UT.
28-cm Sch.-Cass., f/21, 0.10 s.
Col. = 204°.8, Solar Lat. = 0°.0S,
Librations = 3°.7W/6°.8N.

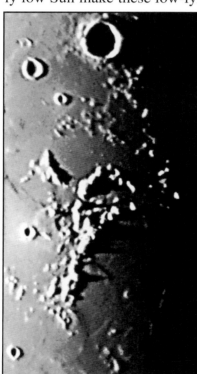

2. Lansberg - Montes Riphaeus.
1995 Oct 18, 12h36m UT.
28-cm Sch.-Cass., f/21, 0.15 s.
Col. = 204°.9, Solar Lat. = 0°.0S,
Librations = 3°.7W/6°.8N.

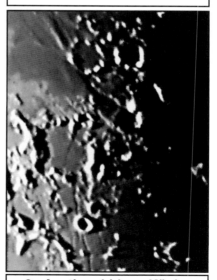

3. Agatharchides - König.
1995 Oct 18, 12h40m UT.
28-cm Sch.-Cass., f/21, 0.15 s.
Col. = 204°.9, Solar Lat. = 0°.0S,
Librations = 3°.7W/6°.8N.

205°–N

90°N to 21°N

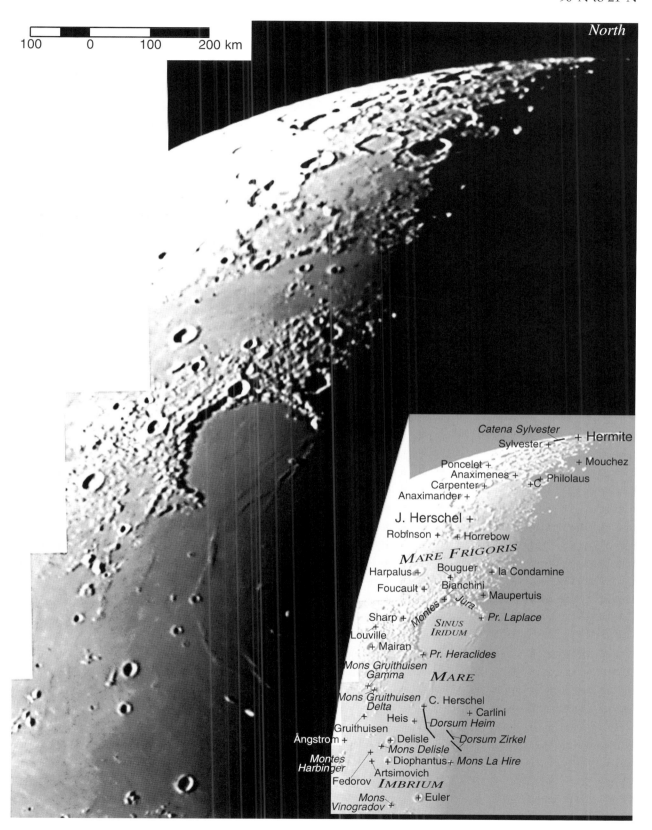

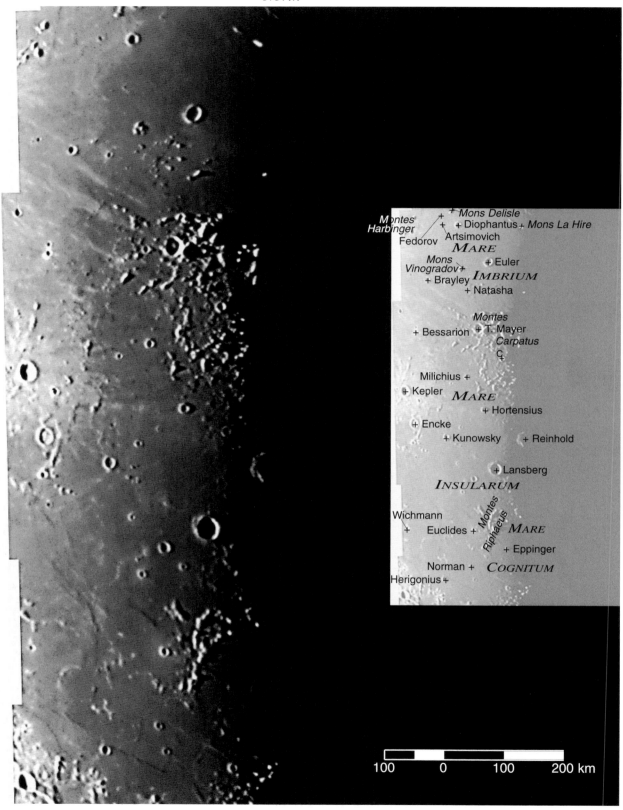

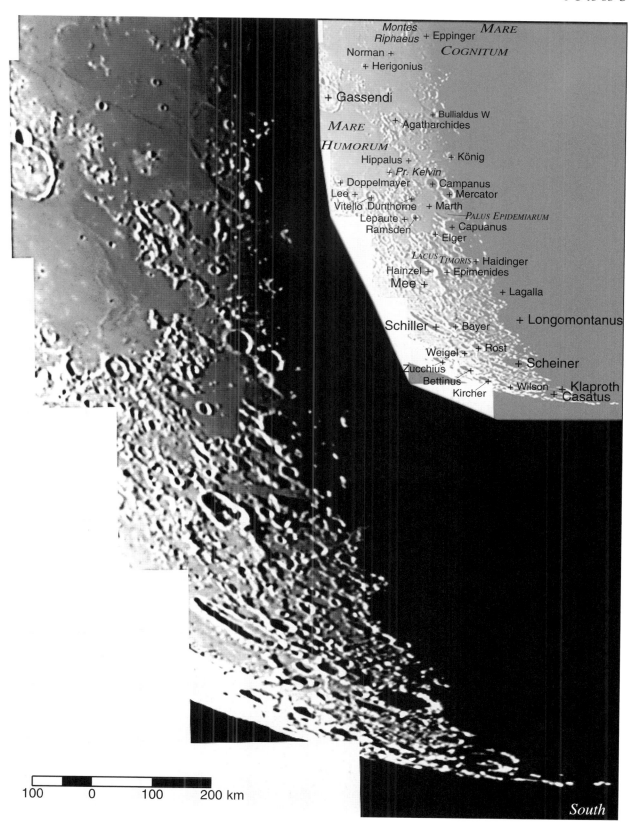

Colongitude 213°

1996 Feb 14, 13h46m UT. 28-cm Sch.-Cass., f/10, 0.25 s, W58 (green) Filter.
Colongitude = 213°.1, Solar latitude = 1°.2N, Librations = 3°.8W/5°.2S. 1.36 km/pixel.

The glancing illumination now brings out the undulating floor of J. Herschel in the Northern Highlands. To the crater's southeast in Mare Frigoris a group of hills can be seen.

The peaks of the Montes Jura are casting long shadows over Sinus Iridum, revealing that feature as a large crater that antedates the Imbrian-Period ejecta that form the Jura-Alpes-Caucasus Highlands. At the southwestern edge of these highlands, the highland domes Mons Gruithuisen Delta and Gamma now stand out dramatically, clearly steeper than mare domes.

In southwestern Mare Imbrium, Mons Vinogradov is now on the terminator, as are the westernmost hills of the Montes Carpatus.

The craters Kepler and Encke now form a striking pair on the eastern margin of Oceanus Procellarum. These two craters lie in an area of scattered highlands features; some consisting of ejecta from the Imbrian Basin, others of the remnants of pre-mare craters. Such low-lying hills and crater rims are also characteristic of southeastern Oceanus Procellarum between Wichmann and Herigonius.

This is also a favorable time to inspect the detail on the floor of the concentric-walled crater Gassendi and the concentric ridge system of eastern Mare Humorum (Figure 1). In the southern mare, note also the flooded pre-mare craters Puiseux, Lee, and Doppelmayer. The latter and nearby Vitello are also concentric-walled craters.

The Southern Highlands contain a distinctive unnamed feature on the terminator north of Hainzel, giving the appearance of an ancient group of craters or a double-rimmed basin.

The librations in this mosaic are favorable for viewing the southwestern limb. Although the sun angle is still high in this area, we are looking at the shaded faces of its relief features, so detail is still evident in such features as the inner and outer rims of the Schiller-Zucchius Basin, as well as the giant crater Bailly (Figure 2), itself constituting a two-rimmed basin. East of Bailly is a chain of three similar craters: Zucchius, Bettinus, and Kircher. Finally, on the limb south of Bailly is the deep terraced crater Drygalski, containing a multiple central peak.

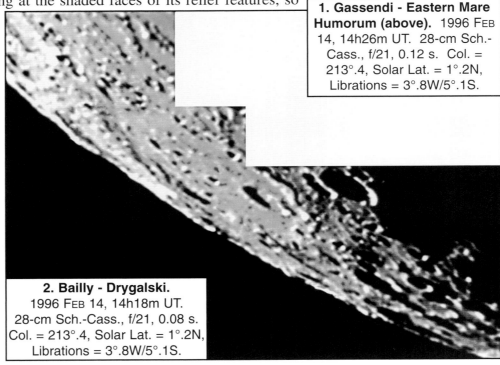

1. **Gassendi - Eastern Mare Humorum (above).** 1996 Feb 14, 14h26m UT. 28-cm Sch.-Cass., f/21, 0.12 s. Col. = 213°.4, Solar Lat. = 1°.2N, Librations = 3°.8W/5°.1S.

2. **Bailly - Drygalski.** 1996 Feb 14, 14h18m UT. 28-cm Sch.-Cass., f/21, 0.08 s. Col. = 213°.4, Solar Lat. = 1°.2N, Librations = 3°.8W/5°.1S.

213°–N

85°N to 4°N

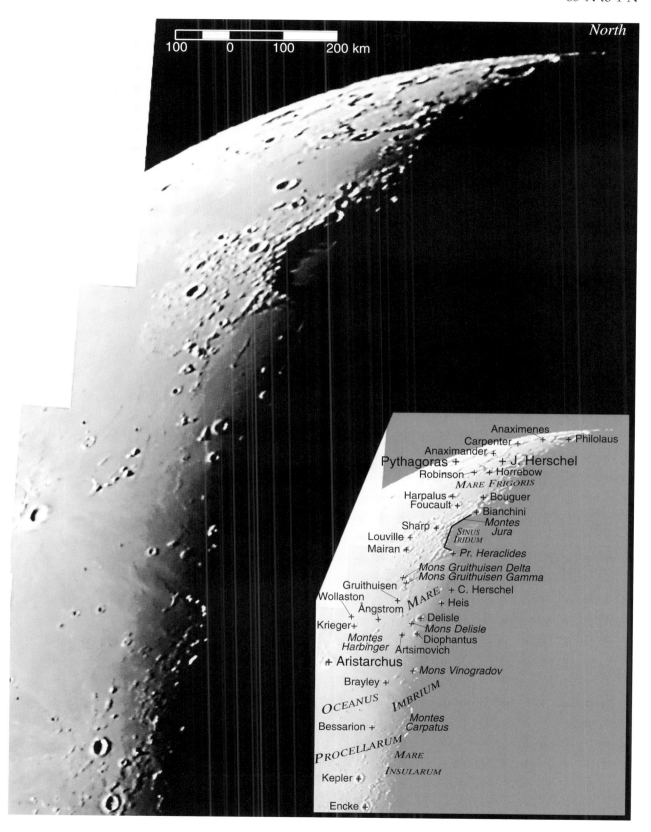

213°–C

20°N to 31°S

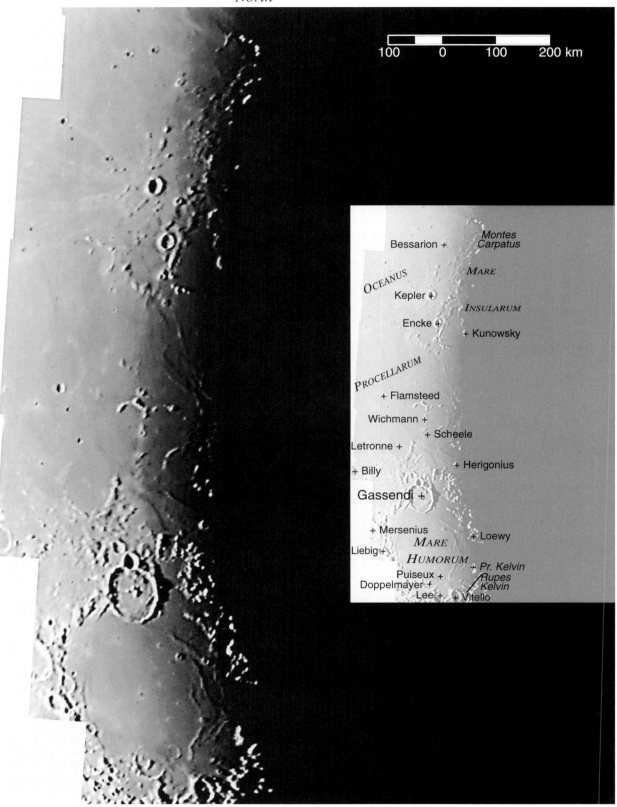

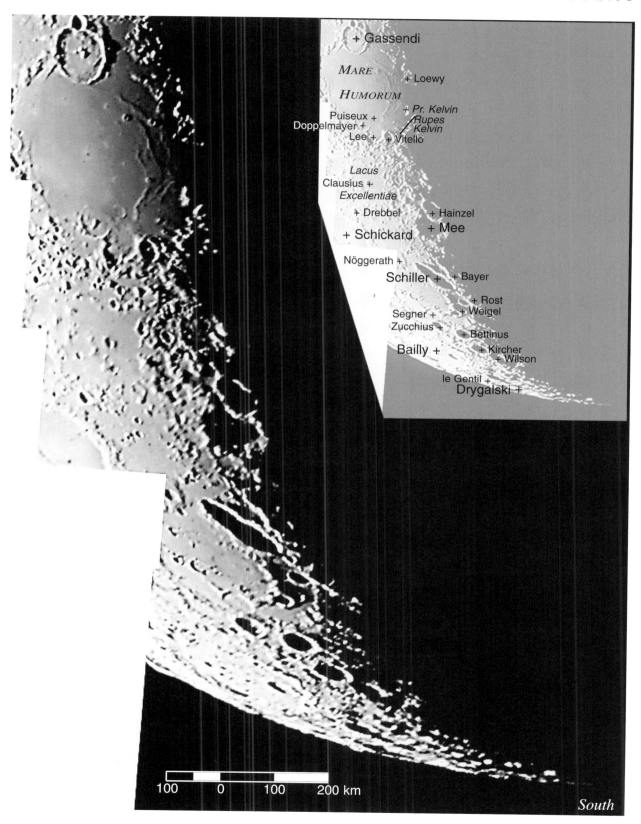

Colongitude 225°

1996 Feb 15, 13h38m UT. 28-cm Sch.-Cass., f/10, 0.08 s.
Colongitude = 225°.2, Solar latitude = 1°.2N, Librations = 2°.2W/5°.7S. 1.34 km/pixel.

With the sunset terminator's longitude midway between the disk's mean center and limb (i.e., longitude 45° west), the lunar crescent is rapidly narrowing and harder to observe in good seeing just before dawn. However, waning-crescent phases such as this are the only way to observe the westernmost quarter of the Earth-facing hemisphere under afternoon lighting.

One area near the terminator contains the westernmost portions of Mare Frigoris and the Jura-Alpes-Caucasus Highlands (Figure 1). Although composed of Imbrian Basin ejecta, this portion of the highlands contains two southwest-northeast valleys that appear radial to the Orientale Basin (see Figure 225°–N). If this is the case, these valleys imply that the Orientale impact occurred later than the Imbrian impact, although both are dated to the Early Imbrian Period. Also note (Figure 1) how brilliant the highlands domes Gruithuisen Delta and Gamma appear when on the terminator.

Most of Oceanus Procellarum can now be seen under afternoon illumination. The most prominent feature in the Oceanus is the diamond-shaped extrusive volcanic feature named the "Aristarchus Plateau." This elevated area is crossed by the Vallis Schröteri. On the plateau's northwest are its highest features, while the prominent crater pair of Aristarchus and Herodotus define its southeastern boundary.

Another volcanic area, farther south in the Oceanus, is the "Marius Hills." About the same size as the Aristarchus Plateau, the Marius Hills are lower and characterized by hummocky terrain with domes and what appear to be cinder cones.

The partial crater Letronne marks the southern edge of Oceanus Procellarum; its floor inundated by mare lava but with a remnant of its central peak surviving (Figure 2). To Letronne's south is Mare Humorum; the ridges near its western edge are evident in Figure 2. The present lighting is also favorable for discerning the convex nature of the floor of the nearby crater Mersenius.

The libration of the present mosaic aids in viewing the Southwestern Highlands, including the large crater Schickard with its level, flooded floor. To Schickard's south is the flooded crater Wargentin and the double feature Nasmyth-Phocylides. To the east of these three craters lie the relatively flat highlands-plains deposits of the floor of the Schiller-Zucchius Basin. Finally, Bailly is well shown in the present mosaic, although the limb areas to its south are becoming lost in shadow.

1. Harpalus - Gruithuisen.
1997 Sep 27, 13h02m UT. 28-cm Sch.-Cass., f/21, 0.40 s. Col. = 220°.6, Solar Lat. = 0°.4N, Librations = 5°.2E/3°.8N.

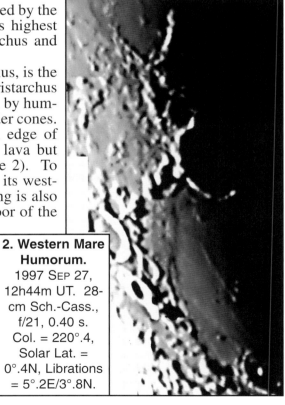

2. Western Mare Humorum.
1997 Sep 27, 12h44m UT. 28-cm Sch.-Cass., f/21, 0.40 s. Col. = 220°.4, Solar Lat. = 0°.4N, Librations = 5°.2E/3°.8N.

225°–N

84°N to 1°S

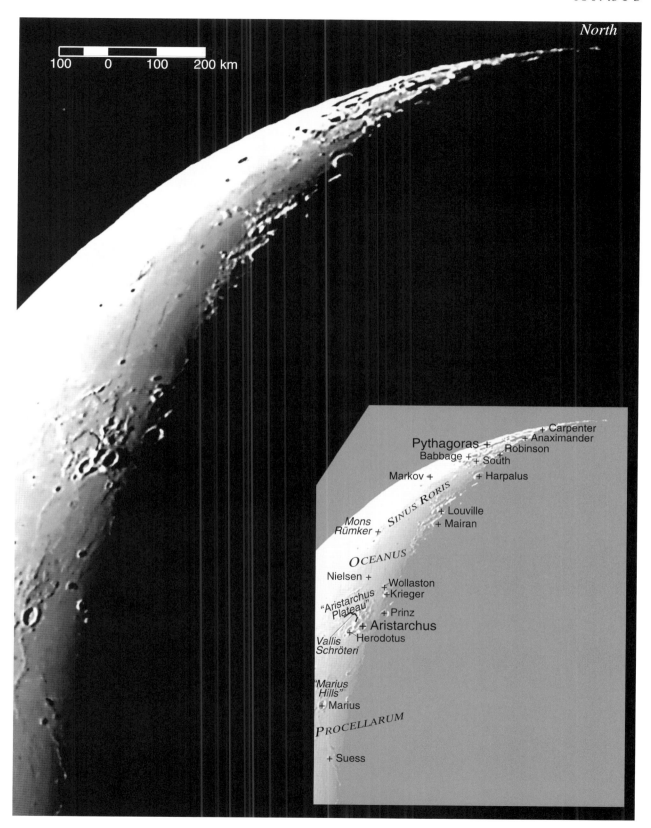

225°–C
23°N to 34°S

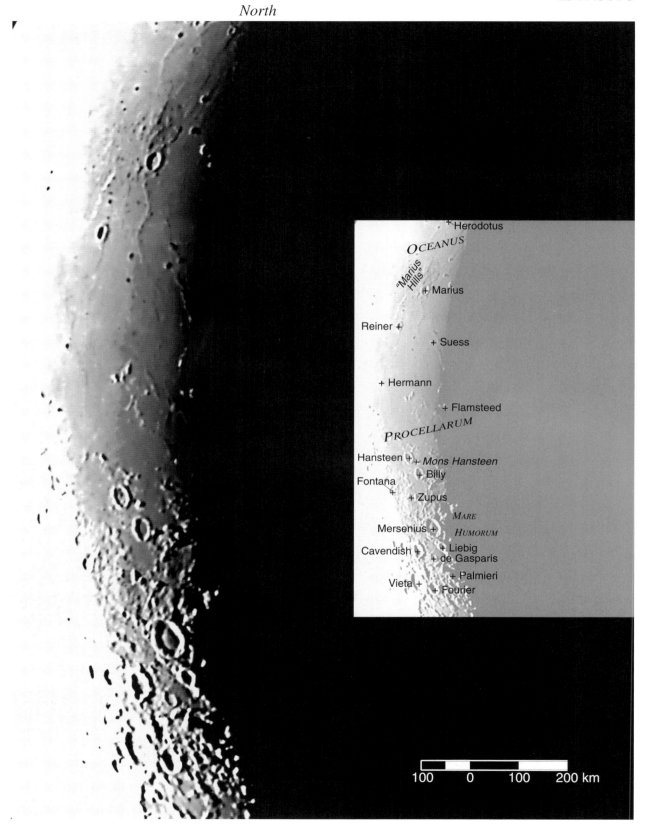

225°–S
10°S to 90°S

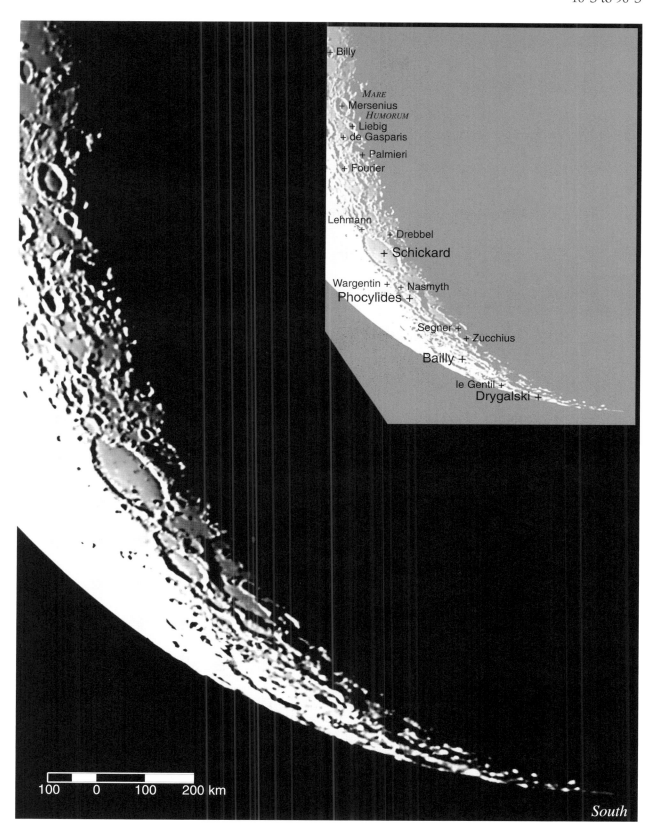

Colongitude 233°

1996 Jan 17, 14h17m UT. 28-cm Sch.-Cass., f/10, 0.35 s, W58 (green) Filter.
Colongitude = 232°.8, Solar latitude = 1°.5N, Librations = 3°.9W/4°.5S. 1.35 km/pixel.

Foreshortening is deceiving; about one-ninth of the Moon's surface, or one-fifth of the surface ever visible from Earth, is compressed into the narrow crescent on the next three pages.

Fortunately, the libration is favorable for the western and southwestern limbs. However, the northwestern limb is not shown well and its most prominent feature is the large, terraced, Eratosthenian-Period crater Pythagoras, with its prominent central peak. The ancient depression Babbage, adjoining Pythagoras on the southeast, is also shown well at this phase.

Sinus Roris, now with a low Sun angle, contains relatively few features, the most obvious being the crater Markov. South of Mons Rümker, Oceanus Procellarum stretches south for 1600 kilometers to the crater Billy (see Figure 1, which shows this area with the Sun 2°.1 higher than in the mosaic).

The afternoon Sun is now low enough to reveal the postmare Mons Rümker volcanic dome complex. Southwest of Mons Rümker the "fresh" Copernican-Period crater Lichtenberg stands out due to its bright ray system. The rays extend to the west of the crater, but not to its east, because mare lavas flooded that area after the crater was formed, thus constituting the Moon's most recent extensive lava flows.

The most prominent feature in Oceanus Procellarum is now the Aristarchus Plateau, bisected by the terminator. This large quadrangle, bounded by ridges and chains of hills on its northwest and southwest, is of Imbrian age and largely covered with "dark mantling material." Farther south, the Marius Hills, like Mons Rümker, are interpreted as composed of Eratosthenian-Period mare lava cones and domes. Just southwest, note how the Reiner Gamma tonal feature is prominent in Figure 1, with a solar altitude of 8°, but is considerably less evident in the mosaic with its 6° Sun angle; an unusually large change in appearance for a small change in lighting.

In the heavily cratered Southwestern Highlands the terminator crosses the floor of Schickard, indicating how shallow this crater is in proportion to its diameter as well as how little relief most of the floor contains. Immediately southwest of Schickard, the present lighting is effective in revealing Wargentin's plateau nature. Finally, to the west and northwest of Schickard, the valleys radial to the Oriental Basin are beginning to appear, including Vallis Inghirami.

The libration and lighting are favorable for viewing the large craters of the southwestern limb. Hausen, with terraced walls and a high central peak, lies on the extreme limb immediately west of the double ring Bailly. To Bailly's south is Drygalski, similar to Hausen in appearance but probably much older. Due to the Sun's northerly selenocentric declination, the area south of Drygalski is in shadow except for a few isolated mountain peaks.

1. Oceanus Procellarum.
1996 Nov 07, 13h28m UT.
28-cm Sch.-Cass., f/10, 0.06 s.
Col. = 230°.7, Solar Lat. = 1°.0N,
Librations = 3°.3W/1°.0N.

233°–N

86°N to 5°N

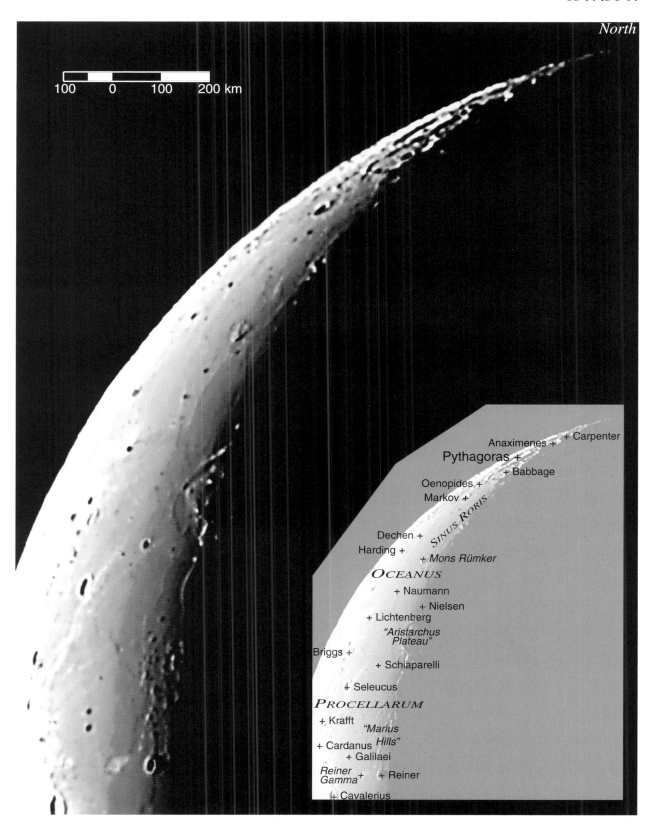

North

Anaximenes +   + Carpenter
Pythagoras +
+ Babbage
Oenopides +
Markov +
Sinus Roris
Dechen +
Harding +
+ Mons Rümker
Oceanus
+ Naumann
+ Nielsen
+ Lichtenberg
"Aristarchus Plateau"
Briggs +
+ Schiaparelli
+ Seleucus
Procellarum
+ Krafft
"Marius Hills"
+ Cardanus
+ Galilaei
Reiner Gamma +   + Reiner
+ Cavalerius

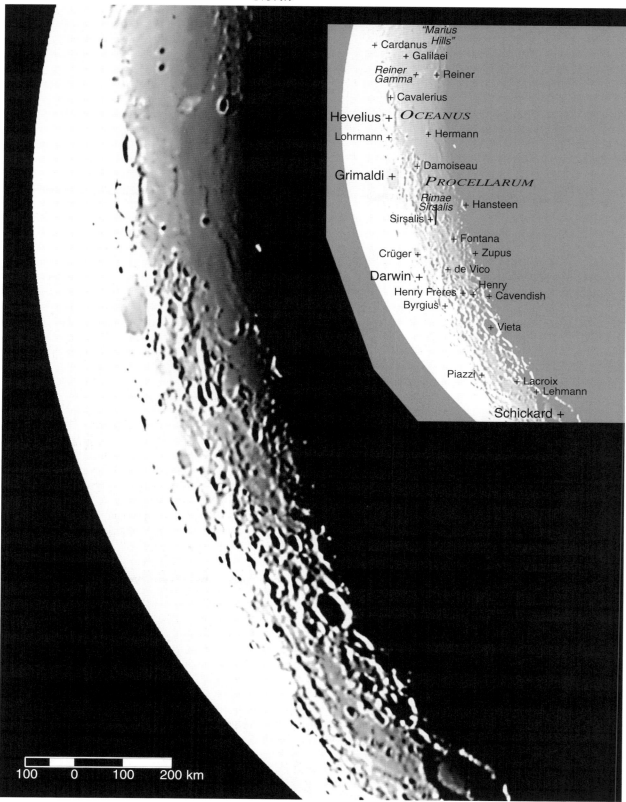

233°–C
16°N to 46°S

233°–S
32°S to 90°s

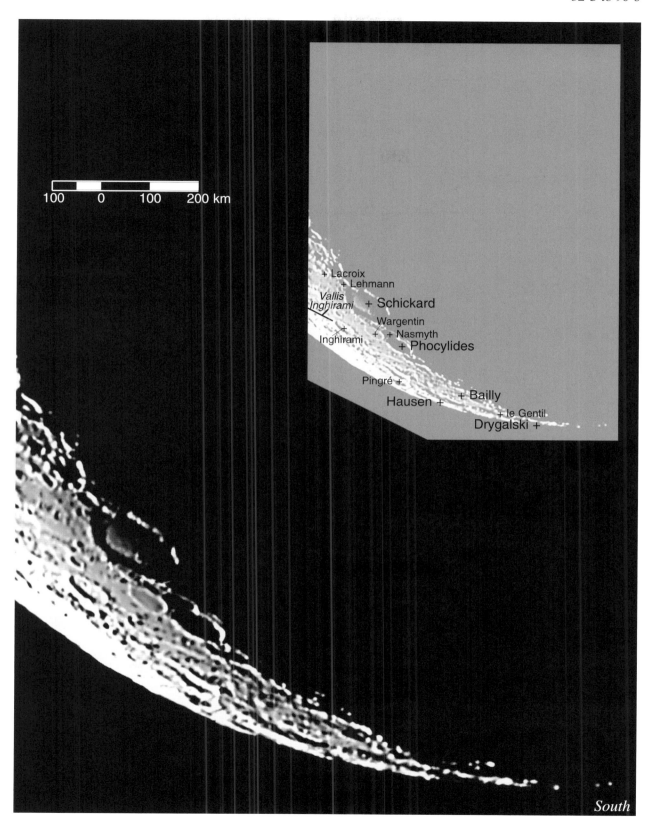

South

Colongitude 237°

1996 OCT 09, 13h00m UT. 28-cm Sch.-Cass., f/10, 0.06 s.
Colongitude = 237°.3, Solar latitude = 0°.2N, Librations = 1°.3W/3°.3N. 1.48 km/pixel.

One of the problems in observing a thin crescent Moon is the same one faced by observers of Mercury and Venus; the object must be viewed at a low elevation, with resulting poor seeing, if one wishes to see it against a dark or even a twilit sky. However, the infrared sensitivity of the CCD camera can darken the sky so much that it is feasible to image the Moon in full daylight, as the experimental image in Figure 1 shows (admittedly taken from a mountain location 1600 meters above sea level).

The larger-scale mosaic on the following pages may resemble the one taken at colongitude 233°, but there are subtle changes, in part due to a libration now 8 degrees more northerly. This last factor shows to advantage such northwest-limb features as Pythagoras, Oenopides, Pascal, and Brianchon.

Oceanus Procellarum continues to span almost half the length of the terminator. Within it, the Sun is setting on Mons Rümker, which is better seen in Figure 1, where it is clearly casting a shadow. The Sun has already set on the Aristarchus Plateau but part of the Dorsa Burnet ridge system to its west is one of the few non-crater features still visible in this portion of the Oceanus. Another such feature is Reiner Gamma, still faintly visible as a light patch even with a solar altitude of only 2°.

The lunar limb itself, although "saturated" (overexposed), is still sharply defined and exhibits several irregularities. For example, the "bump" on the terminator west of Grimaldi represents the northern Montes Cordillera, the main rim of the Orientale Basin, silhouetted on the limb. The extent of this range is apparent when we note another limb protrusion west of Lagrange that marks the southern portion of this range.

The thin crescent still covers some 34° of longitude, so the limb remains overexposed; the Valles Bouvard and Inghirami can be seen better in Figure 1, with its shorter exposure. The larger mosaic does show an unnamed low-lying ring structure on the terminator north of Lacroix and a similar feature north of that, enclosing Vieta (itself invisible).

Schickard remains visible only as a protrusion upon the terminator; however, on its southwest, Wargentin is still apparent as a plateau rather than a conventional crater. Still in this general area, the Mendel-Rydberg Basin makes an obvious indentation in the limb between Inghirami and Pingré. Farther south, the crater Bailly marks the southernmost limit of useful detail in this particular view.

1. Waning Crescent by Daylight.
1997 SEP 28, 18h06m UT. 36-cm Sch.-Cass., f/11, 0.05 s, Infrared Filter. Col. = 235°.4, Solar Lat. = 0°.5N, Librations = 2°.9E/2°.8N.

237°–N

90°N to 27°N

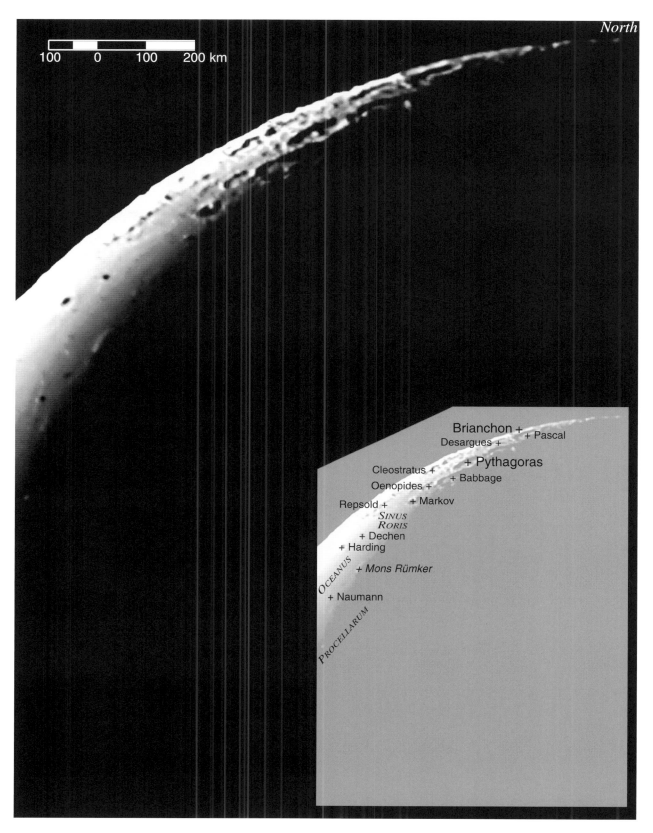

237°–C

37°N to 22°S

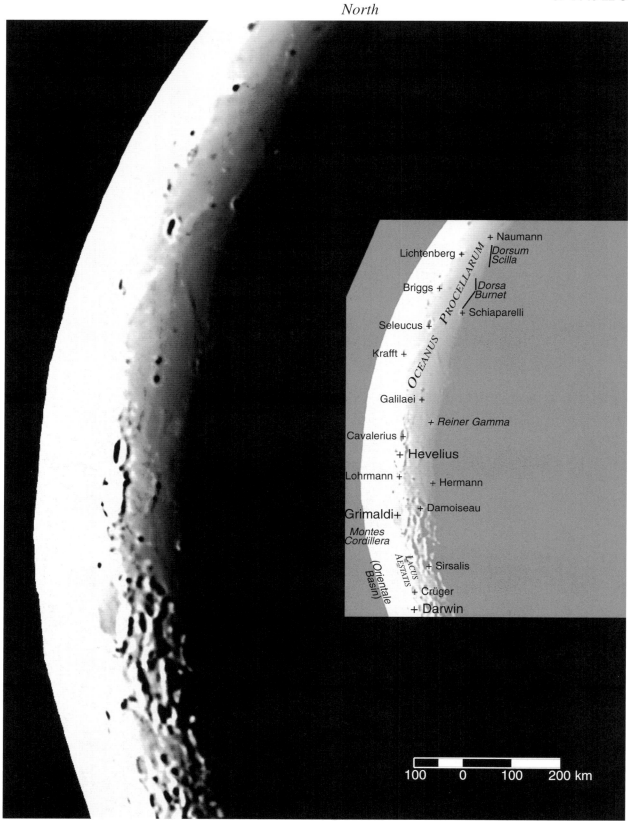

237°–S

3°S to 87°S

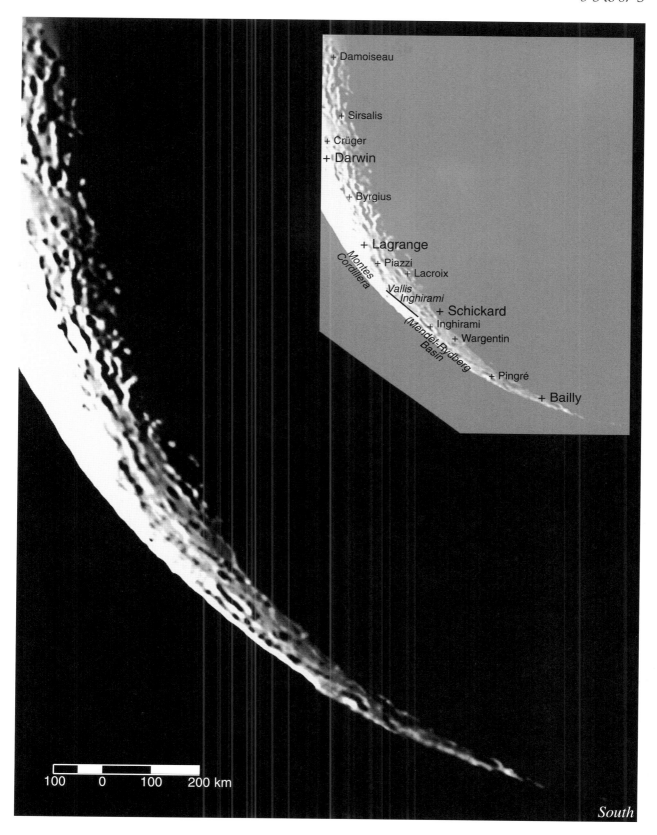

249

Colongitude 247°

1995 Nov 20, 13h59m UT. 28-cm Sch.-Cass., f/10, 0.10 s.
Colongitude = 247°.4, Solar latitude = 0°.8N, Librations = 4°.8W/0°.0S. 1.37 km/pixel.

The mosaic on the following pages shows the oldest Moon to date for which useful detail has been recorded (age 27.4 days). Figure 1 depicts a 28.2-day Moon, with little detail due to a low lunar altitude and increasing twilight. Perhaps an infrared filter and a larger-format CCD camera (allowing the Moon to be imaged more quickly) will prove useful for "old Moons" in the future.

Figure 2, a 27.0-day Moon, helps to bridge the gap between the present mosaic and the previous one. The mosaic on the next three pages shows the Sun setting on Pythagoras and Oenopides in the Northern Highlands. Lichtenberg and Seleucus are also on the terminator, in northwestern Oceanus Procellarum. Near Seleucus are the fresh craters Briggs, Cardanus, and Krafft, as well as the partial rings Struve, Russell, and Eddington.

The interiors of Cavalerius, Hevelius, and Lohrmann are in darkness, but sunlight still grazes the floor of Grimaldi. Flat, with little detail, Grimaldi's interior was flooded after the Orientale impact. Grimaldi itself forms only the inner ring of the large Grimaldi Basin. In contrast to Grimaldi, its western neighbors, Hedin, Riccioli, and Rocca, have extensive ejecta fields on their interiors. Farther south, the hummocky ejecta on Darwin's floor has distorted the shadow of its western wall.

From Oceanus Procellarum almost as far south as Bailly the highlands are scarred and littered as a result of the formation of the Orientale Basin. Large pre-Imbrian craters such as Darwin, Lamarck, Lagrange, and Piazzi are almost lost in the chaotic terrain, with their floors partly covered by mounds and ridges of deposits. As is the case throughout the mosaic, extreme foreshortening adds to the difficulty in interpreting detail. Along with the Grimaldi Basin and the large craters near it, most of the large craters in this region are ancient; Nectarian or pre-Nectarian in date. Three broad valleys radial to the Orientale Basin are also visible in this area: Valles Bouvard, Baade and Inghirami, which appear to be chains of overlapping, elongated secondary craters.

The Mendel-Rydberg Basin can be seen in outline on the limb. The Pingré Basin lies near the terminator, with its eastern rim prominently lit by the setting Sun, along with a ridge parallel to it in the northeastern interior of the basin, representing a partial inner ring.

Bailly is about to disappear, with its interior already in shadow except for a chain of hills that makes up a portion of the inner ring of the Bailly Basin. Finally, to the south of Bailly, Drygalski is directly on the limb, the sunlight still lingering on its walls, if not on its interior. With the Moon now only 6-percent illuminated, the lunation is over in terms of observing significant detail.

1. Montes Cordillera (Lunar Age 28.2 days).
1997 Nov 28, 14h49m UT. 28-cm Sch.-Cass., f/10, 0.05-0.10 s, W29 (deep red) Filter.
Col. = 256°.6, Solar Lat. = 1°.5N, Librations = 3°.2W/5°.4S.

2. Waning Crescent, Age 27.0 days.
1996 Nov 08, 13h44m UT. 28-cm Sch.-Cass., f/10, 0.06 s.
Col. = 243°.0, Solar Lat. = 1°.0N, Librations = 3°.9W/0°.5S.

247°–N
90°N to 14°N

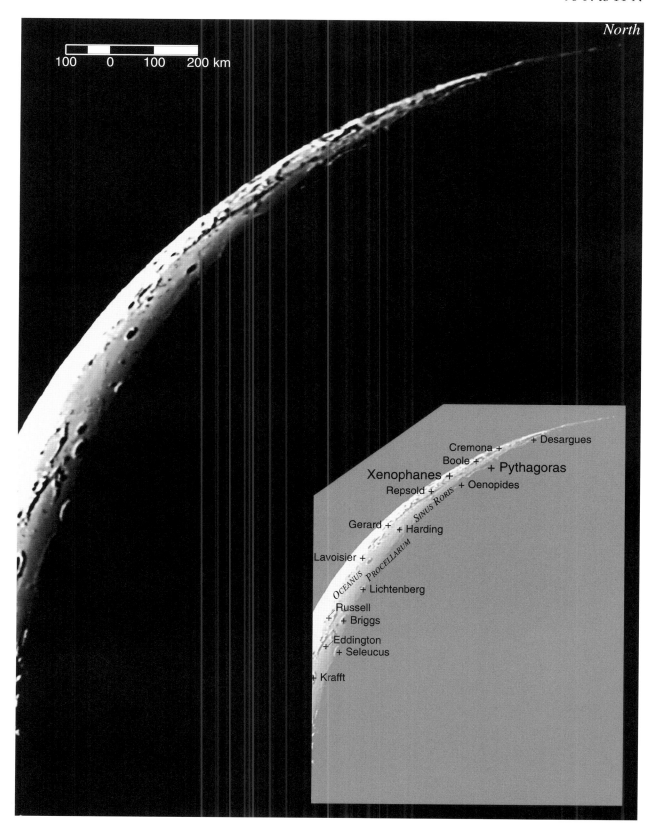

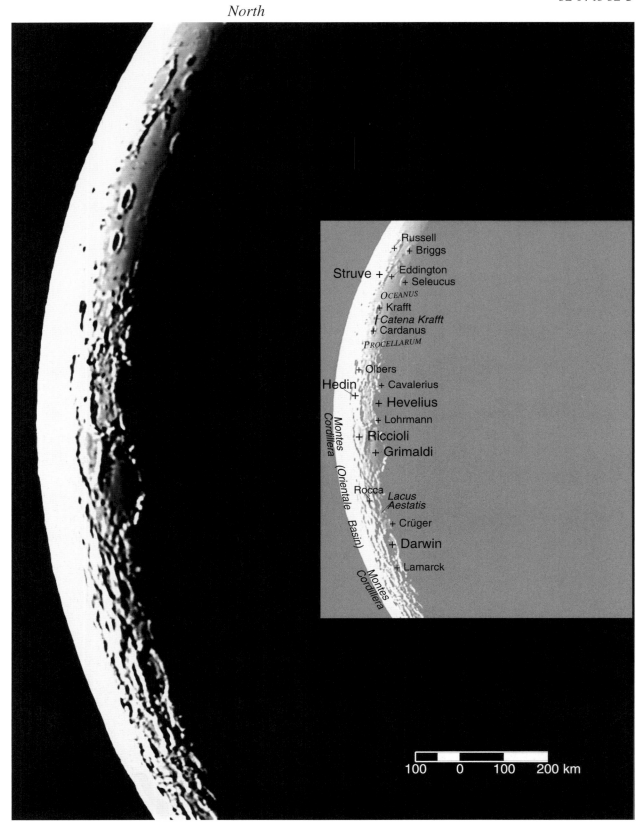

247°–S
19°S to 90°S

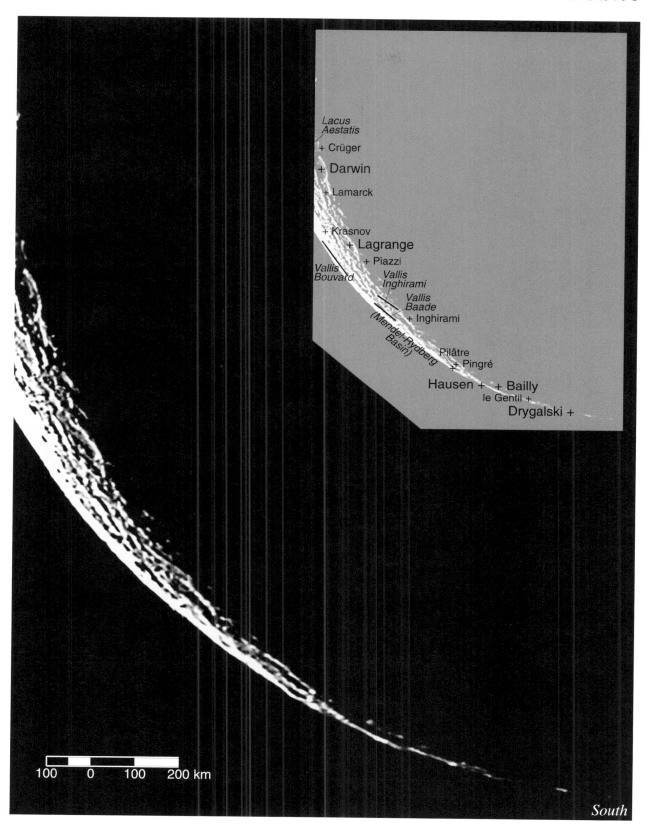

Appendix: Lunar Physical Ephemeris, 2000-2010

The following data are for 00h00m UT for each Gregorian day, 2000-2010. In order, the items given are: UT day; then in degrees: colongitude, solar selenographic latitude, and geocentric longitude and latitude libration.

2000 JANUARY

01	201.7	+0.6	+5.6	-6.8
02	213.9	+0.6	+4.5	-6.6
03	226.1	+0.6	+3.2	-6.1
04	238.3	+0.6	+1.9	-5.4
05	250.5	+0.6	+0.5	-4.4
06	262.6	+0.5	-0.8	-3.2
07	274.8	+0.5	-2.2	-1.9
08	287.0	+0.5	-3.4	-0.4
09	299.2	+0.5	-4.5	+1.0
10	311.4	+0.4	-5.5	+2.5
11	323.6	+0.4	-6.3	+3.8
12	335.7	+0.4	-6.9	+4.9
13	347.9	+0.4	-7.1	+5.9
14	000.1	+0.3	-7.0	+6.5
15	012.2	+0.3	-6.6	+6.8
16	024.4	+0.3	-5.8	+6.7
17	036.5	+0.2	-4.6	+6.2
18	048.6	+0.2	-3.1	+5.3
19	060.8	+0.2	-1.3	+4.0
20	072.9	+0.1	+0.5	+2.4
21	085.0	+0.1	+2.3	+0.7
22	097.1	+0.0	+3.9	-1.1
23	109.3	+0.0	+5.3	-2.7
24	121.4	-0.0	+6.3	-4.2
25	133.5	-0.1	+6.8	-5.3
26	145.7	-0.1	+7.0	-6.1
27	157.8	-0.1	+6.7	-6.6
28	170.0	-0.1	+6.1	-6.8
29	182.1	-0.2	+5.2	-6.7
30	194.3	-0.2	+4.1	-6.3
31	206.5	-0.2	+2.8	-5.6

2000 FEBRUARY

01	218.6	-0.2	+1.4	-4.6
02	230.8	-0.2	+0.1	-3.5
03	243.0	-0.3	-1.3	-2.2
04	255.2	-0.3	-2.5	-0.8
05	267.4	-0.3	-3.6	+0.6
06	279.6	-0.3	-4.5	+2.1
07	291.8	-0.3	-5.2	+3.5
08	304.0	-0.4	-5.6	+4.7
09	316.2	-0.4	-5.9	+5.7
10	328.3	-0.4	-5.8	+6.4
11	340.5	-0.5	-5.6	+6.8
12	352.7	-0.5	-5.1	+6.8
13	004.9	-0.5	-4.3	+6.4
14	017.0	-0.5	-3.4	+5.6
15	029.2	-0.6	-2.3	+4.4
16	041.3	-0.6	-1.0	+3.0
17	053.5	-0.7	+0.4	+1.3
18	065.6	-0.7	+1.7	-0.4
19	077.7	-0.7	+3.1	-2.1
20	089.9	-0.8	+4.2	-3.6
21	102.0	-0.8	+5.1	-4.9
22	114.1	-0.8	+5.6	-5.8
23	126.3	-0.8	+5.9	-6.4
24	138.4	-0.9	+5.7	-6.7
25	150.6	-0.9	+5.1	-6.6
26	162.7	-0.9	+4.3	-6.3
27	174.9	-0.9	+3.2	-5.7
28	187.1	-0.9	+1.9	-4.8
29	199.3	-1.0	+0.6	-3.7

2000 MARCH

01	211.4	-1.0	-0.8	-2.5
02	223.6	-1.0	-2.1	-1.2
03	235.8	-1.0	-3.3	+0.2
04	248.0	-1.0	-4.2	+1.7
05	260.2	-1.0	-4.8	+3.0
06	272.4	-1.0	-5.2	+4.3
07	284.7	-1.1	-5.3	+5.4
08	296.9	-1.1	-5.1	+6.1
09	309.1	-1.1	-4.6	+6.6
10	321.3	-1.1	-4.0	+6.7
11	333.5	-1.1	-3.2	+6.3
12	345.6	-1.1	-2.4	+5.6
13	357.8	-1.2	-1.5	+4.6
14	010.0	-1.2	-0.5	+3.2
15	022.2	-1.2	+0.4	+1.7
16	034.3	-1.2	+1.4	+0.0
17	046.5	-1.3	+2.3	-1.6
18	058.7	-1.3	+3.1	-3.1
19	070.8	-1.3	+3.9	-4.5
20	083.0	-1.3	+4.4	-5.5
21	095.1	-1.4	+4.8	-6.2
22	107.3	-1.4	+4.8	-6.6
23	119.4	-1.4	+4.6	-6.6
24	131.6	-1.4	+4.0	-6.3
25	143.8	-1.4	+3.2	-5.7
26	155.9	-1.4	+2.1	-4.9
27	168.1	-1.4	+0.8	-3.8
28	180.3	-1.4	-0.6	-2.7
29	192.5	-1.4	-2.0	-1.4
30	204.7	-1.4	-3.2	+0.0
31	216.9	-1.4	-4.3	+1.4

2000 APRIL

01	229.1	-1.4	-5.2	+2.7
02	241.3	-1.4	-5.7	+3.9
03	253.5	-1.5	-5.8	+5.0
04	265.8	-1.5	-5.5	+5.9
05	278.0	-1.5	-4.9	+6.4
06	290.2	-1.5	-4.0	+6.5
07	302.4	-1.5	-2.9	+6.3
08	314.6	-1.5	-1.8	+5.6
09	326.9	-1.5	-0.6	+4.6
10	339.1	-1.5	+0.5	+3.3
11	351.3	-1.5	+1.5	+1.8
12	003.5	-1.5	+2.4	+0.2
13	015.7	-1.5	+3.1	-1.4
14	027.8	-1.5	+3.7	-2.9
15	040.0	-1.5	+4.1	-4.3
16	052.2	-1.5	+4.5	-5.3
17	064.4	-1.6	+4.6	-6.1
18	076.5	-1.6	+4.6	-6.5
19	088.7	-1.6	+4.4	-6.6
20	100.9	-1.6	+3.9	-6.3
21	113.1	-1.6	+3.2	-5.7
22	125.2	-1.6	+2.2	-4.9
23	137.4	-1.6	+1.1	-3.9
24	149.6	-1.6	-0.2	-2.8
25	161.8	-1.5	-1.6	-1.5
26	174.0	-1.5	-3.0	-0.1
27	186.2	-1.5	-4.3	+1.2
28	198.4	-1.5	-5.4	+2.5
29	210.6	-1.5	-6.2	+3.8
30	222.8	-1.5	-6.7	+4.8

2000 MAY

01	235.1	-1.5	-6.7	+5.7
02	247.3	-1.5	-6.2	+6.3
03	259.5	-1.5	-5.4	+6.5
04	271.8	-1.5	-4.1	+6.4
05	284.0	-1.4	-2.7	+5.8
06	296.3	-1.4	-1.1	+4.8
07	308.5	-1.4	+0.5	+3.5
08	320.7	-1.4	+1.9	+2.0
09	333.0	-1.4	+3.1	+0.3
10	345.2	-1.4	+4.1	-1.3
11	357.4	-1.4	+4.8	-2.9
12	009.6	-1.4	+5.3	-4.2
13	021.8	-1.4	+5.6	-5.3
14	034.0	-1.4	+5.6	-6.1
15	046.2	-1.4	+5.4	-6.5
16	058.4	-1.4	+5.1	-6.7
17	070.6	-1.4	+4.5	-6.4
18	082.8	-1.4	+3.8	-5.9
19	094.9	-1.3	+2.8	-5.1
20	107.1	-1.3	+1.7	-4.1
21	119.3	-1.3	+0.4	-2.9
22	131.5	-1.3	-0.9	-1.6
23	143.7	-1.3	-2.4	-0.3
24	155.9	-1.3	-3.8	+1.1
25	168.1	-1.2	-5.0	+2.4
26	180.3	-1.2	-6.1	+3.7
27	192.5	-1.2	-7.0	+4.8
28	204.8	-1.2	-7.4	+5.6
29	217.0	-1.2	-7.4	+6.3
30	229.2	-1.1	-7.0	+6.6
31	241.5	-1.1	-6.1	+6.5

2000 JUNE

01	253.7	-1.1	-4.7	+6.1
02	266.0	-1.1	-3.0	+5.2
03	278.2	-1.1	-1.2	+3.9
04	290.5	-1.0	+0.7	+2.4
05	302.7	-1.0	+2.4	+0.7
06	315.0	-1.0	+4.0	-1.0
07	327.2	-1.0	+5.2	-2.6
08	339.4	-1.0	+6.1	-4.1
09	351.7	-0.9	+6.6	-5.2
10	003.9	-0.9	+6.8	-6.1
11	016.1	-0.9	+6.7	-6.6
12	028.3	-0.9	+6.4	-6.8
13	040.5	-0.9	+5.7	-6.6
14	052.7	-0.9	+4.9	-6.1
15	064.9	-0.8	+3.9	-5.4
16	077.1	-0.8	+2.8	-4.4
17	089.3	-0.8	+1.5	-3.2
18	101.5	-0.8	+0.1	-1.9
19	113.7	-0.7	-1.3	-0.5
20	125.9	-0.7	-2.7	+0.9
21	138.1	-0.7	-4.0	+2.3
22	150.3	-0.7	-5.3	+3.5
23	162.5	-0.6	-6.3	+4.7
24	174.7	-0.6	-7.1	+5.6
25	186.9	-0.6	-7.6	+6.3
26	199.1	-0.5	-7.7	+6.7
27	211.4	-0.5	-7.3	+6.7
28	223.6	-0.5	-6.5	+6.4
29	235.8	-0.5	-5.2	+5.6
30	248.1	-0.4	-3.6	+4.5

2000 JULY

01	260.4	-0.4	-1.8	+3.0
02	272.6	-0.4	+0.2	+1.4
03	284.9	-0.4	+2.1	-0.4
04	297.1	-0.3	+3.9	-2.1
05	309.4	-0.3	+5.4	-3.7
06	321.6	-0.3	+6.5	-5.0
07	333.8	-0.3	+7.2	-6.0
08	346.1	-0.2	+7.5	-6.6
09	358.3	-0.2	+7.4	-6.8
10	010.5	-0.2	+7.0	-6.7
11	022.7	-0.2	+6.2	-6.3
12	034.9	-0.1	+5.3	-5.6
13	047.1	-0.1	+4.1	-4.7
14	059.3	-0.1	+2.8	-3.6
15	071.5	-0.1	+1.4	-2.3
16	083.7	-0.0	+0.0	-0.9
17	095.9	+0.0	-1.4	+0.6
18	108.1	+0.0	-2.7	+2.0
19	120.3	+0.1	-4.0	+3.3
20	132.5	+0.1	-5.1	+4.5
21	144.7	+0.1	-6.0	+5.5
22	156.9	+0.1	-6.7	+6.2
23	169.1	+0.2	-7.1	+6.7
24	181.3	+0.2	-7.2	+6.8
25	193.5	+0.2	-6.9	+6.6
26	205.8	+0.2	-6.2	+6.0
27	218.0	+0.3	-5.2	+5.0
28	230.2	+0.3	-3.8	+3.7
29	242.5	+0.3	-2.2	+2.1
30	254.7	+0.4	-0.4	+0.3
31	267.0	+0.4	+1.5	-1.5

2000 AUGUST

01	279.3	+0.4	+3.3	-3.1
02	291.5	+0.4	+4.9	-4.5
03	303.7	+0.5	+6.2	-5.6
04	316.0	+0.5	+7.0	-6.4
05	328.2	+0.5	+7.4	-6.7
06	340.5	+0.5	+7.4	-6.7
07	352.7	+0.6	+7.0	-6.4
08	004.9	+0.6	+6.2	-5.7
09	017.1	+0.6	+5.2	-4.9
10	029.3	+0.6	+4.0	-3.8
11	041.5	+0.7	+2.6	-2.6
12	053.7	+0.7	+1.2	-1.2
13	065.9	+0.7	-0.2	+0.2
14	078.1	+0.7	-1.6	+1.6
15	090.3	+0.8	-2.8	+2.9
16	102.5	+0.8	-3.8	+4.2
17	114.6	+0.8	-4.7	+5.2
18	126.8	+0.8	-5.4	+6.0
19	139.0	+0.8	-5.8	+6.6
20	151.2	+0.9	-6.0	+6.8
21	163.4	+0.9	-6.0	+6.6
22	175.6	+0.9	-5.7	+6.1
23	187.8	+0.9	-5.2	+5.2
24	200.0	+0.9	-4.4	+4.0
25	212.3	+1.0	-3.3	+2.6
26	224.5	+1.0	-2.0	+0.9
27	236.7	+1.0	-0.5	-0.8
28	249.0	+1.0	+1.1	-2.5
29	261.2	+1.1	+2.7	-4.0
30	273.5	+1.1	+4.2	-5.2
31	285.7	+1.1	+5.4	-6.1

2000 SEPTEMBER

01	297.9	+1.1	+6.3	-6.5
02	310.2	+1.1	+6.8	-6.6
03	322.4	+1.2	+6.8	-6.3
04	334.6	+1.2	+6.4	-5.8
05	346.8	+1.2	+5.7	-4.9
06	359.1	+1.2	+4.6	-3.9
07	011.3	+1.2	+3.4	-2.7
08	023.5	+1.2	+2.0	-1.4
09	035.6	+1.3	+0.6	-0.1
10	047.8	+1.3	-0.8	+1.3
11	060.0	+1.3	-2.0	+2.6
12	072.2	+1.3	-3.1	+3.9
13	084.4	+1.3	-3.9	+4.9
14	096.5	+1.3	-4.5	+5.8
15	108.7	+1.4	-4.8	+6.4
16	120.9	+1.4	-4.9	+6.6
17	133.1	+1.4	-4.8	+6.5
18	145.2	+1.4	-4.5	+6.1
19	157.4	+1.4	-4.1	+5.3
20	169.6	+1.4	-3.5	+4.1
21	181.8	+1.4	-2.7	+2.8
22	194.0	+1.4	-1.8	+1.2
23	206.2	+1.4	-0.8	-0.5
24	218.4	+1.4	+0.3	-2.1
25	230.6	+1.4	+1.5	-3.6
26	242.9	+1.4	+2.7	-4.9
27	255.1	+1.5	+3.9	-5.8
28	267.3	+1.5	+4.8	-6.4
29	279.6	+1.5	+5.5	-6.5
30	291.8	+1.5	+5.9	-6.3

2000 OCTOBER

01	304.0	+1.5	+5.9	-5.8
02	316.2	+1.5	+5.5	-5.0
03	328.4	+1.5	+4.7	-4.0
04	340.6	+1.5	+3.6	-2.8
05	352.8	+1.5	+2.4	-1.5
06	005.0	+1.5	+1.0	-0.2
07	017.2	+1.5	-0.4	+1.1
08	029.4	+1.6	-1.8	+2.4
09	041.6	+1.6	-2.9	+3.6
10	053.7	+1.6	-3.8	+4.7
11	065.9	+1.6	-4.5	+5.6
12	078.1	+1.6	-4.8	+6.2
13	090.2	+1.6	-4.8	+6.5
14	102.4	+1.5	-4.5	+6.5
15	114.5	+1.5	-4.0	+6.1
16	126.7	+1.5	-3.3	+5.3
17	138.8	+1.5	-2.5	+4.2
18	151.0	+1.5	-1.6	+2.8
19	163.2	+1.5	-0.8	+1.3
20	175.4	+1.5	+0.1	-0.3
21	187.5	+1.5	+1.0	-2.0
22	199.7	+1.5	+1.9	-3.4
23	211.9	+1.5	+2.7	-4.7
24	224.1	+1.5	+3.5	-5.7
25	236.3	+1.5	+4.3	-6.3
26	248.5	+1.5	+4.8	-6.6
27	260.8	+1.5	+5.2	-6.4
28	273.0	+1.5	+5.2	-6.0
29	285.2	+1.5	+5.0	-5.2
30	297.4	+1.5	+4.5	-4.2
31	309.6	+1.5	+3.6	-3.0

2000 NOVEMBER

01	321.8	+1.5	+2.5	-1.7
02	334.0	+1.4	+1.2	-0.3
03	346.2	+1.4	-0.2	+1.0
04	358.3	+1.4	-1.6	+2.3
05	010.5	+1.4	-2.9	+3.5
06	022.7	+1.4	-4.1	+4.6
07	034.8	+1.4	-5.0	+5.5
08	047.0	+1.4	-5.5	+6.2
09	059.2	+1.4	-5.7	+6.5
10	071.3	+1.4	-5.5	+6.6
11	083.4	+1.3	-4.9	+6.2
12	095.6	+1.3	-4.0	+5.5
13	107.7	+1.3	-2.9	+4.5
14	119.9	+1.3	-1.6	+3.1
15	132.0	+1.3	-0.3	+1.5
16	144.1	+1.2	+0.9	-0.2
17	156.3	+1.2	+2.0	-1.8
18	168.5	+1.2	+3.0	-3.4
19	180.6	+1.2	+3.8	-4.7
20	192.8	+1.1	+4.5	-5.7
21	205.0	+1.1	+5.0	-6.4
22	217.2	+1.1	+5.3	-6.7
23	229.3	+1.1	+5.5	-6.6
24	241.5	+1.1	+5.4	-6.2
25	253.7	+1.1	+5.1	-5.5
26	265.9	+1.0	+4.6	-4.5
27	278.1	+1.0	+3.8	-3.3
28	290.3	+1.0	+2.8	-2.0
29	302.5	+1.0	+1.6	-0.6
30	314.7	+1.0	+0.2	+0.8

2000 DECEMBER

01	326.9	+1.0	-1.2	+2.2
02	339.1	+0.9	-2.6	+3.4
03	351.2	+0.9	-3.9	+4.5
04	003.4	+0.9	-5.1	+5.5
05	015.5	+0.9	-6.0	+6.2
06	027.7	+0.9	-6.6	+6.6
07	039.8	+0.8	-6.8	+6.7
08	052.0	+0.8	-6.6	+6.5
09	064.1	+0.8	-5.9	+5.9
10	076.2	+0.8	-4.8	+4.9
11	088.4	+0.7	-3.5	+3.6
12	100.5	+0.7	-1.9	+2.0
13	112.6	+0.7	-0.1	+0.3
14	124.8	+0.6	+1.5	-1.4
15	136.9	+0.6	+3.1	-3.1
16	149.0	+0.6	+4.4	-4.5
17	161.2	+0.5	+5.4	-5.6
18	173.3	+0.5	+6.1	-6.4
19	185.5	+0.5	+6.5	-6.8
20	197.6	+0.4	+6.7	-6.8
21	209.8	+0.4	+6.5	-6.4
22	222.0	+0.4	+6.1	-5.8
23	234.2	+0.4	+5.4	-4.8
24	246.4	+0.3	+4.6	-3.7
25	258.5	+0.3	+3.5	-2.3
26	270.7	+0.3	+2.3	-0.9
27	282.9	+0.3	+1.0	+0.5
28	295.1	+0.2	-0.4	+1.9
29	307.3	+0.2	-1.8	+3.2
30	319.5	+0.2	-3.2	+4.4
31	331.6	+0.2	-4.5	+5.4

2001 JANUARY

01	343.8	+0.2	-5.7	+6.1
02	356.0	+0.1	-6.6	+6.6
03	008.1	+0.1	-7.3	+6.8
04	020.3	+0.1	-7.6	+6.7
05	032.4	+0.1	-7.5	+6.2
06	044.6	+0.0	-7.0	+5.4
07	056.7	+0.0	-6.0	+4.2
08	068.8	-0.0	-4.6	+2.7
09	080.9	-0.1	-2.9	+1.0
10	093.1	-0.1	-1.0	-0.7
11	105.2	-0.2	+1.0	-2.5
12	117.3	-0.2	+2.9	-4.0
13	129.4	-0.2	+4.6	-5.3
14	141.6	-0.3	+6.0	-6.2
15	153.7	-0.3	+7.0	-6.7
16	165.9	-0.3	+7.5	-6.8
17	178.0	-0.4	+7.6	-6.5
18	190.2	-0.4	+7.4	-5.9
19	202.3	-0.4	+6.8	-5.0
20	214.5	-0.4	+5.9	-3.9
21	226.7	-0.5	+4.8	-2.7
22	238.9	-0.5	+3.5	-1.3
23	251.1	-0.5	+2.2	+0.1
24	263.3	-0.5	+0.8	+1.5
25	275.4	-0.6	-0.6	+2.9
26	287.6	-0.6	-2.0	+4.1
27	299.8	-0.6	-3.3	+5.2
28	312.0	-0.6	-4.5	+6.0
29	324.2	-0.6	-5.6	+6.5
30	336.4	-0.6	-6.5	+6.8
31	348.5	-0.7	-7.2	+6.7

2001 FEBRUARY

01	000.7	-0.7	-7.6	+6.4
02	012.9	-0.7	-7.6	+5.6
03	025.0	-0.7	-7.3	+4.6
04	037.2	-0.8	-6.6	+3.3
05	049.3	-0.8	-5.5	+1.7
06	061.4	-0.8	-3.9	+0.0
07	073.6	-0.9	-2.1	-1.7
08	085.7	-0.9	-0.1	-3.4
09	097.8	-0.9	+2.0	-4.8
10	110.0	-0.9	+3.9	-5.8
11	122.1	-1.0	+5.6	-6.4
12	134.2	-1.0	+6.8	-6.7
13	146.4	-1.0	+7.5	-6.5
14	158.5	-1.1	+7.8	-5.9
15	170.7	-1.1	+7.5	-5.1
16	182.9	-1.1	+6.9	-4.1
17	195.0	-1.1	+6.0	-2.9
18	207.2	-1.1	+4.8	-1.5
19	219.4	-1.2	+3.4	-0.2
20	231.6	-1.2	+2.0	+1.2
21	243.8	-1.2	+0.6	+2.6
22	256.0	-1.2	-0.8	+3.8
23	268.2	-1.2	-2.1	+4.9
24	280.4	-1.2	-3.3	+5.7
25	292.6	-1.2	-4.3	+6.3
26	304.8	-1.2	-5.2	+6.6
27	317.0	-1.3	-5.9	+6.7
28	329.2	-1.3	-6.4	+6.3

2001 MARCH

01	341.4	-1.3	-6.7	+5.7
02	353.5	-1.3	-6.8	+4.7
03	005.7	-1.3	-6.6	+3.5
04	017.9	-1.3	-6.1	+2.0
05	030.0	-1.3	-5.2	+0.4
06	042.2	-1.3	-4.0	-1.2
07	054.4	-1.4	-2.5	-2.8
08	066.5	-1.4	-0.8	-4.3
09	078.6	-1.4	+1.1	-5.4
10	090.8	-1.4	+3.0	-6.2
11	102.9	-1.4	+4.6	-6.5
12	115.1	-1.4	+5.9	-6.4
13	127.2	-1.5	+6.7	-5.9
14	139.4	-1.5	+7.1	-5.2
15	151.6	-1.5	+7.0	-4.1
16	163.7	-1.5	+6.4	-2.9
17	175.9	-1.5	+5.5	-1.6
18	188.1	-1.5	+4.3	-0.3
19	200.3	-1.5	+2.9	+1.1
20	212.5	-1.5	+1.5	+2.4
21	224.7	-1.5	+0.1	+3.6
22	236.9	-1.5	-1.2	+4.7
23	249.1	-1.5	-2.4	+5.5
24	261.3	-1.5	-3.4	+6.2
25	273.5	-1.5	-4.2	+6.5
26	285.8	-1.5	-4.8	+6.6
27	298.0	-1.5	-5.3	+6.3
28	310.2	-1.5	-5.5	+5.7
29	322.4	-1.5	-5.6	+4.8
30	334.6	-1.5	-5.4	+3.6
31	346.8	-1.5	-5.2	+2.1

2001 APRIL

01	359.0	-1.5	-4.7	+0.6
02	011.2	-1.5	-3.9	-1.0
03	023.4	-1.5	-3.0	-2.6
04	035.5	-1.5	-1.8	-4.0
05	047.7	-1.5	-0.4	-5.2
06	059.9	-1.5	+1.1	-6.0
07	072.0	-1.5	+2.6	-6.5
08	084.2	-1.5	+3.9	-6.5
09	096.4	-1.5	+5.0	-6.1
10	108.5	-1.5	+5.7	-5.3
11	120.7	-1.5	+6.0	-4.3
12	132.9	-1.5	+5.9	-3.1
13	145.0	-1.5	+5.4	-1.8
14	157.2	-1.5	+4.5	-0.4
15	169.4	-1.5	+3.3	+1.0
16	181.6	-1.5	+2.0	+2.3
17	193.8	-1.5	+0.6	+3.5
18	206.0	-1.5	-0.7	+4.6
19	218.2	-1.5	-2.0	+5.4
20	230.5	-1.5	-3.0	+6.1
21	242.7	-1.4	-3.9	+6.5
22	254.9	-1.4	-4.5	+6.6
23	267.2	-1.4	-4.9	+6.4
24	279.4	-1.4	-5.0	+5.8
25	291.6	-1.4	-4.9	+4.9
26	303.9	-1.4	-4.6	+3.7
27	316.1	-1.4	-4.1	+2.3
28	328.3	-1.3	-3.5	+0.7
29	340.5	-1.3	-2.8	-0.9
30	352.7	-1.3	-2.0	-2.5

2001 MAY

01	005.0	-1.3	-1.1	-3.9
02	017.2	-1.3	-0.1	-5.1
03	029.3	-1.3	+1.0	-6.0
04	041.5	-1.3	+2.1	-6.5
05	053.7	-1.2	+3.1	-6.6
06	065.9	-1.2	+4.0	-6.3
07	078.1	-1.2	+4.7	-5.6
08	090.2	-1.2	+5.1	-4.6
09	102.4	-1.2	+5.2	-3.4
10	114.6	-1.2	+4.9	-2.1
11	126.8	-1.1	+4.2	-0.6
12	139.0	-1.1	+3.3	+0.8
13	151.2	-1.1	+2.2	+2.1
14	163.4	-1.1	+0.9	+3.4
15	175.6	-1.1	-0.4	+4.5
16	187.8	-1.1	-1.7	+5.4
17	200.0	-1.0	-2.9	+6.1
18	212.2	-1.0	-4.0	+6.5
19	224.5	-1.0	-4.7	+6.7
20	236.7	-1.0	-5.2	+6.5
21	249.0	-1.0	-5.4	+6.0
22	261.2	-0.9	-5.3	+5.2
23	273.4	-0.9	-4.8	+4.1
24	285.7	-0.9	-4.2	+2.7
25	297.9	-0.9	-3.3	+1.1
26	310.2	-0.9	-2.3	-0.6
27	322.4	-0.8	-1.2	-2.3
28	334.6	-0.8	-0.1	-3.8
29	346.9	-0.8	+1.0	-5.1
30	359.1	-0.7	+2.0	-6.0
31	011.3	-0.7	+3.0	-6.6

2001 JUNE

01	023.5	-0.7	+3.8	-6.8
02	035.7	-0.7	+4.4	-6.6
03	047.9	-0.6	+4.9	-6.0
04	060.1	-0.6	+5.1	-5.0
05	072.3	-0.6	+5.1	-3.9
06	084.5	-0.6	+4.8	-2.5
07	096.7	-0.5	+4.2	-1.0
08	108.8	-0.5	+3.4	+0.4
09	121.0	-0.5	+2.4	+1.9
10	133.2	-0.5	+1.2	+3.2
11	145.4	-0.4	-0.1	+4.4
12	157.6	-0.4	-1.4	+5.3
13	169.9	-0.4	-2.7	+6.1
14	182.1	-0.4	-3.9	+6.6
15	194.3	-0.3	-4.9	+6.8
16	206.5	-0.3	-5.7	+6.7
17	218.8	-0.3	-6.1	+6.3
18	231.0	-0.3	-6.3	+5.6
19	243.3	-0.3	-6.0	+4.5
20	255.5	-0.2	-5.4	+3.2
21	267.8	-0.2	-4.4	+1.6
22	280.0	-0.2	-3.2	-0.1
23	292.3	-0.2	-1.8	-1.8
24	304.5	-0.1	-0.3	-3.4
25	316.8	-0.1	+1.3	-4.8
26	329.0	-0.1	+2.7	-5.9
27	341.2	-0.0	+3.9	-6.5
28	353.5	+0.0	+4.9	-6.8
29	005.7	+0.0	+5.6	-6.7
30	017.9	+0.1	+6.0	-6.2

2001 JULY

01	030.1	+0.1	+6.1	-5.3
02	042.3	+0.1	+5.9	-4.2
03	054.5	+0.2	+5.5	-2.9
04	066.7	+0.2	+4.9	-1.5
05	078.9	+0.2	+4.0	+0.0
06	091.1	+0.3	+3.0	+1.5
07	103.3	+0.3	+1.9	+2.9
08	115.5	+0.3	+0.6	+4.1
09	127.7	+0.3	-0.7	+5.1
10	139.9	+0.4	-2.0	+5.9
11	152.1	+0.4	-3.3	+6.5
12	164.3	+0.4	-4.5	+6.8
13	176.5	+0.4	-5.5	+6.8
14	188.7	+0.4	-6.3	+6.5
15	200.9	+0.4	-6.9	+5.8
16	213.2	+0.5	-7.0	+4.9
17	225.4	+0.5	-6.8	+3.7
18	237.7	+0.5	-6.2	+2.2
19	249.9	+0.5	-5.1	+0.6
20	262.2	+0.5	-3.7	-1.1
21	274.4	+0.6	-2.0	-2.8
22	286.7	+0.6	-0.1	-4.3
23	298.9	+0.6	+1.8	-5.5
24	311.2	+0.7	+3.5	-6.3
25	323.4	+0.7	+5.1	-6.7
26	335.6	+0.7	+6.2	-6.7
27	347.9	+0.7	+7.0	-6.2
28	000.1	+0.8	+7.3	-5.5
29	012.3	+0.8	+7.2	-4.4
30	024.5	+0.8	+6.8	-3.2
31	036.7	+0.9	+6.1	-1.8

2001 AUGUST

01	048.9	+0.9	+5.2	-0.3
02	061.1	+0.9	+4.2	+1.1
03	073.3	+0.9	+3.0	+2.5
04	085.5	+1.0	+1.7	+3.8
05	097.7	+1.0	+0.4	+4.9
06	109.9	+1.0	-0.9	+5.7
07	122.1	+1.0	-2.2	+6.3
08	134.3	+1.0	-3.5	+6.7
09	146.5	+1.1	-4.6	+6.7
10	158.7	+1.1	-5.6	+6.5
11	170.9	+1.1	-6.5	+5.9
12	183.1	+1.1	-7.1	+5.1
13	195.3	+1.1	-7.4	+4.0
14	207.5	+1.1	-7.3	+2.7
15	219.8	+1.1	-6.8	+1.1
16	232.0	+1.1	-5.8	-0.5
17	244.2	+1.1	-4.4	-2.2
18	256.5	+1.2	-2.6	-3.7
19	268.7	+1.2	-0.6	-5.0
20	281.0	+1.2	+1.5	-5.9
21	293.2	+1.2	+3.6	-6.5
22	305.5	+1.2	+5.3	-6.5
23	317.7	+1.3	+6.7	-6.2
24	329.9	+1.3	+7.6	-5.5
25	342.2	+1.3	+7.9	-4.5
26	354.4	+1.3	+7.8	-3.3
27	006.6	+1.3	+7.4	-1.9
28	018.8	+1.4	+6.5	-0.5
29	031.0	+1.4	+5.5	+0.9
30	043.2	+1.4	+4.3	+2.3
31	055.4	+1.4	+3.0	+3.5

2001 SEPTEMBER

01	067.5	+1.4	+1.7	+4.6
02	079.7	+1.5	+0.3	+5.5
03	091.9	+1.5	-0.9	+6.2
04	104.1	+1.5	-2.2	+6.5
05	116.3	+1.5	-3.3	+6.6
06	128.4	+1.5	-4.4	+6.4
07	140.6	+1.5	-5.3	+5.9
08	152.8	+1.5	-6.1	+5.1
09	165.0	+1.5	-6.6	+4.1
10	177.2	+1.5	-7.0	+2.8
11	189.4	+1.5	-7.0	+1.4
12	201.6	+1.5	-6.6	-0.2
13	213.8	+1.5	-5.8	-1.8
14	226.1	+1.5	-4.6	-3.3
15	238.3	+1.5	-3.0	-4.6
16	250.5	+1.5	-1.1	-5.6
17	262.8	+1.5	+1.0	-6.3
18	275.0	+1.5	+3.0	-6.5
19	287.2	+1.5	+4.8	-6.2
20	299.5	+1.5	+6.3	-5.6
21	311.7	+1.5	+7.3	-4.6
22	323.9	+1.5	+7.8	-3.4
23	336.1	+1.5	+7.7	-2.0
24	348.3	+1.5	+7.3	-0.6
25	000.5	+1.5	+6.4	+0.8
26	012.7	+1.6	+5.4	+2.2
27	024.9	+1.6	+4.1	+3.4
28	037.1	+1.6	+2.8	+4.5
29	049.3	+1.6	+1.5	+5.4
30	061.4	+1.6	+0.2	+6.1

2001 OCTOBER

01	073.6	+1.6	-1.0	+6.5
02	085.8	+1.6	-2.2	+6.6
03	097.9	+1.6	-3.1	+6.4
04	110.1	+1.6	-4.0	+6.0
05	122.3	+1.5	-4.7	+5.2
06	134.4	+1.5	-5.3	+4.2
07	146.6	+1.5	-5.7	+2.9
08	158.8	+1.5	-5.9	+1.5
09	171.0	+1.5	-5.9	-0.1
10	183.1	+1.5	-5.6	-1.6
11	195.3	+1.4	-5.0	-3.1
12	207.5	+1.4	-4.0	-4.4
13	219.7	+1.4	-2.7	-5.5
14	231.9	+1.4	-1.1	-6.2
15	244.2	+1.4	+0.7	-6.5
16	256.4	+1.4	+2.5	-6.4
17	268.6	+1.4	+4.1	-5.8
18	280.8	+1.4	+5.5	-4.9
19	293.0	+1.4	+6.4	-3.7
20	305.2	+1.3	+6.9	-2.3
21	317.4	+1.3	+6.9	-0.8
22	329.6	+1.3	+6.5	+0.7
23	341.8	+1.3	+5.7	+2.1
24	354.0	+1.3	+4.7	+3.3
25	006.2	+1.3	+3.5	+4.4
26	018.4	+1.3	+2.2	+5.4
27	030.5	+1.3	+0.9	+6.1
28	042.7	+1.3	-0.4	+6.5
29	054.9	+1.3	-1.5	+6.7
30	067.0	+1.3	-2.5	+6.5
31	079.2	+1.3	-3.3	+6.1

2001 NOVEMBER

01	091.3	+1.2	-4.0	+5.4
02	103.5	+1.2	-4.4	+4.4
03	115.6	+1.2	-4.7	+3.1
04	127.8	+1.2	-4.8	+1.7
05	139.9	+1.1	-4.7	+0.1
06	152.1	+1.1	-4.4	-1.5
07	164.2	+1.1	-4.0	-3.0
08	176.4	+1.0	-3.3	-4.4
09	188.6	+1.0	-2.4	-5.5
10	200.7	+1.0	-1.3	-6.3
11	212.9	+1.0	-0.1	-6.7
12	225.1	+0.9	+1.3	-6.6
13	237.3	+0.9	+2.6	-6.2
14	249.5	+0.9	+3.8	-5.3
15	261.7	+0.9	+4.8	-4.1
16	273.9	+0.8	+5.5	-2.7
17	286.1	+0.8	+5.8	-1.2
18	298.3	+0.8	+5.7	+0.3
19	310.5	+0.8	+5.3	+1.8
20	322.7	+0.8	+4.6	+3.1
21	334.9	+0.8	+3.6	+4.3
22	347.0	+0.7	+2.5	+5.3
23	359.2	+0.7	+1.2	+6.0
24	011.4	+0.7	-0.1	+6.5
25	023.5	+0.7	-1.3	+6.8
26	035.7	+0.7	-2.4	+6.7
27	047.8	+0.7	-3.4	+6.3
28	060.0	+0.6	-4.1	+5.7
29	072.1	+0.6	-4.5	+4.7
30	084.2	+0.6	-4.7	+3.5

2001 DECEMBER

01	096.4	+0.5	-4.7	+2.1
02	108.5	+0.5	-4.4	+0.5
03	120.6	+0.5	-3.8	-1.2
04	132.8	+0.4	-3.1	-2.8
05	144.9	+0.4	-2.3	-4.2
06	157.0	+0.4	-1.3	-5.4
07	169.2	+0.3	-0.3	-6.3
08	181.3	+0.3	+0.8	-6.7
09	193.5	+0.3	+1.8	-6.8
10	205.7	+0.2	+2.7	-6.4
11	217.9	+0.2	+3.6	-5.7
12	230.0	+0.2	+4.3	-4.6
13	242.2	+0.1	+4.8	-3.3
14	254.4	+0.1	+5.0	-1.8
15	266.6	+0.1	+5.0	-0.2
16	278.8	+0.0	+4.7	+1.3
17	291.0	+0.0	+4.2	+2.8
18	303.2	+0.0	+3.4	+4.0
19	315.4	-0.0	+2.4	+5.1
20	327.5	-0.0	+1.3	+5.9
21	339.7	-0.1	+0.1	+6.5
22	351.9	-0.1	-1.2	+6.8
23	004.0	-0.1	-2.5	+6.8
24	016.2	-0.1	-3.6	+6.5
25	028.3	-0.1	-4.6	+5.9
26	040.5	-0.2	-5.3	+5.1
27	052.6	-0.2	-5.7	+4.0
28	064.7	-0.2	-5.8	+2.6
29	076.9	-0.2	-5.5	+1.1
30	089.0	-0.3	-4.8	-0.6
31	101.1	-0.3	-3.9	-2.2

2002 JANUARY

01	113.2	-0.3	-2.6	-3.7
02	125.4	-0.4	-1.3	-5.1
03	137.5	-0.4	+0.2	-6.0
04	149.6	-0.5	+1.5	-6.6
05	161.8	-0.5	+2.8	-6.8
06	173.9	-0.5	+3.8	-6.5
07	186.1	-0.6	+4.6	-5.9
08	198.2	-0.6	+5.2	-4.9
09	210.4	-0.6	+5.5	-3.6
10	222.6	-0.7	+5.5	-2.2
11	234.8	-0.7	+5.3	-0.7
12	246.9	-0.7	+4.9	+0.9
13	259.1	-0.7	+4.3	+2.3
14	271.3	-0.8	+3.6	+3.7
15	283.5	-0.8	+2.7	+4.8
16	295.7	-0.8	+1.6	+5.7
17	307.9	-0.8	+0.4	+6.3
18	320.1	-0.8	-0.8	+6.7
19	332.2	-0.8	-2.1	+6.7
20	344.4	-0.9	-3.4	+6.5
21	356.6	-0.9	-4.6	+6.0
22	008.7	-0.9	-5.6	+5.3
23	020.9	-0.9	-6.4	+4.3
24	033.0	-0.9	-6.9	+3.0
25	045.2	-0.9	-7.0	+1.6
26	057.3	-1.0	-6.7	+0.0
27	069.4	-1.0	-5.9	-1.6
28	081.6	-1.0	-4.7	-3.1
29	093.7	-1.0	-3.1	-4.5
30	105.8	-1.1	-1.3	-5.6
31	117.9	-1.1	+0.6	-6.3

2002 FEBRUARY

01	130.1	-1.1	+2.4	-6.6
02	142.2	-1.1	+4.0	-6.5
03	154.4	-1.2	+5.3	-5.9
04	166.5	-1.2	+6.2	-5.0
05	178.7	-1.2	+6.7	-3.8
06	190.8	-1.2	+6.8	-2.4
07	203.0	-1.3	+6.5	-0.9
08	215.2	-1.3	+6.0	+0.6
09	227.4	-1.3	+5.3	+2.0
10	239.6	-1.3	+4.5	+3.4
11	251.8	-1.3	+3.5	+4.5
12	264.0	-1.3	+2.4	+5.5
13	276.2	-1.4	+1.2	+6.1
14	288.4	-1.4	-0.0	+6.5
15	300.6	-1.4	-1.3	+6.7

Appendix 257

16	312.7	-1.4	-2.6	+6.5
17	324.9	-1.4	-3.9	+6.0
18	337.1	-1.4	-5.1	+5.3
19	349.3	-1.4	-6.1	+4.4
20	001.5	-1.4	-7.0	+3.2
21	013.6	-1.4	-7.6	+1.9
22	025.8	-1.4	-7.9	+0.4
23	038.0	-1.4	-7.7	-1.1
24	050.1	-1.4	-7.0	-2.6
25	062.2	-1.4	-5.8	-4.0
26	074.4	-1.4	-4.1	-5.2
27	086.5	-1.5	-2.1	-6.0
28	098.7	-1.5	+0.1	-6.4

2002 MARCH

01	110.8	-1.5	+2.2	-6.4
02	122.9	-1.5	+4.2	-5.9
03	135.1	-1.5	+5.7	-5.0
04	147.2	-1.5	+6.8	-3.8
05	159.4	-1.5	+7.4	-2.5
06	171.6	-1.5	+7.6	-1.0
07	183.8	-1.5	+7.3	+0.5
08	195.9	-1.5	+6.7	+1.9
09	208.1	-1.5	+5.8	+3.2
10	220.3	-1.5	+4.8	+4.4
11	232.5	-1.6	+3.7	+5.3
12	244.7	-1.6	+2.4	+6.0
13	256.9	-1.6	+1.2	+6.5
14	269.2	-1.6	-0.1	+6.6
15	281.4	-1.6	-1.4	+6.5
16	293.6	-1.6	-2.6	+6.0
17	305.8	-1.6	-3.9	+5.3
18	318.0	-1.5	-5.0	+4.4
19	330.2	-1.5	-6.1	+3.2
20	342.4	-1.5	-6.9	+1.9
21	354.6	-1.5	-7.6	+0.5
22	006.8	-1.5	-7.9	-1.0
23	018.9	-1.5	-7.8	-2.4
24	031.1	-1.5	-7.3	-3.8
25	043.3	-1.5	-6.3	-5.0
26	055.5	-1.5	-4.8	-5.9
27	067.6	-1.5	-2.9	-6.4
28	079.8	-1.5	-0.7	-6.4
29	091.9	-1.4	+1.5	-6.1
30	104.1	-1.4	+3.5	-5.2
31	116.2	-1.4	+5.2	-4.1

2002 APRIL

01	128.4	-1.4	+6.5	-2.7
02	140.6	-1.4	+7.2	-1.2
03	152.7	-1.4	+7.5	+0.3
04	164.9	-1.4	+7.3	+1.8
05	177.1	-1.4	+6.8	+3.2
06	189.3	-1.4	+5.9	+4.3
07	201.5	-1.4	+4.9	+5.3
08	213.7	-1.4	+3.7	+6.0
09	225.9	-1.4	+2.4	+6.5
10	238.1	-1.4	+1.1	+6.7
11	250.4	-1.4	-0.2	+6.6
12	262.6	-1.3	-1.4	+6.2
13	274.8	-1.3	-2.6	+5.5
14	287.0	-1.3	-3.7	+4.6
15	299.3	-1.3	-4.7	+3.4
16	311.5	-1.3	-5.6	+2.1
17	323.7	-1.3	-6.3	+0.6
18	335.9	-1.2	-6.8	-0.9
19	348.1	-1.2	-7.1	-2.3
20	000.3	-1.2	-7.0	-3.7
21	012.5	-1.2	-6.5	-4.9
22	024.7	-1.2	-5.6	-5.9
23	036.9	-1.1	-4.4	-6.4
24	049.1	-1.1	-2.8	-6.6
25	061.3	-1.1	-1.0	-6.3
26	073.4	-1.1	+0.9	-5.6
27	085.6	-1.0	+2.7	-4.6
28	097.8	-1.0	+4.3	-3.2
29	110.0	-1.0	+5.6	-1.6
30	122.1	-1.0	+6.4	-0.0

2002 MAY

01	134.3	-0.9	+6.8	+1.5
02	146.5	-0.9	+6.7	+3.0
03	158.7	-0.9	+6.2	+4.2
04	170.9	-0.9	+5.5	+5.2
05	183.1	-0.9	+4.5	+6.0
06	195.3	-0.9	+3.3	+6.5
07	207.5	-0.8	+2.1	+6.7
08	219.8	-0.8	+0.8	+6.7
09	232.0	-0.8	-0.5	+6.3
10	244.2	-0.8	-1.7	+5.7
11	256.5	-0.8	-2.8	+4.8
12	268.7	-0.8	-3.7	+3.7
13	280.9	-0.7	-4.5	+2.4
14	293.2	-0.7	-5.1	+0.9
15	305.4	-0.7	-5.6	-0.6
16	317.7	-0.7	-5.8	-2.1
17	329.9	-0.6	-5.8	-3.6
18	342.1	-0.6	-5.5	-4.8
19	354.3	-0.6	-4.9	-5.8
20	006.5	-0.5	-4.1	-6.5
21	018.8	-0.5	-3.0	-6.8
22	031.0	-0.5	-1.7	-6.6
23	043.1	-0.4	-0.3	-6.0
24	055.3	-0.4	+1.2	-5.1
25	067.5	-0.4	+2.6	-3.8
26	079.7	-0.3	+3.8	-2.2
27	091.9	-0.3	+4.8	-0.6
28	104.1	-0.3	+5.4	+1.0
29	116.2	-0.2	+5.7	+2.6
30	128.4	-0.2	+5.7	+3.9
31	140.6	-0.2	+5.3	+5.0

2002 JUNE

01	152.8	-0.2	+4.6	+5.9
02	165.0	-0.2	+3.6	+6.5
03	177.3	-0.1	+2.5	+6.8
04	189.5	-0.1	+1.3	+6.8
05	201.7	-0.1	+0.0	+6.5
06	213.9	-0.1	-1.2	+5.9
07	226.2	-0.1	-2.4	+5.1
08	238.4	-0.0	-3.4	+4.1
09	250.7	-0.0	-4.2	+2.8
10	262.9	+0.0	-4.7	+1.4
11	275.2	+0.0	-5.1	-0.2
12	287.4	+0.0	-5.1	-1.7
13	299.7	+0.1	-4.9	-3.2
14	311.9	+0.1	-4.5	-4.6
15	324.1	+0.1	-3.8	-5.7
16	336.4	+0.2	-3.0	-6.4
17	348.6	+0.2	-2.0	-6.8
18	000.8	+0.2	-1.0	-6.7
19	013.1	+0.3	+0.2	-6.3
20	025.3	+0.3	+1.2	-5.4
21	037.5	+0.3	+2.3	-4.2
22	049.7	+0.4	+3.2	-2.8
23	061.9	+0.4	+4.0	-1.2
24	074.0	+0.4	+4.5	+0.5
25	086.2	+0.5	+4.9	+2.1
26	098.4	+0.5	+5.0	+3.5
27	110.6	+0.5	+4.8	+4.7
28	122.8	+0.6	+4.3	+5.7
29	135.0	+0.6	+3.6	+6.3
30	147.2	+0.6	+2.7	+6.7

2002 JULY

01	159.4	+0.6	+1.6	+6.8
02	171.6	+0.6	+0.3	+6.5
03	183.9	+0.6	-0.9	+6.1
04	196.1	+0.7	-2.2	+5.3
05	208.3	+0.7	-3.3	+4.3
06	220.6	+0.7	-4.3	+3.2
07	232.8	+0.7	-5.0	+1.8
08	245.1	+0.7	-5.4	+0.3
09	257.3	+0.7	-5.5	-1.2
10	269.6	+0.8	-5.2	-2.7
11	281.8	+0.8	-4.6	-4.1
12	294.1	+0.8	-3.7	-5.3
13	306.3	+0.8	-2.6	-6.2
14	318.6	+0.8	-1.3	-6.6
15	330.8	+0.9	-0.0	-6.7
16	343.0	+0.9	+1.2	-6.3
17	355.3	+0.9	+2.3	-5.5
18	007.5	+1.0	+3.2	-4.4
19	019.7	+1.0	+3.9	-3.1
20	031.9	+1.0	+4.5	-1.6
21	044.1	+1.1	+4.8	+0.0
22	056.3	+1.1	+5.0	+1.6
23	068.5	+1.1	+4.9	+3.1
24	080.7	+1.1	+4.7	+4.4
25	092.9	+1.2	+4.3	+5.4
26	105.1	+1.2	+3.7	+6.1
27	117.3	+1.2	+2.8	+6.5
28	129.5	+1.2	+1.8	+6.7
29	141.7	+1.2	+0.7	+6.5
30	153.9	+1.2	-0.6	+6.1
31	166.1	+1.2	-1.9	+5.4

2002 AUGUST

01	178.3	+1.3	-3.2	+4.5
02	190.5	+1.3	-4.3	+3.4
03	202.7	+1.3	-5.3	+2.1
04	215.0	+1.3	-6.0	+0.7
05	227.2	+1.3	-6.4	-0.8
06	239.5	+1.3	-6.3	-2.3
07	251.7	+1.3	-5.8	-3.7
08	263.9	+1.3	-4.9	-4.9
09	276.2	+1.3	-3.6	-5.8
10	288.5	+1.3	-2.0	-6.4
11	300.7	+1.3	-0.3	-6.5
12	312.9	+1.3	+1.3	-6.3
13	325.2	+1.4	+2.9	-5.6
14	337.4	+1.4	+4.1	-4.5
15	349.6	+1.4	+5.0	-3.2
16	001.9	+1.4	+5.7	-1.7
17	014.1	+1.4	+6.0	-0.2
18	026.3	+1.5	+6.0	+1.4
19	038.5	+1.5	+5.9	+2.8
20	050.7	+1.5	+5.5	+4.1
21	062.8	+1.5	+5.0	+5.2
22	075.0	+1.5	+4.3	+5.9
23	087.2	+1.5	+3.4	+6.4
24	099.4	+1.5	+2.5	+6.6
25	111.6	+1.6	+1.4	+6.5
26	123.8	+1.6	+0.1	+6.1
27	136.0	+1.6	-1.2	+5.4
28	148.1	+1.6	-2.5	+4.5
29	160.3	+1.5	-3.8	+3.5
30	172.5	+1.5	-5.0	+2.2
31	184.8	+1.5	-6.1	+0.9

2002 SEPTEMBER

01	197.0	+1.5	-6.9	-0.6
02	209.2	+1.5	-7.2	-2.0
03	221.4	+1.5	-7.2	-3.3
04	233.7	+1.5	-6.6	-4.6
05	245.9	+1.5	-5.6	-5.5
06	258.1	+1.5	-4.1	-6.2
07	270.4	+1.5	-2.2	-6.5
08	282.6	+1.5	-0.2	-6.3
09	294.8	+1.5	+1.8	-5.6
10	307.1	+1.5	+3.6	-4.6
11	319.3	+1.5	+5.1	-3.3
12	331.5	+1.5	+6.2	-1.8
13	343.7	+1.5	+6.9	-0.3
14	356.0	+1.5	+7.2	+1.3
15	008.2	+1.5	+7.1	+2.7
16	020.3	+1.5	+6.8	+4.0
17	032.5	+1.5	+6.2	+5.1
18	044.7	+1.5	+5.5	+5.9
19	056.9	+1.5	+4.6	+6.4
20	069.1	+1.5	+3.6	+6.6
21	081.2	+1.5	+2.4	+6.5
22	093.4	+1.5	+1.2	+6.2
23	105.6	+1.5	-0.1	+5.5
24	117.7	+1.5	-1.4	+4.6
25	129.9	+1.5	-2.8	+3.6
26	142.1	+1.5	-4.1	+2.3
27	154.3	+1.4	-5.3	+1.0
28	166.5	+1.4	-6.3	-0.4
29	178.7	+1.4	-7.1	-1.9
30	190.8	+1.4	-7.6	-3.2

2002 OCTOBER

01	203.1	+1.4	-7.6	-4.4
02	215.3	+1.3	-7.2	-5.4
03	227.5	+1.3	-6.2	-6.1
04	239.7	+1.3	-4.8	-6.5
05	251.9	+1.3	-2.9	-6.4
06	264.1	+1.3	-0.8	-5.9
07	276.4	+1.3	+1.3	-5.0
08	288.6	+1.2	+3.4	-3.7
09	300.8	+1.2	+5.1	-2.2
10	313.0	+1.2	+6.4	-0.5
11	325.2	+1.2	+7.3	+1.1
12	337.4	+1.2	+7.7	+2.6
13	349.6	+1.2	+7.7	+4.0
14	001.8	+1.2	+7.4	+5.1
15	014.0	+1.2	+6.7	+5.9
16	026.2	+1.2	+5.9	+6.4
17	038.3	+1.2	+4.8	+6.7
18	050.5	+1.2	+3.7	+6.7
19	062.6	+1.1	+2.4	+6.3
20	074.8	+1.1	+1.2	+5.7
21	087.0	+1.1	-0.2	+4.9
22	099.1	+1.1	-1.5	+3.8
23	111.3	+1.1	-2.8	+2.5
24	123.4	+1.0	-4.0	+1.1
25	135.6	+1.0	-5.1	-0.3
26	147.7	+1.0	-6.0	-1.7
27	159.9	+0.9	-6.8	-3.1
28	172.1	+0.9	-7.2	-4.4
29	184.2	+0.9	-7.3	-5.4
30	196.4	+0.8	-6.9	-6.2
31	208.6	+0.8	-6.1	-6.6

2002 NOVEMBER

01	220.8	+0.8	-4.9	-6.7
02	233.0	+0.8	-3.3	-6.3
03	245.2	+0.7	-1.5	-5.5
04	257.4	+0.7	+0.5	-4.3
05	269.6	+0.7	+2.5	-2.8
06	281.8	+0.7	+4.2	-1.1
07	294.0	+0.6	+5.7	+0.6
08	306.2	+0.6	+6.7	+2.2
09	318.4	+0.6	+7.3	+3.7
10	330.6	+0.6	+7.5	+4.9
11	342.8	+0.5	+7.2	+5.8
12	355.0	+0.5	+6.6	+6.4
13	007.1	+0.5	+5.8	+6.8
14	019.3	+0.5	+4.7	+6.8
15	031.5	+0.5	+3.5	+6.5
16	043.6	+0.5	+2.2	+5.9
17	055.8	+0.4	+0.9	+5.1
18	067.9	+0.4	-0.4	+4.1
19	080.0	+0.4	-1.7	+2.9
20	092.2	+0.3	-2.8	+1.5
21	104.3	+0.3	-3.9	+0.0
22	116.4	+0.3	-4.8	-1.5
23	128.6	+0.2	-5.5	-2.9
24	140.7	+0.2	-5.9	-4.2
25	152.9	+0.2	-6.1	-5.3
26	165.0	+0.1	-6.1	-6.2
27	177.2	+0.1	-5.7	-6.7
28	189.3	+0.1	-5.1	-6.8
29	201.5	+0.0	-4.1	-6.6
30	213.7	+0.0	-2.9	-5.9

2002 DECEMBER

01	225.9	-0.0	-1.4	-4.8
02	238.1	-0.1	+0.2	-3.4
03	250.2	-0.1	+1.8	-1.8
04	262.4	-0.1	+3.3	-0.1
05	274.6	-0.1	+4.6	+1.6
06	286.8	-0.2	+5.6	+3.2
07	299.0	-0.2	+6.2	+4.5
08	311.2	-0.2	+6.5	+5.6
09	323.4	-0.2	+6.4	+6.3
10	335.6	-0.3	+5.9	+6.7
11	347.7	-0.3	+5.1	+6.8
12	359.9	-0.3	+4.1	+6.6
13	012.0	-0.3	+2.8	+6.1
14	024.2	-0.4	+1.5	+5.3
15	036.3	-0.4	+0.2	+4.4
16	048.5	-0.4	-1.1	+3.2
17	060.6	-0.4	-2.3	+1.9
18	072.7	-0.5	-3.3	+0.5
19	084.9	-0.5	-4.1	-1.0
20	097.0	-0.5	-4.7	-2.5
21	109.1	-0.5	-5.1	-3.8
22	121.3	-0.6	-5.1	-5.0
23	133.4	-0.6	-5.0	-5.9
24	145.5	-0.6	-4.6	-6.6
25	157.7	-0.7	-4.0	-6.8
26	169.8	-0.7	-3.2	-6.6
27	182.0	-0.7	-2.4	-6.1
28	194.1	-0.7	-1.4	-5.1
29	206.3	-0.8	-0.3	-3.9
30	218.5	-0.8	+0.8	-2.4
31	230.6	-0.8	+1.9	-0.7

2003 JANUARY

01	242.8	-0.9	+3.0	+1.0
02	255.0	-0.9	+3.9	+2.6
03	267.2	-0.9	+4.6	+4.0
04	279.4	-0.9	+5.1	+5.2
05	291.6	-1.0	+5.3	+6.0
06	303.8	-1.0	+5.2	+6.5
07	315.9	-1.0	+4.7	+6.7
08	328.1	-1.0	+4.0	+6.6
09	340.3	-1.0	+3.0	+6.1
10	352.5	-1.0	+1.8	+5.4
11	004.6	-1.1	+0.4	+4.5
12	016.8	-1.1	-0.9	+3.4
13	028.9	-1.1	-2.2	+2.2
14	041.1	-1.1	-3.4	+0.8
15	053.2	-1.1	-4.3	-0.6
16	065.3	-1.1	-5.0	-2.0
17	077.5	-1.2	-5.3	-3.4
18	089.6	-1.2	-5.3	-4.6
19	101.7	-1.2	-4.9	-5.6
20	113.8	-1.2	-4.3	-6.3
21	126.0	-1.2	-3.4	-6.6
22	138.1	-1.3	-2.4	-6.5
23	150.2	-1.3	-1.3	-6.0
24	162.4	-1.3	-0.3	-5.2
25	174.5	-1.3	+0.7	-4.0
26	186.7	-1.3	+1.6	-2.6
27	198.9	-1.3	+2.4	-1.0
28	211.0	-1.4	+3.1	+0.7
29	223.2	-1.4	+3.7	+2.2
30	235.4	-1.4	+4.1	+3.7
31	247.6	-1.4	+4.5	+4.9

2003 FEBRUARY

01	259.8	-1.4	+4.6	+5.8
02	272.0	-1.4	+4.6	+6.3
03	284.2	-1.4	+4.3	+6.6
04	296.4	-1.5	+3.7	+6.5
05	308.6	-1.5	+2.9	+6.1
06	320.7	-1.5	+1.8	+5.5
07	332.9	-1.5	+0.6	+4.6
08	345.1	-1.5	-0.7	+3.5
09	357.3	-1.5	-2.1	+2.3
10	009.4	-1.5	-3.5	+1.0
11	021.6	-1.5	-4.6	-0.4
12	033.8	-1.5	-5.6	-1.7
13	045.9	-1.5	-6.2	-3.1
14	058.0	-1.5	-6.4	-4.3
15	070.2	-1.5	-6.1	-5.3
16	082.3	-1.5	-5.4	-6.0
17	094.5	-1.5	-4.4	-6.5
18	106.6	-1.5	-3.1	-6.5
19	118.7	-1.5	-1.6	-6.0
20	130.9	-1.5	-0.1	-5.2
21	143.0	-1.5	+1.3	-4.1
22	155.2	-1.5	+2.5	-2.6
23	167.3	-1.5	+3.5	-1.1
24	179.5	-1.5	+4.2	+0.5
25	191.7	-1.5	+4.7	+2.1
26	203.8	-1.5	+5.0	+3.5
27	216.0	-1.5	+5.2	+4.7
28	228.2	-1.5	+5.1	+5.7

2003 MARCH

01	240.4	-1.5	+4.9	+6.3
02	252.6	-1.5	+4.5	+6.6
03	264.8	-1.5	+3.9	+6.5
04	277.0	-1.5	+3.2	+6.2
05	289.3	-1.5	+2.2	+5.6
06	301.5	-1.5	+1.0	+4.7
07	313.7	-1.5	-0.3	+3.6
08	325.9	-1.5	-1.7	+2.4
09	338.0	-1.5	-3.1	+1.1
10	350.2	-1.5	-4.4	-0.3
11	002.4	-1.5	-5.6	-1.6
12	014.6	-1.5	-6.6	-2.9
13	026.8	-1.5	-7.2	-4.1
14	038.9	-1.5	-7.5	-5.1
15	051.1	-1.4	-7.2	-5.9
16	063.2	-1.4	-6.4	-6.4
17	075.4	-1.4	-5.2	-6.5
18	087.5	-1.4	-3.6	-6.2
19	099.7	-1.4	-1.8	-5.4
20	111.8	-1.4	+0.1	-4.3
21	124.0	-1.3	+1.9	-2.9
22	136.1	-1.3	+3.4	-1.3
23	148.3	-1.3	+4.7	+0.4
24	160.5	-1.3	+5.6	+2.0
25	172.7	-1.3	+6.2	+3.5
26	184.8	-1.3	+6.4	+4.7
27	197.0	-1.3	+6.4	+5.7
28	209.2	-1.2	+6.1	+6.3
29	221.5	-1.2	+5.6	+6.7
30	233.7	-1.2	+5.0	+6.7
31	245.9	-1.2	+4.1	+6.4

2003 APRIL

01	258.1	-1.2	+3.1	+5.8
02	270.3	-1.2	+1.9	+4.9

258 Appendix

03	282.5	-1.2	+0.7	+3.8
04	294.8	-1.2	-0.7	+2.6
05	307.0	-1.1	-2.1	+1.3
06	319.2	-1.1	-3.6	-0.1
07	331.4	-1.1	-4.9	-1.5
08	343.6	-1.1	-6.1	-2.8
09	355.8	-1.1	-7.1	-4.0
10	008.0	-1.1	-7.7	-5.1
11	020.2	-1.0	-8.0	-5.9
12	032.4	-1.0	-7.8	-6.5
13	044.5	-1.0	-7.1	-6.7
14	056.7	-1.0	-6.0	-6.5
15	068.9	-0.9	-4.4	-5.8
16	081.1	-0.9	-2.5	-4.8
17	093.2	-0.9	-0.5	-3.4
18	105.4	-0.8	+1.5	-1.8
19	117.5	-0.8	+3.3	-0.1
20	129.7	-0.8	+4.9	+1.6
21	141.9	-0.7	+6.1	+3.2
22	154.1	-0.7	+6.9	+4.6
23	166.3	-0.7	+7.2	+5.6
24	178.5	-0.7	+7.2	+6.3
25	190.7	-0.7	+6.9	+6.7
26	202.9	-0.6	+6.3	+6.8
27	215.1	-0.6	+5.5	+6.5
28	227.3	-0.6	+4.4	+6.0
29	239.6	-0.6	+3.2	+5.2
30	251.8	-0.6	+1.9	+4.1

2003 MAY

01	264.0	-0.5	+0.5	+2.9
02	276.3	-0.5	-0.9	+1.6
03	288.5	-0.5	-2.3	+0.2
04	300.7	-0.5	-3.6	-1.3
05	313.0	-0.5	-4.9	-2.6
06	325.2	-0.4	-6.0	-3.9
07	337.4	-0.4	-6.9	-5.0
08	349.6	-0.4	-7.5	-5.9
09	001.8	-0.4	-7.8	-6.5
10	014.0	-0.3	-7.6	-6.8
11	026.2	-0.3	-7.1	-6.7
12	038.4	-0.3	-6.2	-6.2
13	050.6	-0.2	-4.8	-5.3
14	062.8	-0.2	-3.1	-4.1
15	075.0	-0.2	-1.2	-2.5
16	087.2	-0.1	+0.7	-0.8
17	099.3	-0.1	+2.6	+1.0
18	111.5	-0.1	+4.3	+2.7
19	123.7	-0.0	+5.7	+4.2
20	135.9	+0.0	+6.7	+5.3
21	148.1	+0.0	+7.3	+6.2
22	160.3	+0.1	+7.4	+6.7
23	172.5	+0.1	+7.1	+6.8
24	184.7	+0.1	+6.5	+6.6
25	196.9	+0.1	+5.6	+6.1
26	209.1	+0.2	+4.4	+5.4
27	221.4	+0.2	+3.2	+4.4
28	233.6	+0.2	+1.8	+3.3
29	245.9	+0.2	+0.4	+1.9
30	258.1	+0.2	-1.0	+0.5
31	270.4	+0.3	-2.4	-0.9

2003 JUNE

01	282.6	+0.3	-3.6	-2.3
02	294.8	+0.3	-4.7	-3.6
03	307.1	+0.3	-5.5	-4.8
04	319.3	+0.3	-6.2	-5.7
05	331.6	+0.4	-6.6	-6.4
06	343.8	+0.4	-6.8	-6.8
07	356.0	+0.4	-6.6	-6.8
08	008.2	+0.4	-6.2	-6.4
09	020.5	+0.5	-5.4	-5.6
10	032.7	+0.5	-4.3	-4.5
11	044.9	+0.5	-3.0	-3.1
12	057.1	+0.6	-1.4	-1.5
13	069.2	+0.6	+0.3	+0.3
14	081.4	+0.6	+2.0	+2.0
15	093.6	+0.7	+3.6	+3.6
16	105.8	+0.7	+5.0	+4.9
17	118.0	+0.7	+6.0	+5.9
18	130.2	+0.8	+6.7	+6.5
19	142.4	+0.8	+6.9	+6.7
20	154.6	+0.8	+6.7	+6.6
21	166.8	+0.8	+6.1	+6.2
22	179.0	+0.9	+5.2	+5.5
23	191.2	+0.9	+4.0	+4.6
24	203.5	+0.9	+2.7	+3.5
25	215.7	+0.9	+1.3	+2.2
26	228.0	+0.9	-0.1	+0.9
27	240.2	+0.9	-1.4	-0.5
28	252.5	+0.9	-2.6	-1.9
29	264.7	+1.0	-3.7	-3.3
30	277.0	+1.0	-4.5	-4.5

2003 JULY

01	289.2	+1.0	-5.1	-5.5
02	301.5	+1.0	-5.4	-6.2
03	313.7	+1.0	-5.6	-6.6
04	326.0	+1.0	-5.4	-6.7
05	338.2	+1.1	-5.1	-6.4
06	350.4	+1.1	-4.6	-5.8
07	002.7	+1.1	-3.8	-4.7
08	014.9	+1.1	-2.9	-3.4
09	027.1	+1.1	-1.9	-1.9
10	039.3	+1.2	-0.7	-0.2
11	051.5	+1.2	+0.6	+1.5
12	063.7	+1.2	+2.0	+3.1
13	075.9	+1.2	+3.2	+4.4
14	088.1	+1.3	+4.4	+5.5
15	100.3	+1.3	+5.3	+6.2
16	112.4	+1.3	+5.9	+6.6
17	124.6	+1.3	+6.0	+6.5
18	136.8	+1.3	+5.8	+6.2
19	149.0	+1.4	+5.3	+5.5
20	161.3	+1.4	+4.4	+4.6
21	173.5	+1.4	+3.2	+3.6
22	185.7	+1.4	+1.9	+2.4
23	197.9	+1.4	+0.5	+1.1
24	210.2	+1.4	-0.9	-0.3
25	222.4	+1.4	-2.2	-1.6
26	234.6	+1.4	-3.3	-3.0
27	246.9	+1.4	-4.1	-4.1
28	259.1	+1.4	-4.7	-5.2
29	271.4	+1.4	-5.0	-6.0
30	283.7	+1.4	-4.9	-6.4
31	295.9	+1.4	-4.7	-6.6

2003 AUGUST

01	308.2	+1.4	-4.2	-6.4
02	320.4	+1.4	-3.5	-5.8
03	332.6	+1.4	-2.7	-4.8
04	344.9	+1.5	-1.9	-3.5
05	357.1	+1.5	-1.0	-2.1
06	009.3	+1.5	-0.1	-0.4
07	021.5	+1.5	+0.8	+1.2
08	033.7	+1.5	+1.8	+2.8
09	045.9	+1.5	+2.7	+4.2
10	058.1	+1.5	+3.6	+5.3
11	070.3	+1.5	+4.4	+6.1
12	082.5	+1.5	+5.0	+6.5
13	094.7	+1.5	+5.3	+6.5
14	106.9	+1.6	+5.3	+6.2
15	119.0	+1.6	+4.9	+5.6
16	131.2	+1.6	+4.3	+4.7
17	143.4	+1.6	+3.3	+3.7
18	155.6	+1.6	+2.2	+2.5
19	167.8	+1.6	+0.8	+1.2
20	180.1	+1.6	-0.6	-0.2
21	192.3	+1.5	-2.0	-1.5
22	204.5	+1.5	-3.2	-2.8
23	216.7	+1.5	-4.2	-4.0
24	229.0	+1.5	-5.0	-5.0
25	241.2	+1.5	-5.3	-5.8
26	253.4	+1.5	-5.4	-6.3
27	265.7	+1.5	-5.0	-6.5
28	277.9	+1.5	-4.3	-6.4
29	290.2	+1.5	-3.4	-5.8
30	302.4	+1.5	-2.4	-4.9
31	314.6	+1.5	-1.2	-3.7

2003 SEPTEMBER

01	326.9	+1.5	-0.1	-2.2
02	339.1	+1.5	+1.0	-0.6
03	351.3	+1.5	+1.9	+1.1
04	003.5	+1.5	+2.8	+2.7
05	015.7	+1.4	+3.5	+4.1
06	027.9	+1.4	+4.2	+5.2
07	040.1	+1.4	+4.7	+6.0
08	052.3	+1.4	+5.1	+6.5
09	064.5	+1.4	+5.2	+6.6
10	076.6	+1.4	+5.2	+6.4
11	088.8	+1.4	+4.9	+5.8
12	101.0	+1.4	+4.3	+4.9
13	113.2	+1.4	+3.5	+3.9
14	125.4	+1.4	+2.4	+2.6
15	137.5	+1.4	+1.2	+1.3
16	149.7	+1.4	-0.2	-0.0
17	161.9	+1.3	-1.6	-1.4
18	174.1	+1.3	-3.0	-2.7
19	186.3	+1.3	-4.2	-3.9
20	198.5	+1.3	-5.3	-4.9
21	210.7	+1.3	-6.0	-5.7
22	222.9	+1.3	-6.3	-6.3
23	235.2	+1.2	-6.2	-6.6
24	247.4	+1.2	-5.7	-6.5
25	259.6	+1.2	-4.8	-6.0
26	271.8	+1.2	-3.5	-5.2
27	284.1	+1.2	-2.1	-4.0
28	296.3	+1.1	-0.5	-2.5
29	308.5	+1.1	+1.0	-0.9
30	320.7	+1.1	+2.4	+0.9

2003 OCTOBER

01	333.0	+1.1	+3.6	+2.5
02	345.2	+1.1	+4.6	+4.0
03	357.3	+1.0	+5.3	+5.2
04	009.5	+1.0	+5.8	+6.1
05	021.7	+1.0	+6.1	+6.6
06	033.9	+1.0	+6.2	+6.8
07	046.1	+1.0	+6.0	+6.6
08	058.2	+0.9	+5.6	+6.0
09	070.4	+0.9	+4.9	+5.2
10	082.5	+0.9	+4.1	+4.2
11	094.7	+0.9	+3.0	+3.0
12	106.9	+0.9	+1.8	+1.6
13	119.0	+0.8	+0.5	+0.2
14	131.2	+0.8	-0.9	-1.2
15	143.4	+0.8	-2.3	-2.5
16	155.5	+0.8	-3.7	-3.8
17	167.7	+0.7	-5.0	-4.8
18	179.9	+0.7	-6.0	-5.7
19	192.1	+0.7	-6.8	-6.3
20	204.3	+0.7	-7.1	-6.7
21	216.4	+0.6	-7.1	-6.7
22	228.7	+0.6	-6.6	-6.4
23	240.9	+0.6	-5.7	-5.6
24	253.1	+0.6	-4.4	-4.5
25	265.3	+0.5	-2.7	-3.1
26	277.5	+0.5	-0.9	-1.4
27	289.7	+0.5	+1.0	+0.3
28	301.9	+0.4	+2.8	+2.1
29	314.1	+0.4	+4.3	+3.6
30	326.3	+0.4	+5.6	+5.0
31	338.5	+0.4	+6.6	+6.0

2003 NOVEMBER

01	350.7	+0.3	+7.1	+6.6
02	002.9	+0.3	+7.3	+6.8
03	015.0	+0.3	+7.2	+6.7
04	027.2	+0.3	+6.8	+6.3
05	039.3	+0.2	+6.1	+5.5
06	051.5	+0.2	+5.2	+4.5
07	063.6	+0.2	+4.1	+3.3
08	075.8	+0.1	+2.9	+2.0
09	087.9	+0.1	+1.5	+0.6
10	100.1	+0.1	+0.1	-0.9
11	112.2	+0.1	-1.3	-2.3
12	124.4	+0.0	-2.7	-3.5
13	136.5	+0.0	-4.0	-4.7
14	148.7	-0.0	-5.2	-5.6
15	160.8	-0.1	-6.2	-6.3
16	173.0	-0.1	-6.9	-6.7
17	185.1	-0.1	-7.4	-6.8
18	197.3	-0.1	-7.5	-6.6
19	209.5	-0.2	-7.2	-6.0
20	221.7	-0.2	-6.4	-5.0
21	233.9	-0.2	-5.3	-3.7
22	246.0	-0.2	-3.7	-2.2
23	258.2	-0.3	-1.9	-0.4
24	270.4	-0.3	+0.1	+1.3
25	282.6	-0.3	+2.1	+3.0
26	294.8	-0.4	+4.0	+4.5
27	307.0	-0.4	+5.6	+5.6
28	319.2	-0.4	+6.8	+6.4
29	331.4	-0.4	+7.6	+6.7
30	343.6	-0.5	+7.9	+6.7

2003 DECEMBER

01	355.7	-0.5	+7.8	+6.3
02	007.9	-0.5	+7.3	+5.6
03	020.0	-0.6	+6.4	+4.7
04	032.2	-0.6	+5.4	+3.6
05	044.3	-0.6	+4.1	+2.3
06	056.5	-0.6	+2.8	+0.9
07	068.6	-0.7	+1.3	-0.5
08	080.7	-0.7	-0.1	-1.9
09	092.9	-0.7	-1.4	-3.2
10	105.0	-0.8	-2.7	-4.4
11	117.1	-0.8	-3.9	-5.4
12	129.3	-0.8	-4.9	-6.1
13	141.4	-0.8	-5.8	-6.6
14	153.5	-0.8	-6.4	-6.8
15	165.7	-0.9	-6.9	-6.6
16	177.8	-0.9	-7.0	-6.1
17	190.0	-0.9	-6.9	-5.3
18	202.2	-0.9	-6.3	-4.1
19	214.3	-0.9	-5.5	-2.7
20	226.5	-1.0	-4.2	-1.1
21	238.7	-1.0	-2.7	+0.6
22	250.9	-1.0	-0.9	+2.3
23	263.1	-1.0	+1.1	+3.9
24	275.3	-1.0	+3.0	+5.2
25	287.5	-1.1	+4.7	+6.0
26	299.7	-1.1	+6.1	+6.5
27	311.8	-1.1	+7.1	+6.6
28	324.0	-1.1	+7.5	+6.3
29	336.2	-1.2	+7.5	+5.7
30	348.4	-1.2	+7.1	+4.8
31	000.5	-1.2	+6.2	+3.7

2004 JANUARY

01	012.7	-1.2	+5.1	+2.5
02	024.8	-1.2	+3.8	+1.1
03	037.0	-1.3	+2.4	-0.2
04	049.1	-1.3	+0.9	-1.6
05	061.2	-1.3	-0.4	-2.9
06	073.4	-1.3	-1.7	-4.1
07	085.5	-1.3	-2.9	-5.1
08	097.6	-1.4	-3.8	-5.9
09	109.7	-1.4	-4.6	-6.4
10	121.9	-1.4	-5.2	-6.7
11	134.0	-1.4	-5.6	-6.5
12	146.1	-1.4	-5.8	-6.1
13	158.3	-1.4	-5.8	-5.3
14	170.4	-1.4	-5.6	-4.2
15	182.6	-1.4	-5.2	-2.9
16	194.8	-1.4	-4.6	-1.4
17	206.9	-1.4	-3.6	+0.3
18	219.1	-1.4	-2.5	+1.9
19	231.3	-1.4	-1.0	+3.4
20	243.5	-1.4	+0.6	+4.7
21	255.7	-1.4	+2.2	+5.7
22	267.9	-1.5	+3.7	+6.3
23	280.0	-1.5	+5.0	+6.5
24	292.2	-1.5	+6.0	+6.3
25	304.4	-1.5	+6.5	+5.7
26	316.6	-1.5	+6.5	+4.9
27	328.8	-1.5	+6.1	+3.8
28	341.0	-1.5	+5.4	+2.6
29	353.1	-1.5	+4.3	+1.2
30	005.3	-1.5	+3.0	-0.1
31	017.5	-1.5	+1.6	-1.4

2004 FEBRUARY

01	029.6	-1.6	+0.2	-2.7
02	041.7	-1.6	-1.2	-3.9
03	053.9	-1.6	-2.4	-4.9
04	066.0	-1.6	-3.4	-5.7
05	078.2	-1.6	-4.1	-6.3
06	090.3	-1.6	-4.6	-6.6
07	102.4	-1.6	-4.9	-6.5
08	114.6	-1.6	-4.9	-6.1
09	126.7	-1.5	-4.8	-5.3
10	138.8	-1.5	-4.5	-4.3
11	151.0	-1.5	-4.1	-3.0
12	163.1	-1.5	-3.5	-1.5
13	175.3	-1.5	-2.8	+0.1
14	187.5	-1.5	-2.0	+1.7
15	199.6	-1.5	-1.1	+3.3
16	211.8	-1.5	+0.0	+4.6
17	224.0	-1.5	+1.2	+5.6
18	236.2	-1.5	+2.4	+6.3
19	248.4	-1.5	+3.5	+6.5
20	260.6	-1.4	+4.4	+6.4
21	272.8	-1.4	+5.1	+5.9
22	285.0	-1.4	+5.4	+5.1
23	297.2	-1.4	+5.4	+4.0
24	309.4	-1.4	+4.9	+2.7
25	321.6	-1.4	+4.1	+1.4
26	333.8	-1.4	+3.1	+0.0
27	346.0	-1.4	+1.8	-1.3
28	358.1	-1.4	+0.4	-2.6
29	010.3	-1.4	-0.9	-3.8

2004 MARCH

01	022.5	-1.4	-2.2	-4.8
02	034.6	-1.4	-3.4	-5.7
03	046.8	-1.4	-4.3	-6.3
04	058.9	-1.4	-4.9	-6.6
05	071.1	-1.4	-5.2	-6.6
06	083.2	-1.3	-5.2	-6.2
07	095.4	-1.3	-4.9	-5.5
08	107.5	-1.3	-4.4	-4.5
09	119.7	-1.3	-3.6	-3.2
10	131.8	-1.2	-2.8	-1.7
11	144.0	-1.2	-1.9	-0.0
12	156.1	-1.2	-0.9	+1.6
13	168.3	-1.2	+0.0	+3.2
14	180.5	-1.1	+1.0	+4.5
15	192.7	-1.1	+1.9	+5.6
16	204.9	-1.1	+2.8	+6.3
17	217.1	-1.1	+3.5	+6.7
18	229.3	-1.1	+4.2	+6.6
19	241.5	-1.0	+4.7	+6.2
20	253.7	-1.0	+4.9	+5.4
21	265.9	-1.0	+4.9	+4.3
22	278.1	-1.0	+4.5	+3.1
23	290.3	-1.0	+3.9	+1.7
24	302.5	-1.0	+3.0	+0.3
25	314.7	-1.0	+1.9	-1.1
26	326.9	-0.9	+0.7	-2.5
27	339.1	-0.9	-0.7	-3.7
28	351.3	-0.9	-2.1	-4.8
29	003.5	-0.9	-3.3	-5.6
30	015.7	-0.9	-4.5	-6.3
31	027.9	-0.9	-5.3	-6.6

2004 APRIL

01	040.1	-0.8	-5.9	-6.7
02	052.2	-0.8	-6.2	-6.5
03	064.4	-0.8	-6.0	-5.9
04	076.6	-0.8	-5.5	-4.9
05	088.7	-0.7	-4.7	-3.7
06	100.9	-0.7	-3.6	-2.1
07	113.0	-0.7	-2.3	-0.5
08	125.2	-0.6	-1.0	+1.3
09	137.4	-0.6	+0.4	+2.9
10	149.5	-0.6	+1.7	+4.4
11	161.7	-0.5	+2.9	+5.5
12	173.9	-0.5	+3.9	+6.3
13	186.1	-0.5	+4.6	+6.8
14	198.3	-0.4	+5.2	+6.8
15	210.5	-0.4	+5.5	+6.4
16	222.7	-0.4	+5.6	+5.7
17	235.0	-0.4	+5.4	+4.7
18	247.2	-0.3	+5.0	+3.5
19	259.4	-0.3	+4.3	+2.1
20	271.6	-0.3	+3.4	+0.7
21	283.9	-0.3	+2.4	-0.8
22	296.1	-0.2	+1.2	-2.2
23	308.3	-0.2	-0.2	-3.5
24	320.5	-0.2	-1.5	-4.6
25	332.8	-0.2	-2.9	-5.5
26	345.0	-0.2	-4.1	-6.2
27	357.2	-0.2	-5.3	-6.7
28	009.4	-0.1	-6.2	-6.8
29	021.6	-0.1	-6.8	-6.6
30	033.8	-0.1	-7.1	-6.2

2004 MAY

01	046.0	-0.1	-7.0	-5.3
02	058.1	-0.0	-6.5	-4.2
03	070.3	+0.0	-5.6	-2.8
04	082.5	+0.0	-4.3	-1.1
05	094.7	+0.1	-2.7	+0.6
06	106.8	+0.1	-1.0	+2.3
07	119.0	+0.2	+0.8	+3.9
08	131.2	+0.2	+2.6	+5.2
09	143.4	+0.2	+4.1	+6.1
10	155.6	+0.3	+5.3	+6.7
11	167.8	+0.3	+6.2	+6.8
12	180.0	+0.3	+6.7	+6.5
13	192.2	+0.4	+6.8	+5.9
14	204.4	+0.4	+6.6	+5.0
15	216.6	+0.4	+6.2	+3.8
16	228.9	+0.4	+5.4	+2.5
17	241.1	+0.5	+4.5	+1.1
18	253.4	+0.5	+3.3	-0.4
19	265.6	+0.5	+2.1	-1.8

Appendix 259

20	277.8	+0.5	+0.8	-3.2
21	290.1	+0.5	-0.5	-4.3
22	302.3	+0.6	-1.9	-5.3
23	314.6	+0.6	-3.2	-6.1
24	326.8	+0.6	-4.5	-6.6
25	339.0	+0.6	-5.6	-6.8
26	351.3	+0.6	-6.5	-6.7
27	003.5	+0.6	-7.2	-6.3
28	015.7	+0.7	-7.5	-5.6
29	027.9	+0.7	-7.6	-4.6
30	040.1	+0.7	-7.2	-3.3
31	052.3	+0.7	-6.3	-1.7

2004 JUNE

01	064.5	+0.8	-5.1	-0.1
02	076.7	+0.8	-3.4	+1.6
03	088.8	+0.8	-1.5	+3.3
04	101.0	+0.9	+0.5	+4.7
05	113.2	+0.9	+2.6	+5.8
06	125.4	+0.9	+4.4	+6.4
07	137.6	+0.9	+5.9	+6.7
08	149.8	+1.0	+7.0	+6.5
09	162.0	+1.0	+7.6	+5.9
10	174.2	+1.0	+7.7	+5.1
11	186.4	+1.0	+7.4	+4.0
12	198.7	+1.1	+6.8	+2.7
13	210.9	+1.1	+5.8	+1.3
14	223.1	+1.1	+4.7	-0.1
15	235.4	+1.1	+3.4	-1.5
16	247.6	+1.1	+2.1	-2.8
17	259.9	+1.2	+0.7	-4.0
18	272.1	+1.2	-0.6	-5.1
19	284.4	+1.2	-1.9	-5.9
20	296.6	+1.2	-3.2	-6.4
21	308.9	+1.2	-4.3	-6.7
22	321.1	+1.2	-5.4	-6.6
23	333.4	+1.2	-6.2	-6.3
24	345.6	+1.2	-6.9	-5.7
25	357.8	+1.2	-7.3	-4.7
26	010.1	+1.2	-7.4	-3.5
27	022.3	+1.3	-7.1	-2.1
28	034.5	+1.3	-6.5	-0.5
29	046.7	+1.3	-5.4	+1.1
30	058.9	+1.3	-3.9	+2.7

2004 JULY

01	071.1	+1.3	-2.1	+4.2
02	083.3	+1.3	+0.0	+5.3
03	095.5	+1.4	+2.1	+6.1
04	107.6	+1.4	+4.0	+6.5
05	119.8	+1.4	+5.7	+6.4
06	132.0	+1.4	+6.9	+5.9
07	144.2	+1.4	+7.6	+5.1
08	156.4	+1.4	+7.8	+4.0
09	168.7	+1.4	+7.5	+2.8
10	180.9	+1.5	+6.9	+1.4
11	193.1	+1.5	+5.9	+0.0
12	205.3	+1.5	+4.7	-1.3
13	217.6	+1.5	+3.4	-2.6
14	229.8	+1.5	+2.0	-3.8
15	242.1	+1.5	+0.6	-4.9
16	254.3	+1.5	-0.7	-5.7
17	266.6	+1.5	-1.9	-6.3
18	278.8	+1.5	-3.0	-6.6
19	291.1	+1.5	-4.0	-6.6
20	303.3	+1.5	-4.9	-6.3
21	315.6	+1.5	-5.6	-5.7
22	327.8	+1.5	-6.1	-4.8
23	340.1	+1.5	-6.4	-3.6
24	352.3	+1.5	-6.5	-2.2
25	004.5	+1.5	-6.3	-0.7
26	016.7	+1.5	-5.8	+0.9
27	029.0	+1.5	-4.9	+2.4
28	041.2	+1.5	-3.6	+3.9
29	053.4	+1.5	-2.0	+5.1
30	065.5	+1.5	-0.2	+5.9
31	077.7	+1.5	+1.7	+6.4

2004 AUGUST

01	089.9	+1.5	+3.5	+6.4
02	102.1	+1.5	+5.1	+6.0
03	114.3	+1.5	+6.3	+5.3
04	126.5	+1.5	+7.0	+4.2
05	138.7	+1.5	+7.2	+2.9
06	150.9	+1.5	+7.0	+1.6
07	163.1	+1.5	+6.4	+0.2
08	175.3	+1.5	+5.4	-1.2
09	187.5	+1.5	+4.2	-2.5
10	199.8	+1.5	+2.9	-3.7
11	212.0	+1.5	+1.6	-4.8
12	224.2	+1.5	+0.3	-5.6
13	236.5	+1.5	-1.0	-6.2
14	248.7	+1.5	-2.1	-6.5
15	261.0	+1.5	-3.0	-6.6
16	273.2	+1.5	-3.8	-6.3
17	285.5	+1.5	-4.4	-5.7
18	297.7	+1.5	-4.9	-4.9
19	309.9	+1.4	-5.1	-3.7
20	322.2	+1.4	-5.2	-2.4
21	334.4	+1.4	-5.1	-0.8
22	346.6	+1.4	-4.8	+0.7
23	358.9	+1.4	-4.3	+2.3
24	011.1	+1.4	-3.4	+3.7
25	023.3	+1.4	-2.4	+5.0
26	035.5	+1.3	-1.1	+5.9
27	047.6	+1.3	+0.4	+6.4
28	059.8	+1.3	+1.9	+6.6
29	072.0	+1.3	+3.3	+6.3
30	084.2	+1.3	+4.6	+5.6
31	096.4	+1.3	+5.5	+4.5

2004 SEPTEMBER

01	108.5	+1.2	+6.1	+3.3
02	120.7	+1.2	+6.3	+1.8
03	132.9	+1.2	+6.0	+0.4
04	145.1	+1.2	+5.4	-1.1
05	157.3	+1.2	+4.5	-2.4
06	169.5	+1.2	+3.4	-3.7
07	181.7	+1.2	+2.1	-4.7
08	193.9	+1.2	+0.8	-5.6
09	206.1	+1.1	-0.5	-6.2
10	218.3	+1.1	-1.7	-6.6
11	230.6	+1.1	-2.7	-6.7
12	242.8	+1.1	-3.5	-6.5
13	255.0	+1.1	-4.1	-5.9
14	267.3	+1.1	-4.4	-5.1
15	279.5	+1.0	-4.6	-4.0
16	291.7	+1.0	-4.5	-2.6
17	304.0	+1.0	-4.2	-1.1
18	316.2	+1.0	-3.7	+0.5
19	328.4	+0.9	-3.1	+2.1
20	340.6	+0.9	-2.4	+3.6
21	352.8	+0.9	-1.5	+4.9
22	005.0	+0.9	-0.5	+5.9
23	017.2	+0.8	+0.6	+6.5
24	029.4	+0.8	+1.7	+6.7
25	041.6	+0.8	+2.8	+6.5
26	053.7	+0.8	+3.8	+5.9
27	065.9	+0.7	+4.6	+5.0
28	078.1	+0.7	+5.2	+3.7
29	090.2	+0.7	+5.4	+2.3
30	102.4	+0.6	+5.4	+0.8

2004 OCTOBER

01	114.6	+0.6	+5.0	-0.7
02	126.7	+0.6	+4.3	-2.2
03	138.9	+0.6	+3.4	-3.5
04	151.1	+0.5	+2.3	-4.6
05	163.2	+0.5	+1.0	-5.5
06	175.4	+0.5	-0.3	-6.2
07	187.6	+0.5	-1.5	-6.6
08	199.8	+0.5	-2.7	-6.8
09	212.0	+0.5	-3.7	-6.6
10	224.2	+0.4	-4.4	-6.2
11	236.4	+0.4	-4.9	-5.5
12	248.7	+0.4	-5.1	-4.4
13	260.9	+0.4	-5.0	-3.1
14	273.1	+0.3	-4.5	-1.6
15	285.3	+0.3	-3.8	+0.1
16	297.5	+0.3	-2.9	+1.7
17	309.7	+0.3	-1.8	+3.3
18	321.9	+0.2	-0.7	+4.7
19	334.1	+0.2	+0.5	+5.8
20	346.3	+0.2	+1.7	+6.5
21	358.5	+0.1	+2.7	+6.8
22	010.7	+0.1	+3.7	+6.7
23	022.9	+0.1	+4.4	+6.2
24	035.0	+0.0	+5.0	+5.3
25	047.2	-0.0	+5.3	+4.2
26	059.3	-0.1	+5.4	+2.8
27	071.5	-0.1	+5.2	+1.3
28	083.6	-0.1	+4.9	-0.2
29	095.8	-0.2	+4.2	-1.7
30	107.9	-0.2	+3.4	-3.1
31	120.1	-0.2	+2.4	-4.3

2004 NOVEMBER

01	132.2	-0.2	+1.3	-5.3
02	144.4	-0.3	+0.0	-6.1
03	156.5	-0.3	-1.3	-6.6
04	168.7	-0.3	-2.5	-6.8
05	180.9	-0.3	-3.7	-6.7
06	193.0	-0.3	-4.8	-6.4
07	205.2	-0.4	-5.6	-5.7
08	217.4	-0.4	-6.1	-4.8
09	229.6	-0.4	-6.2	-3.6
10	241.8	-0.4	-6.0	-2.2
11	254.0	-0.4	-5.4	-0.6
12	266.2	-0.5	-4.4	+1.1
13	278.4	-0.5	-3.1	+2.7
14	290.6	-0.5	-1.5	+4.2
15	302.8	-0.5	+0.2	+5.4
16	315.0	-0.6	+1.8	+6.3
17	327.2	-0.6	+3.3	+6.7
18	339.4	-0.6	+4.6	+6.7
19	351.5	-0.7	+5.5	+6.3
20	003.7	-0.7	+6.1	+5.5
21	015.9	-0.7	+6.4	+4.4
22	028.0	-0.8	+6.4	+3.1
23	040.2	-0.8	+6.1	+1.7
24	052.3	-0.8	+5.6	+0.2
25	064.4	-0.9	+4.9	-1.3
26	076.6	-0.9	+4.0	-2.7
27	088.7	-0.9	+3.0	-4.0
28	100.8	-1.0	+1.9	-5.0
29	113.0	-1.0	+0.7	-5.9
30	125.1	-1.0	-0.6	-6.4

2004 DECEMBER

01	137.3	-1.0	-1.9	-6.7
02	149.4	-1.0	-3.2	-6.7
03	161.6	-1.0	-4.4	-6.4
04	173.7	-1.0	-5.5	-5.8
05	185.9	-1.1	-6.4	-5.0
06	198.0	-1.1	-7.0	-3.9
07	210.2	-1.1	-7.3	-2.6
08	222.4	-1.1	-7.2	-1.1
09	234.6	-1.1	-6.6	+0.5
10	246.8	-1.1	-5.5	+2.1
11	258.9	-1.1	-4.0	+3.6
12	271.1	-1.1	-2.1	+4.9
13	283.3	-1.2	-0.1	+5.9
14	295.5	-1.2	+2.0	+6.4
15	307.7	-1.2	+3.9	+6.6
16	319.9	-1.2	+5.5	+6.2
17	332.1	-1.3	+6.6	+5.5
18	344.2	-1.3	+7.3	+4.5
19	356.4	-1.3	+7.5	+3.2
20	008.6	-1.3	+7.4	+1.8
21	020.7	-1.3	+6.9	+0.4
22	032.9	-1.4	+6.1	-1.1
23	045.0	-1.4	+5.2	-2.5
24	057.1	-1.4	+4.1	-3.7
25	069.3	-1.4	+2.9	-4.8
26	081.4	-1.5	+1.7	-5.7
27	093.5	-1.5	+0.4	-6.3
28	105.6	-1.5	-0.8	-6.6
29	117.8	-1.5	-2.1	-6.6
30	129.9	-1.5	-3.4	-6.4
31	142.1	-1.5	-4.6	-5.8

2005 JANUARY

01	154.2	-1.5	-5.6	-5.0
02	166.3	-1.5	-6.6	-4.0
03	178.5	-1.5	-7.3	-2.7
04	190.6	-1.5	-7.7	-1.4
05	202.8	-1.5	-7.7	+0.2
06	215.0	-1.5	-7.3	+1.7
07	227.2	-1.5	-6.4	+3.2
08	239.3	-1.5	-5.0	+4.5
09	251.5	-1.5	-3.1	+5.5
10	263.7	-1.5	-1.0	+6.2
11	275.9	-1.5	+1.2	+6.4
12	288.1	-1.5	+3.4	+6.2
13	300.3	-1.5	+5.2	+5.6
14	312.5	-1.5	+6.6	+4.6
15	324.7	-1.5	+7.5	+3.3
16	336.8	-1.5	+7.8	+1.9
17	349.0	-1.5	+7.7	+0.5
18	001.2	-1.5	+7.2	-1.0
19	013.3	-1.5	+6.4	-2.4
20	025.5	-1.6	+5.4	-3.6
21	037.6	-1.6	+4.2	-4.7
22	049.8	-1.6	+2.9	-5.6
23	061.9	-1.6	+1.6	-6.2
24	074.0	-1.6	+0.3	-6.5
25	086.2	-1.6	-0.9	-6.6
26	098.3	-1.6	-2.1	-6.4
27	110.4	-1.6	-3.3	-5.9
28	122.6	-1.5	-4.3	-5.1
29	134.7	-1.5	-5.3	-4.0
30	146.8	-1.5	-6.1	-2.8
31	159.0	-1.5	-6.8	-1.4

2005 FEBRUARY

01	171.1	-1.5	-7.2	+0.0
02	183.3	-1.5	-7.3	+1.6
03	195.5	-1.4	-7.0	+3.0
04	207.6	-1.4	-6.3	+4.3
05	219.8	-1.4	-5.1	+5.4
06	232.0	-1.4	-3.6	+6.1
07	244.2	-1.4	-1.7	+6.5
08	256.4	-1.4	+0.4	+6.4
09	268.6	-1.4	+2.4	+5.8
10	280.8	-1.3	+4.2	+4.9
11	293.0	-1.3	+5.7	+3.6
12	305.2	-1.3	+6.7	+2.2
13	317.4	-1.3	+7.2	+0.7
14	329.5	-1.3	+7.2	-0.8
15	341.7	-1.3	+6.8	-2.3
16	353.9	-1.3	+6.1	-3.6
17	006.1	-1.3	+5.1	-4.7
18	018.2	-1.3	+3.9	-5.5
19	030.4	-1.3	+2.6	-6.2
20	042.5	-1.3	+1.3	-6.6
21	054.7	-1.3	+0.1	-6.7
22	066.8	-1.3	-1.2	-6.5
23	079.0	-1.2	-2.3	-6.0
24	091.1	-1.2	-3.3	-5.3
25	103.2	-1.2	-4.2	-4.2
26	115.4	-1.2	-4.9	-3.0
27	127.5	-1.2	-5.5	-1.6
28	139.7	-1.1	-5.9	-0.1

2005 MARCH

01	151.8	-1.1	-6.1	+1.4
02	164.0	-1.1	-6.1	+2.9
03	176.2	-1.0	-5.7	+4.3
04	188.3	-1.0	-5.1	+5.4
05	200.5	-1.0	-4.1	+6.2
06	212.7	-0.9	-2.9	+6.6
07	224.9	-0.9	-1.4	+6.6
08	237.1	-0.9	+0.3	+6.2
09	249.3	-0.9	+1.9	+5.3
10	261.5	-0.8	+3.4	+4.2
11	273.7	-0.8	+4.6	+2.7
12	285.9	-0.8	+5.5	+1.1
13	298.1	-0.8	+6.0	-0.4
14	310.3	-0.8	+6.1	-2.0
15	322.5	-0.7	+5.8	-3.4
16	334.7	-0.7	+5.2	-4.5
17	346.9	-0.7	+4.3	-5.5
18	359.1	-0.7	+3.1	-6.2
19	011.3	-0.7	+1.9	-6.6
20	023.5	-0.7	+0.6	-6.8
21	035.6	-0.7	-0.6	-6.6
22	047.8	-0.6	-1.8	-6.2
23	060.0	-0.6	-2.8	-5.5
24	072.1	-0.6	-3.7	-4.6
25	084.3	-0.6	-4.4	-3.4
26	096.4	-0.5	-4.9	-2.0
27	108.6	-0.5	-5.1	-0.4
28	120.7	-0.5	-5.2	+1.1
29	132.9	-0.4	-5.0	+2.7
30	145.1	-0.4	-4.6	+4.1
31	157.2	-0.4	-4.0	+5.3

2005 APRIL

01	169.4	-0.3	-3.2	+6.2
02	181.6	-0.3	-2.2	+6.7
03	193.8	-0.2	-1.1	+6.8
04	206.0	-0.2	+0.0	+6.5
05	218.2	-0.2	+1.2	+5.8
06	230.4	-0.1	+2.4	+4.7
07	242.6	-0.1	+3.4	+3.3
08	254.8	-0.1	+4.2	+1.7
09	267.1	-0.1	+4.8	+0.1
10	279.3	-0.0	+5.0	-1.5
11	291.5	-0.0	+5.0	-3.0
12	303.7	+0.0	+4.7	-4.2
13	316.0	+0.0	+4.0	-5.3
14	328.2	+0.0	+3.2	-6.1
15	340.4	+0.1	+2.1	-6.6
16	352.6	+0.1	+0.9	-6.8
17	004.8	+0.1	-0.4	-6.7
18	017.0	+0.1	-1.6	-6.4
19	029.2	+0.1	-2.8	-5.8
20	041.3	+0.2	-3.8	-4.9
21	053.5	+0.2	-4.6	-3.8
22	065.7	+0.2	-5.1	-2.4
23	077.9	+0.2	-5.4	-1.0
24	090.0	+0.3	-5.3	+0.6
25	102.2	+0.3	-5.0	+2.2
26	114.4	+0.3	-4.3	+3.7
27	126.5	+0.4	-3.5	+5.0
28	138.7	+0.4	-2.4	+6.0
29	150.9	+0.4	-1.3	+6.6
30	163.1	+0.5	-0.1	+6.8

2005 MAY

01	175.3	+0.5	+1.0	+6.6
02	187.5	+0.5	+2.1	+6.0
03	199.7	+0.6	+3.0	+5.0
04	211.9	+0.6	+3.7	+3.7
05	224.2	+0.6	+4.2	+2.2
06	236.4	+0.7	+4.6	+0.6
07	248.6	+0.7	+4.7	-1.0
08	260.9	+0.7	+4.6	-2.5
09	273.1	+0.7	+4.4	-3.8
10	285.3	+0.7	+3.8	-5.0
11	297.6	+0.8	+3.1	-5.8
12	309.8	+0.8	+2.2	-6.4
13	322.1	+0.8	+1.1	-6.7
14	334.3	+0.8	-0.1	-6.7
15	346.5	+0.8	-1.4	-6.4
16	358.7	+0.8	-2.6	-5.9
17	010.9	+0.9	-3.8	-5.1
18	023.1	+0.9	-4.8	-4.0
19	035.3	+0.9	-5.7	-2.8
20	047.5	+0.9	-6.2	-1.4
21	059.7	+0.9	-6.3	+0.1
22	071.9	+1.0	-6.0	+1.6
23	084.1	+1.0	-5.4	+3.2
24	096.3	+1.0	-4.3	+4.5
25	108.4	+1.0	-3.0	+5.6
26	120.6	+1.1	-1.5	+6.3
27	132.8	+1.1	+0.1	+6.6
28	145.0	+1.1	+1.6	+6.5
29	157.2	+1.1	+3.0	+6.0
30	169.4	+1.1	+4.0	+5.1
31	181.6	+1.2	+4.8	+3.9

2005 JUNE

01	193.9	+1.2	+5.3	+2.5
02	206.1	+1.2	+5.6	+0.9
03	218.3	+1.2	+5.5	-0.6
04	230.6	+1.3	+5.3	-2.1
05	242.8	+1.3	+4.9	-3.5
06	255.1	+1.3	+4.3	-4.7
07	267.3	+1.3	+3.6	-5.6
08	279.6	+1.3	+2.7	-6.2
09	291.8	+1.3	+1.6	-6.6
10	304.1	+1.3	+0.5	-6.6
11	316.3	+1.4	-0.8	-6.4
12	328.5	+1.4	-2.1	-5.9
13	340.8	+1.4	-3.4	-5.1
14	353.0	+1.4	-4.6	-4.2
15	005.2	+1.4	-5.7	-3.0
16	017.5	+1.4	-6.6	-1.7
17	029.7	+1.4	-7.1	-0.2
18	041.9	+1.4	-7.2	+1.2
19	054.1	+1.4	-6.9	+2.7
20	066.3	+1.4	-6.1	+4.1
21	078.5	+1.4	-4.8	+5.2
22	090.6	+1.4	-3.2	+6.0
23	102.8	+1.4	-1.3	+6.4
24	115.0	+1.4	+0.7	+6.4
25	127.2	+1.5	+2.5	+6.0
26	139.4	+1.5	+4.2	+5.1
27	151.6	+1.5	+5.4	+3.9
28	163.8	+1.5	+6.3	+2.6
29	176.0	+1.5	+6.7	+1.0
30	188.3	+1.5	+6.8	-0.5

2005 JULY

01	200.5	+1.5	+6.6	-2.0
02	212.7	+1.5	+6.2	-3.3
03	225.0	+1.5	+5.5	-4.5

04	237.2	+1.5	+4.7	-5.4
05	249.5	+1.6	+3.7	-6.1
06	261.7	+1.6	+2.6	-6.5
07	274.0	+1.6	+1.5	-6.6
08	286.3	+1.6	+0.3	-6.4
09	298.5	+1.6	-1.0	-5.9
10	310.8	+1.6	-2.4	-5.2
11	323.0	+1.6	-3.7	-4.2
12	335.2	+1.5	-4.9	-3.1
13	347.5	+1.5	-6.1	-1.8
14	359.7	+1.5	-7.0	-0.4
15	011.9	+1.5	-7.6	+1.0
16	024.1	+1.5	-7.8	+2.5
17	036.4	+1.5	-7.5	+3.8
18	048.6	+1.5	-6.7	+4.9
19	060.8	+1.5	-5.4	+5.8
20	072.9	+1.5	-3.7	+6.3
21	085.1	+1.5	-1.7	+6.4
22	097.3	+1.5	+0.5	+6.1
23	109.5	+1.5	+2.6	+5.3
24	121.7	+1.5	+4.4	+4.2
25	133.9	+1.5	+5.9	+2.8
26	146.1	+1.5	+6.9	+1.2
27	158.3	+1.4	+7.5	-0.4
28	170.5	+1.4	+7.6	-1.9
29	182.7	+1.4	+7.4	-3.2
30	195.0	+1.4	+6.8	-4.4
31	207.2	+1.4	+6.0	-5.4

2005 AUGUST

01	219.4	+1.4	+5.1	-6.1
02	231.7	+1.4	+4.0	-6.5
03	243.9	+1.4	+2.8	-6.7
04	256.2	+1.4	+1.5	-6.5
05	268.4	+1.4	+0.2	-6.0
06	280.7	+1.4	-1.1	-5.3
07	292.9	+1.4	-2.4	-4.4
08	305.2	+1.4	-3.7	-3.2
09	317.4	+1.4	-4.8	-1.9
10	329.7	+1.3	-5.9	-0.5
11	341.9	+1.3	-6.8	+0.9
12	354.1	+1.3	-7.3	+2.4
13	006.3	+1.3	-7.6	+3.7
14	018.5	+1.3	-7.4	+4.9
15	030.7	+1.2	-6.7	+5.8
16	042.9	+1.2	-5.6	+6.4
17	055.1	+1.2	-4.0	+6.6
18	067.3	+1.2	-2.1	+6.3
19	079.5	+1.2	-0.0	+5.7
20	091.7	+1.1	+2.0	+4.6
21	103.8	+1.1	+4.0	+3.2
22	116.0	+1.1	+5.5	+1.6
23	128.2	+1.1	+6.7	+0.0
24	140.4	+1.0	+7.4	-1.6
25	152.6	+1.0	+7.6	-3.1
26	164.8	+1.0	+7.5	-4.4
27	177.0	+1.0	+7.0	-5.4
28	189.2	+1.0	+6.2	-6.1
29	201.4	+1.0	+5.2	-6.6
30	213.7	+1.0	+4.0	-6.7
31	225.9	+0.9	+2.8	-6.6

2005 SEPTEMBER

01	238.1	+0.9	+1.5	-6.2
02	250.4	+0.9	+0.2	-5.5
03	262.6	+0.9	-1.1	-4.6
04	274.8	+0.9	-2.3	-3.5
05	287.1	+0.9	-3.5	-2.2
06	299.3	+0.8	-4.5	-0.7
07	311.5	+0.8	-5.4	+0.7
08	323.8	+0.8	-6.1	+2.2
09	336.0	+0.8	-6.5	+3.6
10	348.2	+0.7	-6.7	+4.8
11	000.4	+0.7	-6.5	+5.8
12	012.6	+0.7	-5.9	+6.4
13	024.8	+0.6	-4.9	+6.7
14	037.0	+0.6	-3.6	+6.6
15	049.2	+0.6	-2.1	+6.1
16	061.3	+0.5	-0.3	+5.1
17	073.5	+0.5	+1.5	+3.8
18	085.7	+0.5	+3.2	+2.2
19	097.8	+0.4	+4.7	+0.6
20	110.0	+0.4	+5.9	-1.1
21	122.2	+0.4	+6.6	-2.7
22	134.4	+0.4	+7.0	-4.1
23	146.5	+0.3	+6.9	-5.2
24	158.7	+0.3	+6.5	-6.0
25	170.9	+0.3	+5.8	-6.6
26	183.1	+0.3	+4.8	-6.8
27	195.3	+0.3	+3.6	-6.7
28	207.5	+0.2	+2.4	-6.4
29	219.7	+0.2	+1.0	-5.8
30	231.9	+0.2	-0.3	-4.9

2005 OCTOBER

01	244.2	+0.2	-1.5	-3.8
02	256.4	+0.2	-2.6	-2.5
03	268.6	+0.1	-3.6	-1.1
04	280.8	+0.1	-4.3	+0.4
05	293.0	+0.1	-4.9	+1.9
06	305.3	+0.1	-5.3	+3.3
07	317.5	+0.0	-5.4	+4.6
08	329.7	+0.0	-5.3	+5.6
09	341.9	-0.0	-4.9	+6.4
10	354.1	-0.1	-4.3	+6.8
11	006.3	-0.1	-3.4	+6.8
12	018.4	-0.1	-2.3	+6.4
13	030.6	-0.2	-1.1	+5.5
14	042.8	-0.2	+0.2	+4.3
15	054.9	-0.2	+1.6	+2.9
16	067.1	-0.3	+2.9	+1.2
17	079.2	-0.3	+4.0	-0.5
18	091.4	-0.4	+4.9	-2.1
19	103.5	-0.4	+5.6	-3.6
20	115.7	-0.4	+5.9	-4.9
21	127.9	-0.4	+5.9	-5.8
22	140.0	-0.5	+5.5	-6.4
23	152.2	-0.5	+4.8	-6.7
24	164.3	-0.5	+3.9	-6.7
25	176.5	-0.5	+2.8	-6.5
26	188.7	-0.6	+1.5	-5.9
27	200.9	-0.6	+0.2	-5.1
28	213.1	-0.6	-1.1	-4.1
29	225.3	-0.6	-2.3	-2.9
30	237.5	-0.6	-3.3	-1.5
31	249.7	-0.6	-4.1	-0.1

2005 NOVEMBER

01	261.9	-0.7	-4.7	+1.4
02	274.1	-0.7	-4.9	+2.9
03	286.3	-0.7	-4.9	+4.2
04	298.5	-0.7	-4.5	+5.3
05	310.7	-0.7	-4.0	+6.2
06	322.9	-0.8	-3.2	+6.6
07	335.1	-0.8	-2.3	+6.7
08	347.3	-0.8	-1.3	+6.4
09	359.5	-0.8	-0.3	+5.7
10	011.6	-0.9	+0.6	+4.6
11	023.8	-0.9	+1.6	+3.3
12	035.9	-0.9	+2.5	+1.7
13	048.1	-1.0	+3.3	+0.0
14	060.2	-1.0	+4.0	-1.6
15	072.4	-1.0	+4.5	-3.1
16	084.5	-1.1	+4.9	-4.4
17	096.6	-1.1	+5.0	-5.5
18	108.8	-1.1	+4.9	-6.2
19	120.9	-1.1	+4.4	-6.6
20	133.1	-1.2	+3.7	-6.7
21	145.2	-1.2	+2.8	-6.4
22	157.4	-1.2	+1.6	-5.9
23	169.5	-1.2	+0.4	-5.2
24	181.7	-1.2	-1.0	-4.2
25	193.9	-1.2	-2.3	-3.1
26	206.0	-1.2	-3.5	-1.8
27	218.2	-1.2	-4.5	-0.4
28	230.4	-1.2	-5.2	+1.0
29	242.6	-1.3	-5.6	+2.4
30	254.8	-1.3	-5.6	+3.8

2005 DECEMBER

01	267.0	-1.3	-5.2	+4.9
02	279.2	-1.3	-4.4	+5.8
03	291.4	-1.3	-3.3	+6.4
04	303.6	-1.3	-2.1	+6.6
05	315.7	-1.3	-0.7	+6.3
06	327.9	-1.3	+0.5	+5.7
07	340.1	-1.4	+1.7	+4.7
08	352.3	-1.4	+2.7	+3.4
09	004.4	-1.4	+3.5	+1.9
10	016.6	-1.4	+4.1	+0.3
11	028.7	-1.4	+4.5	-1.3
12	040.9	-1.5	+4.8	-2.8
13	053.0	-1.5	+4.9	-4.1
14	065.1	-1.5	+4.9	-5.2
15	077.3	-1.5	+4.7	-6.0
16	089.4	-1.5	+4.3	-6.5
17	101.5	-1.5	+3.7	-6.6
18	113.7	-1.5	+2.9	-6.4
19	125.8	-1.5	+1.8	-6.0
20	137.9	-1.5	+0.6	-5.2
21	150.1	-1.5	-0.7	-4.3
22	162.2	-1.5	-2.1	-3.2
23	174.4	-1.5	-3.4	-2.0
24	186.5	-1.5	-4.7	-0.6
25	198.7	-1.5	-5.7	+0.8
26	210.9	-1.5	-6.5	+2.1
27	223.1	-1.5	-6.8	+3.4
28	235.2	-1.5	-6.7	+4.6
29	247.4	-1.5	-6.1	+5.6
30	259.6	-1.5	-5.1	+6.2
31	271.8	-1.5	-3.6	+6.5

2006 JANUARY

01	284.0	-1.5	-2.0	+6.3
02	296.2	-1.5	-0.2	+5.7
03	308.4	-1.5	+1.5	+4.8
04	320.6	-1.5	+3.0	+3.5
05	332.7	-1.5	+4.2	+2.0
06	344.9	-1.5	+5.1	+0.4
07	357.1	-1.5	+5.7	-1.2
08	009.2	-1.5	+6.0	-2.7
09	021.4	-1.5	+6.1	-4.0
10	033.5	-1.5	+5.9	-5.1
11	045.6	-1.5	+5.5	-5.9
12	057.8	-1.5	+5.0	-6.4
13	069.9	-1.5	+4.3	-6.6
14	082.0	-1.5	+3.4	-6.5
15	094.2	-1.5	+2.4	-6.0
16	106.3	-1.5	+1.3	-5.3
17	118.4	-1.5	-0.0	-4.4
18	130.6	-1.5	-1.4	-3.3
19	142.7	-1.5	-2.8	-2.0
20	154.8	-1.4	-4.1	-0.7
21	167.0	-1.4	-5.4	+0.7
22	179.2	-1.4	-6.5	+2.0
23	191.3	-1.4	-7.3	+3.3
24	203.5	-1.4	-7.7	+4.5
25	215.7	-1.4	-7.7	+5.4
26	227.8	-1.3	-7.2	+6.1
27	240.0	-1.3	-6.1	+6.5
28	252.2	-1.3	-4.6	+6.5
29	264.4	-1.3	-2.8	+6.0
30	276.6	-1.3	-0.7	+5.1
31	288.8	-1.2	+1.3	+3.8

2006 FEBRUARY

01	301.0	-1.2	+3.2	+2.3
02	313.2	-1.2	+4.7	+0.6
03	325.4	-1.2	+5.9	-1.0
04	337.5	-1.2	+6.7	-2.6
05	349.7	-1.2	+7.1	-4.0
06	001.9	-1.2	+7.1	-5.1
07	014.0	-1.2	+6.8	-6.0
08	026.2	-1.2	+6.3	-6.5
09	038.3	-1.1	+5.5	-6.7
10	050.5	-1.1	+4.6	-6.6
11	062.6	-1.1	+3.5	-6.2
12	074.7	-1.1	+2.4	-5.6
13	086.9	-1.1	+1.1	-4.7
14	099.0	-1.1	-0.3	-3.5
15	111.2	-1.0	-1.7	-2.3
16	123.3	-1.0	-3.0	-0.9
17	135.4	-1.0	-4.4	+0.5
18	147.6	-0.9	-5.6	+1.9
19	159.7	-0.9	-6.6	+3.2
20	171.9	-0.9	-7.4	+4.4
21	184.1	-0.9	-7.9	+5.4
22	196.2	-0.8	-8.0	+6.2
23	208.4	-0.8	-7.6	+6.6
24	220.6	-0.8	-6.7	+6.7
25	232.8	-0.7	-5.4	+6.3
26	245.0	-0.7	-3.7	+5.6
27	257.2	-0.7	-1.7	+4.4
28	269.4	-0.7	+0.4	+2.9

2006 MARCH

01	281.6	-0.6	+2.4	+1.2
02	293.8	-0.6	+4.1	-0.5
03	306.0	-0.6	+5.6	-2.2
04	318.2	-0.6	+6.6	-3.7
05	330.4	-0.5	+7.2	-5.0
06	342.6	-0.5	+7.4	-5.9
07	354.8	-0.5	+7.2	-6.5
08	006.9	-0.5	+6.7	-6.8
09	019.1	-0.5	+5.8	-6.8
10	031.3	-0.4	+4.8	-6.4
11	043.4	-0.4	+3.6	-5.8
12	055.6	-0.4	+2.3	-4.9
13	067.7	-0.4	+0.9	-3.9
14	079.9	-0.3	-0.5	-2.6
15	092.0	-0.3	-1.8	-1.2
16	104.2	-0.3	-3.1	+0.2
17	116.3	-0.3	-4.3	+1.6
18	128.5	-0.2	-5.4	+3.0
19	140.6	-0.2	-6.2	+4.3
20	152.8	-0.2	-6.9	+5.3
21	165.0	-0.1	-7.2	+6.2
22	177.1	-0.1	-7.3	+6.7
23	189.3	-0.1	-7.0	+6.8
24	201.5	-0.0	-6.3	+6.6
25	213.7	+0.0	-5.2	+6.0
26	225.9	+0.0	-3.8	+4.9
27	238.1	+0.1	-2.2	+3.6
28	250.3	+0.1	-0.4	+2.0
29	262.6	+0.1	+1.4	+0.2
30	274.8	+0.1	+3.2	-1.5
31	287.0	+0.2	+4.6	-3.2

2006 APRIL

01	299.2	+0.2	+5.8	-4.6
02	311.4	+0.2	+6.6	-5.6
03	323.6	+0.3	+6.9	-6.4
04	335.8	+0.3	+6.8	-6.7
05	348.0	+0.3	+6.4	-6.8
06	000.2	+0.3	+5.6	-6.5
07	012.4	+0.3	+4.5	-6.0
08	024.6	+0.4	+3.3	-5.2
09	036.8	+0.4	+1.9	-4.1
10	049.0	+0.4	+0.5	-2.9
11	061.1	+0.4	-0.8	-1.6
12	073.3	+0.5	-2.1	-0.2
13	085.5	+0.5	-3.3	+1.2
14	097.6	+0.5	-4.3	+2.7
15	109.8	+0.6	-5.1	+4.0
16	122.0	+0.6	-5.6	+5.1
17	134.1	+0.6	-6.0	+6.0
18	146.3	+0.6	-6.1	+6.6
19	158.5	+0.7	-5.9	+6.8
20	170.7	+0.7	-5.5	+6.7
21	182.9	+0.7	-4.9	+6.1
22	195.1	+0.7	-4.0	+5.2
23	207.3	+0.8	-2.9	+4.0
24	219.5	+0.8	-1.7	+2.5
25	231.7	+0.8	-0.3	+0.8
26	244.0	+0.8	+1.1	-0.9
27	256.2	+0.9	+2.6	-2.6
28	268.4	+0.9	+3.8	-4.0
29	280.7	+0.9	+4.9	-5.2
30	292.9	+0.9	+5.6	-6.1

2006 MAY

01	305.1	+1.0	+6.0	-6.5
02	317.4	+1.0	+6.0	-6.7
03	329.6	+1.0	+5.6	-6.5
04	341.8	+1.0	+4.8	-6.0
05	354.0	+1.0	+3.8	-5.2
06	006.2	+1.0	+2.5	-4.3
07	018.4	+1.1	+1.2	-3.1
08	030.6	+1.1	-0.2	-1.9
09	042.8	+1.1	-1.6	-0.5
10	055.0	+1.1	-2.8	+0.9
11	067.2	+1.1	-3.8	+2.3
12	079.4	+1.2	-4.6	+3.6
13	091.6	+1.2	-5.1	+4.7
14	103.7	+1.2	-5.3	+5.7
15	115.9	+1.2	-5.2	+6.3
16	128.1	+1.2	-4.9	+6.6
17	140.3	+1.2	-4.4	+6.6
18	152.5	+1.3	-3.8	+6.1
19	164.7	+1.3	-3.0	+5.3
20	176.9	+1.3	-2.1	+4.2
21	189.1	+1.3	-1.2	+2.8
22	201.3	+1.3	-0.2	+1.2
23	213.6	+1.3	+0.8	-0.5
24	225.8	+1.3	+1.8	-2.1
25	238.0	+1.4	+2.8	-3.6
26	250.3	+1.4	+3.7	-4.9
27	262.5	+1.4	+4.5	-5.8
28	274.8	+1.4	+5.0	-6.4
29	287.0	+1.4	+5.2	-6.6
30	299.3	+1.4	+5.1	-6.4
31	311.5	+1.4	+4.6	-6.0

2006 JUNE

01	323.8	+1.4	+3.8	-5.3
02	336.0	+1.5	+2.8	-4.4
03	348.2	+1.5	+1.5	-3.2
04	000.4	+1.5	+0.1	-2.0
05	012.7	+1.5	-1.2	-0.7
06	024.9	+1.5	-2.6	+0.7
07	037.1	+1.5	-3.7	+2.0
08	049.3	+1.5	-4.6	+3.3
09	061.5	+1.5	-5.2	+4.4
10	073.7	+1.5	-5.5	+5.4
11	085.9	+1.5	-5.4	+6.1
12	098.0	+1.5	-4.9	+6.5
13	110.2	+1.5	-4.2	+6.5
14	122.4	+1.5	-3.3	+6.1
15	134.6	+1.5	-2.2	+5.4
16	146.8	+1.5	-1.2	+4.2
17	159.0	+1.5	-0.1	+2.9
18	171.2	+1.5	+0.9	+1.3
19	183.5	+1.5	+1.8	-0.3
20	195.7	+1.5	+2.6	-1.9
21	207.9	+1.5	+3.3	-3.4
22	220.2	+1.5	+3.9	-4.7
23	232.4	+1.5	+4.4	-5.6
24	244.7	+1.5	+4.8	-6.3
25	256.9	+1.5	+4.9	-6.6
26	269.2	+1.5	+4.8	-6.5
27	281.4	+1.5	+4.5	-6.1
28	293.7	+1.5	+3.8	-5.4
29	305.9	+1.5	+2.9	-4.5
30	318.2	+1.5	+1.8	-3.4

2006 JULY

01	330.4	+1.5	+0.5	-2.1
02	342.7	+1.5	-0.9	-0.8
03	354.9	+1.5	-2.3	+0.5
04	007.1	+1.5	-3.6	+1.9
05	019.3	+1.5	-4.8	+3.1
06	031.5	+1.5	-5.6	+4.3
07	043.7	+1.5	-6.1	+5.3
08	055.9	+1.5	-6.3	+6.0
09	068.1	+1.5	-5.9	+6.4
10	080.3	+1.5	-5.2	+6.5
11	092.5	+1.5	-4.2	+6.2
12	104.7	+1.4	-2.8	+5.5
13	116.9	+1.4	-1.4	+4.4
14	129.1	+1.4	+0.1	+3.1
15	141.3	+1.4	+1.4	+1.5
16	153.5	+1.4	+2.6	-0.2
17	165.7	+1.4	+3.6	-1.8
18	177.9	+1.3	+4.5	-3.3
19	190.2	+1.3	+5.0	-4.6
20	202.4	+1.3	+5.4	-5.6
21	214.6	+1.3	+5.6	-6.3
22	226.9	+1.3	+5.6	-6.6
23	239.1	+1.3	+5.4	-6.6
24	251.4	+1.3	+5.0	-6.3
25	263.6	+1.3	+4.3	-5.6
26	275.9	+1.3	+3.4	-4.7
27	288.1	+1.3	+2.4	-3.6
28	300.4	+1.2	+1.1	-2.3
29	312.6	+1.2	-0.2	-1.0
30	324.9	+1.2	-1.7	+0.4
31	337.1	+1.2	-3.1	+1.8

2006 AUGUST

01	349.3	+1.2	-4.4	+3.0
02	001.6	+1.2	-5.6	+4.2
03	013.8	+1.2	-6.4	+5.2
04	026.0	+1.1	-7.0	+6.0
05	038.2	+1.1	-7.1	+6.5
06	050.4	+1.1	-6.7	+6.6
07	062.6	+1.1	-5.9	+6.4
08	074.8	+1.0	-4.7	+5.8
09	086.9	+1.0	-3.2	+4.8
10	099.1	+1.0	-1.4	+3.5
11	111.3	+1.0	+0.4	+1.9
12	123.5	+0.9	+2.1	+0.2
13	135.7	+0.9	+3.7	-1.5
14	147.9	+0.9	+4.9	-3.1
15	160.1	+0.8	+5.9	-4.5
16	172.3	+0.8	+6.5	-5.6
17	184.5	+0.8	+6.8	-6.3
18	196.7	+0.8	+6.9	-6.7
19	209.0	+0.8	+6.6	-6.8
20	221.2	+0.7	+6.1	-6.5

Appendix

```
21  233.4  +0.7  +5.4  -5.9
22  245.7  +0.7  +4.5  -5.0
23  257.9  +0.7  +3.4  -3.9
24  270.1  +0.7  +2.1  -2.6
25  282.4  +0.6  +0.8  -1.3
26  294.6  +0.6  -0.6  +0.1
27  306.9  +0.6  -2.1  +1.5
28  319.1  +0.6  -3.5  +2.9
29  331.3  +0.6  -4.8  +4.1
30  343.6  +0.5  -5.9  +5.1
31  355.8  +0.5  -6.8  +6.0
```

2006 SEPTEMBER
```
01  008.0  +0.5  -7.3  +6.5
02  020.2  +0.5  -7.5  +6.8
03  032.4  +0.4  -7.3  +6.7
04  044.6  +0.4  -6.6  +6.2
05  056.7  +0.4  -5.4  +5.3
06  068.9  +0.4  -3.9  +4.1
07  081.1  +0.3  -2.1  +2.6
08  093.3  +0.3  -0.1  +0.9
09  105.4  +0.2  +1.9  -0.9
10  117.6  +0.2  +3.7  -2.6
11  129.8  +0.2  +5.3  -4.1
12  142.0  +0.1  +6.5  -5.4
13  154.1  +0.1  +7.3  -6.2
14  166.3  +0.1  +7.7  -6.7
15  178.5  +0.1  +7.7  -6.8
16  190.7  +0.0  +7.4  -6.6
17  203.0  +0.0  +6.7  -6.1
18  215.2  -0.0  +5.8  -5.2
19  227.4  -0.0  +4.7  -4.2
20  239.6  -0.1  +3.4  -3.0
21  251.8  -0.1  +2.0  -1.6
22  264.1  -0.1  +0.6  -0.2
23  276.3  -0.1  -0.8  +1.2
24  288.5  -0.1  -2.2  +2.6
25  300.7  -0.2  -3.5  +3.9
26  313.0  -0.2  -4.7  +5.0
27  325.2  -0.2  -5.7  +5.8
28  337.4  -0.2  -6.5  +6.5
29  349.6  -0.3  -7.1  +6.8
30  001.8  -0.3  -7.3  +6.8
```

2006 OCTOBER
```
01  014.0  -0.3  -7.1  +6.4
02  026.2  -0.3  -6.6  +5.7
03  038.3  -0.4  -5.7  +4.6
04  050.5  -0.4  -4.4  +3.2
05  062.7  -0.4  -2.8  +1.6
06  074.8  -0.5  -0.9  -0.2
07  087.0  -0.5  +1.0  -1.9
08  099.1  -0.5  +3.0  -3.6
09  111.3  -0.6  +4.7  -4.9
10  123.5  -0.6  +6.2  -5.9
11  135.6  -0.6  +7.2  -6.5
12  147.8  -0.7  +7.8  -6.7
13  160.0  -0.7  +7.9  -6.6
14  172.1  -0.7  +7.5  -6.1
15  184.3  -0.7  +6.8  -5.4
16  196.5  -0.8  +5.8  -4.4
17  208.7  -0.8  +4.6  -3.2
18  220.9  -0.8  +3.2  -1.9
19  233.1  -0.8  +1.8  -0.6
20  245.3  -0.8  +0.4  +0.9
21  257.5  -0.9  -1.0  +2.2
22  269.8  -0.9  -2.3  +3.5
23  282.0  -0.9  -3.4  +4.7
24  294.2  -0.9  -4.4  +5.6
25  306.4  -0.9  -5.2  +6.3
26  318.6  -0.9  -5.8  +6.7
27  330.8  -1.0  -6.2  +6.7
28  343.0  -1.0  -6.3  +6.5
29  355.2  -1.0  -6.2  +5.8
30  007.3  -1.0  -5.8  +4.9
31  019.5  -1.0  -5.1  +3.6
```

2006 NOVEMBER
```
01  031.7  -1.1  -4.1  +2.1
02  043.8  -1.1  -2.8  +0.4
03  056.0  -1.1  -1.3  -1.3
04  068.1  -1.1  +0.4  -2.9
05  080.3  -1.2  +2.1  -4.4
06  092.4  -1.2  +3.8  -5.5
07  104.5  -1.2  +5.2  -6.2
08  116.7  -1.2  +6.3  -6.6
09  128.8  -1.3  +7.0  -6.5
10  141.0  -1.3  +7.2  -6.1
11  153.1  -1.3  +6.9  -5.4
12  165.3  -1.3  +6.3  -4.5
13  177.5  -1.3  +5.3  -3.3
14  189.6  -1.3  +4.0  -2.1
15  201.8  -1.4  +2.7  -0.8
16  214.0  -1.4  +1.2  +0.6
17  226.2  -1.4  -0.2  +1.9
18  238.4  -1.4  -1.5  +3.2
19  250.6  -1.4  -2.6  +4.4
20  262.8  -1.4  -3.5  +5.3
21  275.0  -1.4  -4.3  +6.1
22  287.2  -1.4  -4.7  +6.5
23  299.4  -1.4  -5.0  +6.6
24  311.6  -1.4  -5.1  +6.4
25  323.8  -1.4  -5.0  +5.8
26  335.9  -1.4  -4.7  +4.9
27  348.1  -1.4  -4.2  +3.7
28  000.3  -1.4  -3.6  +2.3
29  012.5  -1.5  -2.8  +0.7
30  024.6  -1.5  -1.9  -1.0
```

2006 DECEMBER
```
01  036.8  -1.5  -0.7  -2.6
02  048.9  -1.5  +0.6  -4.0
03  061.0  -1.5  +1.9  -5.2
04  073.2  -1.5  +3.2  -6.0
05  085.3  -1.5  +4.4  -6.4
06  097.4  -1.5  +5.4  -6.5
07  109.6  -1.5  +5.9  -6.1
08  121.7  -1.5  +6.1  -5.5
09  133.8  -1.5  +5.8  -4.5
10  146.0  -1.5  +5.2  -3.4
11  158.1  -1.6  +4.2  -2.2
12  170.3  -1.6  +3.0  -0.9
13  182.4  -1.6  +1.6  +0.5
14  194.6  -1.5  +0.2  +1.8
15  206.8  -1.5  -1.2  +3.1
16  218.9  -1.5  -2.4  +4.2
17  231.1  -1.5  -3.4  +5.2
18  243.3  -1.5  -4.2  +5.9
19  255.5  -1.5  -4.7  +6.4
20  267.7  -1.5  -4.8  +6.6
21  279.9  -1.5  -4.7  +6.4
22  292.1  -1.5  -4.4  +5.8
23  304.3  -1.5  -3.9  +5.0
24  316.5  -1.5  -3.2  +3.8
25  328.6  -1.5  -2.5  +2.3
26  340.8  -1.5  -1.7  +0.8
27  353.0  -1.5  -0.9  -0.9
28  005.1  -1.5  +0.0  -2.5
29  017.3  -1.5  +0.9  -3.9
30  029.4  -1.4  +1.8  -5.1
31  041.6  -1.4  +2.7  -6.0
```

2007 JANUARY
```
01  053.7  -1.4  +3.6  -6.5
02  065.8  -1.4  +4.3  -6.6
03  078.0  -1.4  +4.9  -6.3
04  090.1  -1.4  +5.1  -5.7
05  102.2  -1.4  +5.1  -4.8
06  114.3  -1.4  +4.7  -3.6
07  126.5  -1.4  +4.0  -2.4
08  138.6  -1.4  +3.0  -1.0
09  150.7  -1.3  +1.8  +0.4
10  162.9  -1.3  +0.4  +1.7
11  175.0  -1.3  -1.0  +3.0
12  187.2  -1.3  -2.4  +4.1
13  199.4  -1.3  -3.6  +5.1
14  211.5  -1.3  -4.6  +5.9
15  223.7  -1.3  -5.3  +6.4
16  235.9  -1.2  -5.6  +6.6
17  248.1  -1.2  -5.6  +6.5
18  260.3  -1.2  -5.2  +6.1
19  272.5  -1.2  -4.6  +5.2
20  284.7  -1.2  -3.6  +4.1
21  296.9  -1.1  -2.5  +2.6
22  309.0  -1.1  -1.4  +1.0
23  321.2  -1.1  -0.2  -0.7
24  333.4  -1.1  +0.9  -2.3
25  345.6  -1.1  +2.0  -3.8
26  357.7  -1.0  +2.9  -5.1
27  009.9  -1.0  +3.7  -6.0
28  022.0  -1.0  +4.3  -6.6
29  034.2  -1.0  +4.8  -6.7
30  046.3  -0.9  +5.1  -6.5
31  058.5  -0.9  +5.2  -6.0
```

2007 FEBRUARY
```
01  070.6  -0.9  +5.0  -5.1
02  082.7  -0.9  +4.6  -4.0
03  094.8  -0.9  +3.9  -2.7
04  107.0  -0.8  +3.0  -1.3
05  119.1  -0.8  +2.0  +0.1
06  131.3  -0.8  +0.7  +1.5
07  143.4  -0.8  -0.7  +2.8
08  155.5  -0.7  -2.1  +4.0
09  167.7  -0.7  -3.4  +5.0
10  179.9  -0.7  -4.6  +5.9
11  192.0  -0.7  -5.6  +6.4
12  204.2  -0.6  -6.4  +6.7
13  216.4  -0.6  -6.7  +6.7
14  228.6  -0.6  -6.7  +6.3
15  240.7  -0.6  -6.3  +5.6
16  252.9  -0.5  -5.5  +4.6
17  265.1  -0.5  -4.3  +3.2
18  277.3  -0.5  -2.9  +1.6
19  289.5  -0.5  -1.3  -0.2
20  301.7  -0.4  +0.4  -1.9
21  313.9  -0.4  +1.9  -3.5
22  326.1  -0.4  +3.3  -4.9
23  338.3  -0.3  +4.5  -5.9
24  350.5  -0.3  +5.4  -6.6
25  002.6  -0.3  +6.0  -6.8
26  014.8  -0.3  +6.3  -6.6
27  027.0  -0.2  +6.2  -6.2
28  039.1  -0.2  +6.0  -5.4
```

2007 MARCH
```
01  051.3  -0.2  +5.4  -4.4
02  063.4  -0.1  +4.7  -3.1
03  075.6  -0.1  +3.7  -1.7
04  087.7  -0.1  +2.6  -0.3
05  099.8  -0.0  +1.4  +1.1
06  112.0  -0.0  +0.1  +2.5
07  124.1  +0.0  -1.3  +3.8
08  136.3  +0.0  -2.7  +4.9
09  148.4  +0.1  -4.0  +5.7
10  160.6  +0.1  -5.2  +6.4
11  172.8  +0.1  -6.2  +6.7
12  185.0  +0.1  -7.0  +6.8
13  197.1  +0.2  -7.5  +6.5
14  209.3  +0.2  -7.6  +6.0
15  221.5  +0.2  -7.2  +5.0
16  233.7  +0.2  -6.5  +3.8
17  245.9  +0.3  -5.3  +2.3
18  258.1  +0.3  -3.8  +0.6
19  270.4  +0.3  -2.0  -1.2
20  282.6  +0.3  -0.1  -2.9
21  294.8  +0.4  +1.9  -4.4
22  307.0  +0.4  +3.7  -5.6
23  319.2  +0.4  +5.2  -6.4
24  331.4  +0.5  +6.3  -6.7
25  343.6  +0.5  +7.0  -6.7
26  355.8  +0.5  +7.3  -6.3
27  008.0  +0.5  +7.2  -5.6
28  020.1  +0.6  +6.7  -4.6
29  032.3  +0.6  +6.0  -3.4
30  044.5  +0.6  +5.0  -2.1
31  056.6  +0.7  +3.8  -0.7
```

2007 APRIL
```
01  068.8  +0.7  +2.5  +0.8
02  081.0  +0.7  +1.2  +2.2
03  093.1  +0.8  -0.2  +3.5
04  105.3  +0.8  -1.6  +4.6
05  117.4  +0.8  -2.9  +5.5
06  129.6  +0.8  -4.1  +6.2
07  141.8  +0.8  -5.2  +6.6
08  154.0  +0.9  -6.2  +6.8
09  166.1  +0.9  -6.9  +6.6
10  178.3  +0.9  -7.4  +6.1
11  190.5  +0.9  -7.6  +5.3
12  202.7  +0.9  -7.5  +4.2
13  214.9  +0.9  -6.9  +2.8
14  227.1  +1.0  -6.0  +1.2
15  239.4  +1.0  -4.6  -0.5
16  251.6  +1.0  -2.9  -2.2
17  263.8  +1.0  -0.9  -3.8
18  276.1  +1.0  +1.2  -5.1
19  288.3  +1.1  +3.2  -6.0
20  300.5  +1.1  +5.0  -6.5
21  312.7  +1.1  +6.3  -6.6
22  325.0  +1.1  +7.2  -6.3
23  337.2  +1.1  +7.6  -5.6
24  349.4  +1.2  +7.5  -4.7
25  001.6  +1.2  +7.0  -3.5
26  013.8  +1.2  +6.2  -2.3
27  026.0  +1.2  +5.1  -0.9
28  038.2  +1.3  +3.8  +0.5
29  050.4  +1.3  +2.4  +1.9
30  062.5  +1.3  +1.0  +3.2
```

2007 MAY
```
01  074.7  +1.3  -0.3  +4.3
02  086.9  +1.3  -1.6  +5.3
03  099.1  +1.4  -2.8  +6.0
04  111.2  +1.4  -3.9  +6.5
05  123.4  +1.4  -4.9  +6.7
06  135.6  +1.4  -5.7  +6.5
07  147.8  +1.4  -6.3  +6.1
08  160.0  +1.4  -6.7  +5.3
09  172.2  +1.4  -6.9  +4.3
10  184.4  +1.4  -6.8  +3.0
11  196.6  +1.4  -6.4  +1.5
12  208.8  +1.4  -5.7  -0.1
13  221.1  +1.4  -4.5  -1.7
14  233.3  +1.4  -3.1  -3.3
15  245.5  +1.4  -1.3  -4.6
16  257.8  +1.4  +0.6  -5.7
17  270.0  +1.4  +2.6  -6.3
18  282.3  +1.4  +4.3  -6.5
19  294.5  +1.5  +5.7  -6.3
20  306.8  +1.5  +6.7  -5.7
21  319.0  +1.5  +7.2  -4.8
22  331.2  +1.5  +7.1  -3.6
23  343.5  +1.5  +6.7  -2.3
24  355.7  +1.5  +5.9  -1.0
25  007.9  +1.5  +4.8  +0.4
26  020.1  +1.5  +3.5  +1.7
27  032.3  +1.6  +2.1  +3.0
28  044.5  +1.6  +0.7  +4.1
29  056.7  +1.6  -0.6  +5.1
30  068.9  +1.6  -1.8  +5.9
31  081.1  +1.6  -2.9  +6.4
```

2007 JUNE
```
01  093.3  +1.6  -3.8  +6.6
02  105.5  +1.6  -4.5  +6.5
03  117.6  +1.6  -5.1  +6.1
04  129.8  +1.6  -5.5  +5.4
05  142.0  +1.6  -5.7  +4.3
06  154.2  +1.5  -5.7  +3.1
07  166.4  +1.5  -5.5  +1.6
08  178.7  +1.5  -5.1  +0.0
09  190.9  +1.5  -4.4  -1.6
10  203.1  +1.5  -3.5  -3.1
11  215.3  +1.5  -2.3  -4.4
12  227.6  +1.5  -0.8  -5.5
13  239.8  +1.5  +0.8  -6.2
14  252.1  +1.5  +2.4  -6.5
15  264.3  +1.5  +3.8  -6.4
16  276.6  +1.5  +5.1  -5.8
17  288.9  +1.5  +5.9  -5.0
18  301.1  +1.5  +6.3  -3.8
19  313.4  +1.5  +6.3  -2.5
20  325.6  +1.5  +5.8  -1.1
21  337.8  +1.5  +5.0  +0.3
22  350.1  +1.5  +4.0  +1.6
23  002.3  +1.5  +2.7  +2.9
24  014.5  +1.5  +1.4  +4.1
25  026.7  +1.5  +0.0  +5.0
26  038.9  +1.4  -1.2  +5.8
27  051.1  +1.4  -2.3  +6.3
28  063.3  +1.4  -3.3  +6.6
29  075.5  +1.4  -4.0  +6.6
30  087.7  +1.4  -4.5  +6.2
```

2007 JULY
```
01  099.8  +1.4  -4.7  +5.5
02  112.1  +1.4  -4.8  +4.5
03  124.3  +1.3  -4.7  +3.2
04  136.5  +1.3  -4.4  +1.8
05  148.7  +1.3  -3.9  +0.2
06  160.9  +1.3  -3.3  -1.4
07  173.1  +1.2  -2.6  -3.0
08  185.3  +1.2  -1.6  -4.4
09  197.6  +1.2  -0.6  -5.5
10  209.8  +1.2  +0.6  -6.2
11  222.0  +1.2  +1.8  -6.6
12  234.3  +1.1  +3.0  -6.6
13  246.5  +1.1  +4.0  -6.1
14  258.8  +1.1  +4.8  -5.3
15  271.1  +1.1  +5.3  -4.2
16  283.3  +1.1  +5.5  -2.9
17  295.6  +1.1  +5.3  -1.5
18  307.8  +1.1  +4.8  -0.0
19  320.0  +1.0  +4.0  +1.4
20  332.3  +1.0  +2.9  +2.8
21  344.5  +1.0  +1.7  +4.0
22  356.7  +1.0  +0.4  +5.0
23  009.0  +1.0  -0.9  +5.8
24  021.2  +1.0  -2.1  +6.4
25  033.4  +1.0  -3.2  +6.7
26  045.6  +1.0  -4.0  +6.7
27  057.8  +0.9  -4.6  +6.4
28  070.0  +0.9  -4.9  +5.8
29  082.2  +0.9  -5.0  +4.8
30  094.4  +0.9  -4.7  +3.6
31  106.5  +0.8  -4.2  +2.2
```

2007 AUGUST
```
01  118.7  +0.8  -3.5  +0.5
02  130.9  +0.8  -2.6  -1.1
03  143.1  +0.7  -1.7  -2.8
04  155.3  +0.7  -0.6  -4.2
05  167.5  +0.7  +0.5  -5.4
06  179.7  +0.6  +1.5  -6.3
07  192.0  +0.6  +2.5  -6.7
08  204.2  +0.6  +3.5  -6.8
09  216.4  +0.5  +4.2  -6.4
10  228.7  +0.5  +4.8  -5.7
11  240.9  +0.5  +5.2  -4.6
12  253.2  +0.5  +5.3  -3.3
13  265.4  +0.4  +5.1  -1.9
14  277.6  +0.4  +4.7  -0.4
15  289.9  +0.4  +4.0  +1.1
16  302.1  +0.4  +3.1  +2.5
17  314.4  +0.4  +1.9  +3.8
18  326.6  +0.3  +0.7  +4.9
19  338.8  +0.3  -0.6  +5.7
20  351.1  +0.3  -1.9  +6.3
21  003.3  +0.3  -3.1  +6.7
22  015.5  +0.3  -4.2  +6.8
23  027.7  +0.3  -5.0  +6.6
24  039.9  +0.2  -5.6  +6.0
25  052.1  +0.2  -5.8  +5.2
26  064.2  +0.2  -5.7  +4.1
27  076.4  +0.1  -5.3  +2.7
28  088.6  +0.1  -4.4  +1.1
29  100.8  +0.1  -3.3  -0.6
30  113.0  +0.0  -2.0  -2.3
31  125.1  +0.0  -0.6  -3.8
```

2007 SEPTEMBER
```
01  137.3  -0.0  +0.9  -5.1
02  149.5  -0.1  +2.3  -6.1
03  161.7  -0.1  +3.5  -6.8
04  173.9  -0.1  +4.5  -6.8
05  186.1  -0.2  +5.3  -6.4
06  198.3  -0.2  +5.8  -5.9
07  210.5  -0.2  +6.1  -5.0
08  222.8  -0.3  +6.0  -3.7
09  235.0  -0.3  +5.7  -2.3
10  247.2  -0.3  +5.2  -0.9
11  259.5  -0.3  +4.4  +0.7
12  271.7  -0.4  +3.5  +2.1
13  283.9  -0.4  +2.5  +3.4
14  296.2  -0.4  +1.3  +4.6
15  308.4  -0.4  +0.0  +5.5
16  320.6  -0.4  -1.3  +6.2
17  332.8  -0.5  -2.6  +6.7
18  345.0  -0.5  -3.9  +6.8
19  357.2  -0.5  -5.0  +6.7
20  009.4  -0.5  -5.9  +6.2
21  021.6  -0.5  -6.5  +5.5
22  033.8  -0.6  -6.8  +4.5
23  046.0  -0.6  -6.7  +3.2
24  058.2  -0.6  -6.2  +1.7
25  070.3  -0.6  -5.3  +0.1
26  082.5  -0.7  -4.0  -1.6
27  094.7  -0.7  -2.3  -3.2
28  106.8  -0.7  -0.5  -4.6
29  119.0  -0.8  +1.4  -5.7
30  131.2  -0.8  +3.2  -6.4
```

2007 OCTOBER
```
01  143.3  -0.8  +4.8  -6.7
02  155.5  -0.9  +6.0  -6.5
03  167.7  -0.9  +6.8  -6.0
```

04	179.9	-0.9	+7.3	-5.1
05	192.1	-0.9	+7.3	-3.9
06	204.3	-1.0	+7.0	-2.6
07	216.5	-1.0	+6.4	-1.2
08	228.7	-1.0	+5.6	+0.3
09	240.9	-1.0	+4.6	+1.7
10	253.1	-1.1	+3.4	+3.1
11	265.3	-1.1	+2.2	+4.3
12	277.6	-1.1	+1.0	+5.3
13	289.8	-1.1	-0.3	+6.0
14	302.0	-1.1	-1.6	+6.5
15	314.2	-1.1	-2.9	+6.7
16	326.4	-1.1	-4.1	+6.6
17	338.6	-1.1	-5.3	+6.2
18	350.8	-1.2	-6.2	+5.6
19	003.0	-1.2	-6.9	+4.7
20	015.2	-1.2	-7.4	+3.5
21	027.3	-1.2	-7.4	+2.1
22	039.5	-1.2	-7.1	+0.6
23	051.7	-1.2	-6.2	-1.0
24	063.8	-1.2	-4.9	-2.6
25	076.0	-1.3	-3.2	-4.1
26	088.1	-1.3	-1.1	-5.3
27	100.3	-1.3	+1.0	-6.1
28	112.4	-1.3	+3.2	-6.5
29	124.6	-1.3	+5.0	-6.4
30	136.7	-1.4	+6.5	-6.0
31	148.9	-1.4	+7.5	-5.1

2007 NOVEMBER

01	161.0	-1.4	+8.0	-4.0
02	173.2	-1.4	+8.0	-2.7
03	185.4	-1.4	+7.6	-1.3
04	197.6	-1.4	+6.9	+0.1
05	209.8	-1.5	+5.9	+1.5
06	221.9	-1.5	+4.8	+2.9
07	234.1	-1.5	+3.5	+4.1
08	246.3	-1.5	+2.2	+5.1
09	258.6	-1.5	+0.9	+5.8
10	270.8	-1.5	-0.4	+6.4
11	283.0	-1.7	-1.7	+6.6
12	295.2	-1.5	-2.9	+6.6
13	307.4	-1.5	-4.0	+6.2
14	319.6	-1.5	-5.1	+5.6
15	331.7	-1.5	-6.0	+4.7
16	343.9	-1.5	-6.7	+3.6
17	356.1	-1.5	-7.2	+2.3
18	008.3	-1.5	-7.4	+0.8
19	020.4	-1.5	-7.2	-0.7
20	032.6	-1.5	-6.6	-2.3
21	044.7	-1.5	-5.4	-3.7
22	056.9	-1.5	-3.9	-4.9
23	069.0	-1.5	-2.0	-5.9
24	081.2	-1.5	+0.2	-6.4
25	093.3	-1.5	+2.3	-6.4
26	105.4	-1.5	+4.3	-6.0
27	117.5	-1.5	+6.0	-5.2
28	129.7	-1.5	+7.1	-4.1
29	141.8	-1.5	+7.8	-2.8
30	154.0	-1.5	+7.9	-1.4

2007 DECEMBER

01	166.1	-1.5	+7.6	+0.0
02	178.3	-1.5	+6.8	+1.5
03	190.5	-1.5	+5.9	+2.8
04	202.6	-1.5	+4.7	+4.0
05	214.8	-1.5	+3.4	+5.0
06	227.0	-1.5	+2.1	+5.8
07	239.2	-1.5	+0.7	+6.3
08	251.4	-1.5	-0.5	+6.6
09	263.6	-1.5	-1.7	+6.6
10	275.8	-1.5	-2.8	+6.3
11	288.0	-1.5	-3.8	+5.6
12	300.1	-1.5	-4.6	+4.8
13	312.3	-1.5	-5.4	+3.6
14	324.5	-1.5	-6.0	+2.3
15	336.7	-1.4	-6.3	+0.9
16	348.9	-1.4	-6.5	-0.6
17	001.0	-1.4	-6.3	-2.2
18	013.2	-1.4	-5.8	-3.6
19	025.3	-1.4	-4.9	-4.8
20	037.5	-1.4	-3.7	-5.8
21	049.6	-1.3	-2.1	-6.4
22	061.7	-1.3	-0.3	-6.5
23	073.9	-1.3	+1.6	-6.3
24	086.0	-1.3	+3.4	-5.6
25	098.1	-1.3	+4.9	-4.5
26	110.2	-1.3	+6.0	-3.2
27	122.4	-1.2	+6.7	-1.7
28	134.5	-1.2	+6.9	-0.2
29	146.6	-1.2	+6.7	+1.3
30	158.8	-1.2	+6.1	+2.7
31	170.9	-1.2	+5.2	+3.9

2008 JANUARY

01	183.1	-1.2	+4.1	+4.9
02	195.3	-1.2	+2.8	+5.8
03	207.4	-1.2	+1.5	+6.3
04	219.6	-1.1	+0.2	+6.6
05	231.8	-1.1	-1.0	+6.7
06	244.0	-1.1	-2.1	+6.4
07	256.2	-1.1	-3.1	+5.8
08	268.3	-1.1	-3.8	+5.0
09	280.5	-1.1	-4.5	+3.9
10	292.7	-1.0	-4.9	+2.6
11	304.9	-1.0	-5.2	+1.1
12	317.1	-1.0	-5.2	-0.5
13	329.3	-1.0	-5.1	-2.0
14	341.4	-0.9	-4.8	-3.5
15	353.6	-0.9	-4.2	-4.8
16	005.8	-0.9	-3.4	-5.8
17	017.9	-0.8	-2.3	-6.5
18	030.1	-0.8	-1.1	-6.7
19	042.2	-0.8	+0.3	-6.6
20	054.3	-0.8	+1.6	-6.0
21	066.5	-0.7	+3.0	-5.0
22	078.6	-0.7	+4.1	-3.7
23	090.7	-0.7	+5.0	-2.2
24	102.8	-0.6	+5.5	-0.7
25	115.0	-0.6	+5.6	+0.9
26	127.1	-0.6	+5.4	+2.4
27	139.2	-0.6	+4.9	+3.7
28	151.4	-0.5	+4.1	+4.8
29	163.5	-0.5	+3.0	+5.7
30	175.7	-0.5	+1.8	+6.3
31	187.8	-0.5	+0.5	+6.7

2008 FEBRUARY

01	200.0	-0.5	-0.7	+6.8
02	212.2	-0.4	-1.9	+6.6
03	224.4	-0.4	-3.0	+6.1
04	236.6	-0.4	-3.8	+5.3
05	248.7	-0.4	-4.5	+4.3
06	260.9	-0.3	-4.9	+3.0
07	273.1	-0.3	-5.0	+1.5
08	285.3	-0.3	-4.9	-0.1
09	297.5	-0.3	-4.5	-1.7
10	309.7	-0.2	-4.0	-3.2
11	321.9	-0.2	-3.2	-4.6
12	334.1	-0.2	-2.3	-5.7
13	346.2	-0.1	-1.3	-6.4
14	358.4	-0.1	-0.3	-6.8
15	010.6	-0.1	+0.8	-6.7
16	022.7	-0.0	+1.8	-6.3
17	034.9	+0.0	+2.8	-5.4
18	047.0	+0.0	+3.6	-4.2
19	059.2	+0.1	+4.2	-2.8
20	071.3	+0.1	+4.6	-1.2
21	083.4	+0.1	+4.8	+0.4
22	095.6	+0.2	+4.7	+1.9
23	107.7	+0.2	+4.3	+3.3
24	119.9	+0.2	+3.7	+4.5
25	132.0	+0.3	+2.9	+5.5
26	144.1	+0.3	+1.8	+6.2
27	156.3	+0.3	+0.7	+6.6
28	168.5	+0.3	-0.6	+6.8
29	180.6	+0.4	-1.8	+6.7

2008 MARCH

01	192.8	+0.4	-3.0	+6.3
02	205.0	+0.4	-4.1	+5.6
03	217.2	+0.4	-5.0	+4.6
04	229.4	+0.4	-5.6	+3.4
05	241.6	+0.4	-5.9	+2.1
06	253.8	+0.5	-5.8	+0.5
07	266.0	+0.5	-5.4	-1.1
08	278.2	+0.5	-4.6	-2.7
09	290.4	+0.5	-3.5	-4.1
10	302.6	+0.6	-2.3	-5.3
11	314.8	+0.6	-0.9	-6.2
12	327.0	+0.6	+0.5	-6.7
13	339.2	+0.6	+1.8	-6.7
14	351.4	+0.7	+2.9	-6.3
15	003.6	+0.7	+3.8	-5.6
16	015.7	+0.7	+4.5	-4.5
17	027.9	+0.8	+4.9	-3.1
18	040.1	+0.8	+5.1	-1.7
19	052.2	+0.9	+5.0	-0.1
20	064.4	+0.9	+4.8	+1.4
21	076.5	+0.9	+4.4	+2.9
22	088.7	+0.9	+3.8	+4.2
23	100.8	+1.0	+3.0	+5.2
24	113.0	+1.0	+2.0	+6.0
25	125.1	+1.0	+1.0	+6.5
26	137.3	+1.0	-0.2	+6.7
27	149.5	+1.0	-1.5	+6.6
28	161.7	+1.1	-2.8	+6.3
29	173.8	+1.1	-4.0	+5.7
30	186.0	+1.1	-5.1	+4.8
31	198.2	+1.1	-6.0	+3.7

2008 APRIL

01	210.4	+1.1	-6.7	+2.4
02	222.6	+1.1	-7.0	+1.0
03	234.8	+1.1	-6.9	-0.5
04	247.1	+1.1	-6.3	-2.1
05	259.3	+1.1	-5.3	-3.6
06	271.5	+1.2	-3.9	-4.9
07	283.7	+1.2	-2.2	-5.8
08	296.0	+1.2	-0.4	-6.4
09	308.2	+1.2	+1.5	-6.6
10	320.4	+1.2	+3.1	-6.3
11	332.6	+1.2	+4.4	-5.6
12	344.8	+1.3	+5.4	-4.6
13	357.0	+1.3	+6.0	-3.3
14	009.2	+1.3	+6.3	-1.8
15	021.4	+1.3	+6.2	-0.3
16	033.6	+1.4	+5.9	+1.2
17	045.8	+1.4	+5.3	+2.6
18	057.9	+1.4	+4.6	+3.9
19	070.1	+1.4	+3.8	+5.0
20	082.3	+1.5	+2.8	+5.8
21	094.4	+1.5	+1.7	+6.3
22	106.6	+1.5	+0.6	+6.6
23	118.8	+1.5	-0.7	+6.6
24	131.0	+1.5	-2.0	+6.3
25	143.2	+1.5	-3.3	+5.7
26	155.3	+1.5	-4.5	+4.9
27	167.5	+1.5	-5.7	+3.8
28	179.7	+1.5	-6.6	+2.6
29	191.9	+1.5	-7.4	+1.2
30	204.2	+1.5	-7.7	-0.2

2008 MAY

01	216.4	+1.5	-7.7	-1.7
02	228.6	+1.5	-7.2	-3.2
03	240.8	+1.5	-6.1	-4.5
04	253.1	+1.5	-4.6	-5.5
05	265.3	+1.5	-2.7	-6.2
06	277.6	+1.5	-0.6	-6.5
07	289.8	+1.5	+1.5	-6.3
08	302.0	+1.5	+3.5	-5.6
09	314.3	+1.5	+5.1	-4.7
10	326.5	+1.5	+6.3	-3.4
11	338.7	+1.5	+7.0	-1.9
12	350.9	+1.5	+7.2	-0.4
13	003.2	+1.5	+7.1	+1.1
14	015.4	+1.5	+6.7	+2.5
15	027.6	+1.6	+6.0	+3.8
16	039.8	+1.6	+5.1	+4.8
17	051.9	+1.6	+4.1	+5.7
18	064.1	+1.6	+2.9	+6.3
19	076.3	+1.6	+1.7	+6.6
20	088.5	+1.6	+0.5	+6.6
21	100.7	+1.6	-0.8	+6.3
22	112.9	+1.6	-2.1	+5.7
23	125.1	+1.6	-3.3	+4.9
24	137.3	+1.5	-4.5	+3.9
25	149.5	+1.5	-5.7	+2.7
26	161.7	+1.5	-6.6	+1.3
27	173.9	+1.5	-7.3	-0.1
28	186.1	+1.5	-7.8	-1.6
29	198.3	+1.5	-7.8	-3.0
30	210.5	+1.4	-7.4	-4.3
31	222.8	+1.4	-6.4	-5.3

2008 JUNE

01	235.0	+1.4	-5.0	-6.1
02	247.3	+1.4	-3.2	-6.5
03	259.5	+1.4	-1.1	-6.4
04	271.8	+1.4	+1.0	-5.9
05	284.0	+1.4	+3.1	-4.9
06	296.3	+1.4	+4.8	-3.7
07	308.5	+1.4	+6.2	-2.2
08	320.8	+1.4	+7.1	-0.6
09	333.0	+1.4	+7.4	+0.9
10	345.2	+1.3	+7.4	+2.4
11	357.5	+1.3	+7.0	+3.7
12	009.7	+1.3	+6.3	+4.8
13	021.9	+1.3	+5.3	+5.7
14	034.1	+1.3	+4.2	+6.3
15	046.3	+1.3	+3.0	+6.6
16	058.5	+1.3	+1.8	+6.7
17	070.7	+1.3	+0.5	+6.4
18	082.9	+1.3	-0.8	+5.9
19	095.1	+1.3	-2.0	+5.1
20	107.3	+1.3	-3.2	+4.1
21	119.5	+1.2	-4.3	+2.8
22	131.7	+1.2	-5.3	+1.5
23	143.9	+1.2	-6.1	+0.0
24	156.1	+1.2	-6.7	-1.5
25	168.3	+1.1	-7.1	-2.9
26	180.5	+1.1	-7.1	-4.2
27	192.7	+1.1	-6.7	-5.3
28	204.9	+1.0	-5.9	-6.1
29	217.2	+1.0	-4.7	-6.6
30	229.4	+1.0	-3.1	-6.6

2008 JULY

01	241.7	+1.0	-1.3	-6.2
02	253.9	+0.9	+0.7	-5.4
03	266.2	+0.9	+2.5	-4.2
04	278.4	+0.9	+4.2	-2.7
05	290.7	+0.9	+5.5	-1.1
06	302.9	+0.9	+6.5	+0.6
07	315.2	+0.9	+6.9	+2.1
08	327.4	+0.8	+7.0	+3.5
09	339.7	+0.8	+6.7	+4.7
10	351.9	+0.8	+6.0	+5.6
11	004.1	+0.8	+5.1	+6.3
12	016.3	+0.8	+4.0	+6.7
13	028.5	+0.8	+2.8	+6.8
14	040.8	+0.8	+1.5	+6.6
15	053.0	+0.7	+0.2	+6.1
16	065.1	+0.7	-1.0	+5.3
17	077.3	+0.7	-2.2	+4.4
18	089.5	+0.7	-3.2	+3.2
19	101.7	+0.6	-4.1	+1.8
20	113.9	+0.6	-4.9	+0.3
21	126.1	+0.6	-5.4	-1.2
22	138.3	+0.5	-5.8	-2.7
23	150.5	+0.5	-5.9	-4.1
24	162.7	+0.5	-5.7	-5.2
25	174.9	+0.4	-5.2	-6.1
26	187.1	+0.4	-4.4	-6.6
27	199.4	+0.4	-3.4	-6.8
28	211.6	+0.3	-2.1	-6.5
29	223.8	+0.3	-0.7	-5.8
30	236.1	+0.3	+0.9	-4.7
31	248.3	+0.3	+2.3	-3.3

2008 AUGUST

01	260.6	+0.2	+3.7	-1.7
02	272.8	+0.2	+4.8	-0.0
03	285.1	+0.2	+5.6	+1.6
04	297.3	+0.2	+6.1	+3.1
05	309.6	+0.1	+6.1	+4.4
06	321.8	+0.1	+5.9	+5.5
07	334.0	+0.1	+5.3	+6.2
08	346.3	+0.1	+4.4	+6.7
09	358.5	+0.1	+3.3	+6.8
10	010.7	+0.0	+2.1	+6.7
11	022.9	+0.0	+0.8	+6.3
12	035.1	+0.0	-0.4	+5.6
13	047.3	-0.0	-1.6	+4.7
14	059.5	-0.0	-2.7	+3.5
15	071.7	-0.1	-3.6	+2.2
16	083.9	-0.1	-4.3	+0.7
17	096.1	-0.1	-4.8	-0.8
18	108.2	-0.2	-5.0	-2.3
19	120.4	-0.2	-4.9	-3.7
20	132.6	-0.2	-4.6	-5.0
21	144.8	-0.3	-4.1	-5.9
22	157.0	-0.3	-3.4	-6.6
23	169.2	-0.3	-2.5	-6.8
24	181.4	-0.4	-1.5	-6.6
25	193.6	-0.4	-0.4	-6.1
26	205.8	-0.4	+0.7	-5.1
27	218.1	-0.5	+1.8	-3.8
28	230.3	-0.5	+2.9	-2.3
29	242.5	-0.5	+3.8	-0.6
30	254.8	-0.5	+4.5	+1.0
31	267.0	-0.6	+5.0	+2.6

2008 SEPTEMBER

01	279.3	-0.6	+5.3	+4.0
02	291.5	-0.6	+5.2	+5.1
03	303.7	-0.6	+4.9	+6.0
04	316.0	-0.7	+4.3	+6.5
05	328.2	-0.7	+3.4	+6.7
06	340.4	-0.7	+2.4	+6.7
07	352.6	-0.7	+1.2	+6.3
08	004.8	-0.7	-0.1	+5.7
09	017.0	-0.7	-1.4	+4.9
10	029.2	-0.8	-2.6	+3.8
11	041.4	-0.8	-3.7	+2.6
12	053.6	-0.8	-4.5	+1.2
13	065.8	-0.8	-5.0	-0.3
14	077.9	-0.8	-5.2	-1.8
15	090.1	-0.9	-5.1	-3.3
16	102.3	-0.9	-4.6	-4.6
17	114.4	-0.9	-3.8	-5.6
18	126.6	-0.9	-2.8	-6.3
19	138.8	-1.0	-1.7	-6.7
20	151.0	-1.0	-0.5	-6.6
21	163.2	-1.0	+0.7	-6.1
22	175.3	-1.0	+1.7	-5.2
23	187.5	-1.1	+2.7	-4.0
24	199.7	-1.1	+3.5	-2.6
25	212.0	-1.1	+4.1	-1.0
26	224.2	-1.1	+4.6	+0.6
27	236.4	-1.2	+4.8	+2.2
28	248.6	-1.2	+5.0	+3.6
29	260.8	-1.2	+4.9	+4.8
30	273.1	-1.2	+4.6	+5.7

2008 OCTOBER

01	285.3	-1.2	+4.1	+6.3
02	297.5	-1.3	+3.4	+6.6
03	309.7	-1.3	+2.5	+6.6
04	321.9	-1.3	+1.4	+6.3
05	334.2	-1.3	+0.2	+5.8
06	346.4	-1.3	-1.2	+4.9
07	358.6	-1.3	-2.5	+3.9
08	010.7	-1.3	-3.7	+2.8
09	022.9	-1.3	-4.8	+1.4
10	035.1	-1.3	-5.6	+0.0
11	047.3	-1.3	-6.1	-1.4
12	059.4	-1.4	-6.2	-2.8
13	071.6	-1.4	-5.8	-4.1
14	083.7	-1.4	-5.0	-5.2
15	095.9	-1.4	-3.8	-6.0
16	108.1	-1.4	-2.3	-6.5
17	120.2	-1.4	-0.7	-6.5
18	132.4	-1.4	+0.9	-6.1
19	144.5	-1.4	+2.4	-5.2
20	156.7	-1.4	+3.6	-4.1
21	168.9	-1.5	+4.6	-2.7
22	181.0	-1.5	+5.2	-1.2
23	193.2	-1.5	+5.6	+0.4
24	205.4	-1.5	+5.8	+1.9
25	217.6	-1.5	+5.8	+3.4
26	229.8	-1.5	+5.5	+4.6
27	242.0	-1.5	+5.1	+5.5
28	254.2	-1.5	+4.6	+6.2
29	266.5	-1.5	+3.9	+6.5
30	278.7	-1.5	+3.0	+6.6
31	290.9	-1.5	+1.9	+6.3

2008 NOVEMBER

01	303.1	-1.6	+0.7	+5.8
02	315.3	-1.6	-0.6	+5.0
03	327.5	-1.6	-2.0	+4.0
04	339.7	-1.5	-3.3	+2.8
05	351.8	-1.5	-4.6	+1.6
06	004.0	-1.5	-5.8	+0.2
07	016.2	-1.5	-6.7	-1.2
08	028.4	-1.5	-7.2	-2.6
09	040.5	-1.5	-7.3	-3.8
10	052.7	-1.5	-6.9	-5.0
11	064.8	-1.5	-5.9	-5.8
12	077.0	-1.5	-4.5	-6.3
13	089.1	-1.5	-2.8	-6.5
14	101.2	-1.5	-0.8	-6.1
15	113.4	-1.5	+1.2	-5.4
16	125.5	-1.5	+3.0	-4.3
17	137.6	-1.5	+4.6	-2.9
18	149.8	-1.5	+5.8	-1.3
19	162.0	-1.5	+6.6	+0.3

20	174.1	-1.5	+7.0	+1.9
21	186.3	-1.5	+7.0	+3.3
22	198.5	-1.5	+6.8	+4.5
23	210.6	-1.5	+6.4	+5.5
24	222.8	-1.4	+5.7	+6.2
25	235.0	-1.4	+4.9	+6.6
26	247.2	-1.4	+3.9	+6.6
27	259.4	-1.4	+2.9	+6.4
28	271.6	-1.4	+1.7	+5.9
29	283.8	-1.4	+0.4	+5.1
30	296.0	-1.4	-1.0	+4.1

2008 DECEMBER

01	308.2	-1.4	-2.3	+3.0
02	320.4	-1.4	-3.7	+1.7
03	332.6	-1.4	-5.0	+0.3
04	344.7	-1.4	-6.2	-1.1
05	356.9	-1.3	-7.1	-2.5
06	009.1	-1.3	-7.7	-3.7
07	021.2	-1.3	-7.9	-4.9
08	033.4	-1.3	-7.6	-5.8
09	045.5	-1.3	-6.8	-6.3
10	057.6	-1.2	-5.5	-6.6
11	069.8	-1.2	-3.7	-6.4
12	081.9	-1.2	-1.7	-5.7
13	094.0	-1.2	+0.4	-4.7
14	106.1	-1.1	+2.5	-3.3
15	118.3	-1.1	+4.4	-1.7
16	130.4	-1.1	+5.8	+0.0
17	142.5	-1.1	+6.9	+1.7
18	154.7	-1.1	+7.5	+3.2
19	166.8	-1.0	+7.7	+4.4
20	179.0	-1.0	+7.5	+5.5
21	191.2	-1.0	+7.0	+6.2
22	203.3	-1.0	+6.2	+6.6
23	215.5	-1.0	+5.2	+6.8
24	227.7	-1.0	+4.1	+6.6
25	239.9	-0.9	+2.9	+6.1
26	252.0	-0.9	+1.6	+5.4
27	264.2	-0.9	+0.3	+4.4
28	276.4	-0.9	-1.1	+3.2
29	288.6	-0.9	-2.5	+1.9
30	300.8	-0.9	-3.8	+0.5
31	313.0	-0.8	-5.0	-0.9

2009 JANUARY

01	325.2	-0.8	-6.0	-2.3
02	337.3	-0.8	-6.9	-3.7
03	349.5	-0.8	-7.4	-4.8
04	001.7	-0.7	-7.7	-5.8
05	013.8	-0.7	-7.5	-6.4
06	026.0	-0.7	-6.9	-6.7
07	038.1	-0.6	-5.8	-6.6
08	050.2	-0.6	-4.3	-6.1
09	062.4	-0.6	-2.6	-5.2
10	074.5	-0.5	-0.6	-3.9
11	086.6	-0.5	+1.4	-2.4
12	098.7	-0.5	+3.3	-0.6
13	110.9	-0.4	+4.9	+1.1
14	123.0	-0.4	+6.2	+2.7
15	135.1	-0.4	+7.0	+4.2
16	147.3	-0.3	+7.4	+5.3
17	159.4	-0.3	+7.3	+6.1
18	171.5	-0.3	+6.9	+6.6
19	183.7	-0.3	+6.2	+6.8
20	195.9	-0.2	+5.2	+6.7
21	208.0	-0.2	+4.0	+6.3
22	220.2	-0.2	+2.7	+5.6
23	232.4	-0.2	+1.4	+4.7
24	244.6	-0.2	+0.0	+3.6
25	256.8	-0.1	-1.3	+2.3
26	269.0	-0.1	-2.6	+0.9
27	281.1	-0.1	-3.7	-0.6
28	293.3	-0.1	-4.7	-2.0
29	305.5	-0.0	-5.6	-3.4
30	317.7	-0.0	-6.1	-4.6
31	329.9	+0.0	-6.5	-5.6

2009 FEBRUARY

01	342.1	+0.0	-6.6	-6.4
02	354.2	+0.1	-6.4	-6.8
03	006.4	+0.1	-5.8	-6.8
04	018.6	+0.1	-5.0	-6.4
05	030.7	+0.2	-3.9	-5.6
06	042.8	+0.2	-2.5	-4.5
07	055.0	+0.2	-1.0	-3.0
08	067.1	+0.3	+0.7	-1.4
09	079.2	+0.3	+2.3	+0.4
10	091.4	+0.4	+3.7	+2.1
11	103.5	+0.4	+5.0	+3.6
12	115.6	+0.4	+5.9	+4.9
13	127.8	+0.5	+6.4	+5.9
14	139.9	+0.5	+6.5	+6.5
15	152.1	+0.5	+6.2	+6.8
16	164.2	+0.5	+5.5	+6.7
17	176.4	+0.6	+4.6	+6.4
18	188.5	+0.6	+3.4	+5.8
19	200.7	+0.6	+2.1	+4.9
20	212.9	+0.6	+0.7	+3.8
21	225.1	+0.6	-0.7	+2.6
22	237.3	+0.6	-1.9	+1.3
23	249.5	+0.7	-3.1	-0.2
24	261.7	+0.7	-4.0	-1.6
25	273.9	+0.7	-4.8	-3.0
26	286.1	+0.7	-5.2	-4.3
27	298.3	+0.7	-5.5	-5.4
28	310.5	+0.8	-5.4	-6.2

2009 MARCH

01	322.7	+0.8	-5.2	-6.7
02	334.9	+0.8	-4.7	-6.8
03	347.1	+0.8	-4.1	-6.5
04	359.2	+0.9	-3.3	-5.8
05	011.4	+0.9	-2.4	-4.8
06	023.6	+0.9	-1.3	-3.4
07	035.7	+1.0	-0.2	-1.9
08	047.9	+1.0	+1.0	-0.2
09	060.0	+1.0	+2.1	+1.5
10	072.2	+1.1	+3.2	+3.1
11	084.3	+1.1	+4.1	+4.4
12	096.5	+1.1	+4.8	+5.5
13	108.6	+1.1	+5.3	+6.2
14	120.8	+1.2	+5.3	+6.6
15	132.9	+1.2	+5.1	+6.6
16	145.1	+1.2	+4.4	+6.3
17	157.2	+1.2	+3.5	+5.8
18	169.4	+1.2	+2.4	+5.0
19	181.6	+1.2	+1.1	+4.0
20	193.8	+1.2	-0.3	+2.8
21	206.0	+1.2	-1.7	+1.5
22	218.2	+1.3	-2.9	+0.2
23	230.4	+1.3	-4.0	-1.2
24	242.6	+1.3	-4.8	-2.6
25	254.8	+1.3	-5.3	-3.9
26	267.0	+1.3	-5.4	-5.0
27	279.2	+1.3	-5.3	-5.9
28	291.5	+1.3	-4.8	-6.4
29	303.7	+1.3	-4.1	-6.6
30	315.9	+1.3	-3.2	-6.4
31	328.1	+1.3	-2.3	-5.8

2009 APRIL

01	340.3	+1.4	-1.3	-4.8
02	352.5	+1.4	-0.3	-3.5
03	004.7	+1.4	+0.6	-2.1
04	016.9	+1.4	+1.5	-0.4
05	029.1	+1.4	+2.3	+1.2
06	041.2	+1.4	+3.0	+2.7
07	053.4	+1.5	+3.6	+4.1
08	065.6	+1.5	+4.2	+5.2
09	077.7	+1.5	+4.5	+6.0
10	089.9	+1.5	+4.6	+6.5
11	102.1	+1.5	+4.5	+6.6
12	114.2	+1.5	+4.1	+6.3
13	126.4	+1.5	+3.4	+5.8
14	138.6	+1.5	+2.4	+5.0
15	150.8	+1.5	+1.3	+4.0
16	162.9	+1.5	-0.1	+2.9
17	175.1	+1.5	-1.4	+1.6
18	187.3	+1.5	-2.8	+0.3
19	199.5	+1.5	-4.1	-1.0
20	211.8	+1.5	-5.1	-2.4
21	224.0	+1.5	-5.8	-3.6
22	236.2	+1.5	-6.2	-4.7
23	248.4	+1.5	-6.1	-5.6
24	260.7	+1.5	-5.6	-6.2
25	272.9	+1.5	-4.8	-6.5
26	285.1	+1.5	-3.7	-6.4
27	297.4	+1.5	-2.4	-5.8
28	309.6	+1.5	-1.0	-4.9
29	321.8	+1.5	+0.3	-3.6
30	334.1	+1.5	+1.5	-2.2

2009 MAY

01	346.3	+1.5	+2.5	-0.6
02	358.5	+1.5	+3.4	+1.1
03	010.7	+1.5	+4.0	+2.6
04	022.9	+1.5	+4.5	+4.0
05	035.1	+1.5	+4.8	+5.1
06	047.3	+1.5	+5.0	+6.0
07	059.4	+1.5	+4.9	+6.5
08	071.6	+1.5	+4.7	+6.6
09	083.8	+1.5	+4.3	+6.4
10	096.0	+1.5	+3.6	+5.9
11	108.2	+1.5	+2.7	+5.2
12	120.4	+1.5	+1.7	+4.2
13	132.5	+1.5	+0.4	+3.0
14	144.7	+1.5	-1.0	+1.8
15	156.9	+1.5	-2.4	+0.4
16	169.1	+1.4	-3.7	-0.9
17	181.3	+1.4	-5.0	-2.3
18	193.6	+1.4	-6.0	-3.5
19	205.8	+1.4	-6.7	-4.6
20	218.0	+1.4	-7.1	-5.5
21	230.3	+1.4	-6.9	-6.2
22	242.5	+1.3	-6.4	-6.5
23	254.7	+1.3	-5.3	-6.5
24	267.0	+1.3	-3.9	-6.0
25	279.2	+1.3	-2.3	-5.2
26	291.5	+1.3	-0.6	-4.0
27	303.7	+1.3	+1.1	-2.5
28	316.0	+1.3	+2.6	-0.8
29	328.2	+1.2	+3.9	+0.9
30	340.4	+1.2	+4.9	+2.5
31	352.7	+1.2	+5.6	+3.9

2009 JUNE

01	004.9	+1.2	+6.0	+5.1
02	017.1	+1.2	+6.2	+6.0
03	029.3	+1.2	+6.1	+6.5
04	041.5	+1.2	+5.8	+6.7
05	053.7	+1.2	+5.2	+6.6
06	065.9	+1.1	+4.5	+6.1
07	078.1	+1.1	+3.6	+5.4
08	090.3	+1.1	+2.5	+4.4
09	102.4	+1.1	+1.3	+3.3
10	114.6	+1.1	-0.1	+2.0
11	126.8	+1.0	-1.5	+0.6
12	139.0	+1.0	-2.9	-0.8
13	151.2	+1.0	-4.3	-2.1
14	163.4	+1.0	-5.5	-3.4
15	175.7	+1.0	-6.5	-4.6
16	187.9	+0.9	-7.3	-5.5
17	200.1	+0.9	-7.6	-6.2
18	212.3	+0.9	-7.5	-6.6
19	224.6	+0.9	-7.0	-6.7
20	236.8	+0.8	-6.0	-6.3
21	249.1	+0.8	-4.6	-5.6
22	261.3	+0.8	-2.8	-4.5
23	273.6	+0.8	-0.9	-3.0
24	285.8	+0.7	+1.0	-1.3
25	298.1	+0.7	+2.8	+0.4
26	310.3	+0.7	+4.4	+2.1
27	322.6	+0.7	+5.7	+3.7
28	334.8	+0.6	+6.6	+4.9
29	347.1	+0.6	+7.1	+5.9
30	359.3	+0.6	+7.2	+6.5

2009 JULY

01	011.5	+0.6	+7.0	+6.8
02	023.7	+0.6	+6.5	+6.7
03	035.9	+0.5	+5.8	+6.3
04	048.1	+0.5	+4.9	+5.7
05	060.3	+0.5	+3.7	+4.7
06	072.5	+0.5	+2.5	+3.6
07	084.7	+0.4	+1.1	+2.3
08	096.9	+0.4	-0.3	+0.9
09	109.1	+0.4	-1.7	-0.5
10	121.3	+0.4	-3.1	-1.9
11	133.5	+0.3	-4.4	-3.2
12	145.7	+0.3	-5.5	-4.4
13	157.9	+0.3	-6.5	-5.4
14	170.1	+0.2	-7.1	-6.2
15	182.3	+0.2	-7.5	-6.7
16	194.5	+0.2	-7.5	-6.8
17	206.8	+0.2	-7.0	-6.6
18	219.0	+0.1	-6.2	-6.0
19	231.2	+0.1	-4.9	-5.0
20	243.5	+0.1	-3.3	-3.6
21	255.7	+0.0	-1.5	-2.0
22	268.0	+0.0	+0.4	-0.3
23	280.3	-0.0	+2.4	+1.5
24	292.5	-0.0	+4.1	+3.2
25	304.8	-0.1	+5.6	+4.6
26	317.0	-0.1	+6.7	+5.7
27	329.2	-0.1	+7.4	+6.4
28	341.5	-0.1	+7.6	+6.8
29	353.7	-0.2	+7.4	+6.8
30	005.9	-0.2	+6.9	+6.4
31	018.1	-0.2	+6.1	+5.8

2009 AUGUST

01	030.3	-0.2	+5.0	+5.0
02	042.5	-0.3	+3.7	+3.9
03	054.7	-0.3	+2.4	+2.6
04	066.9	-0.3	+1.0	+1.3
05	079.1	-0.3	-0.4	-0.1
06	091.3	-0.4	-1.8	-1.5
07	103.5	-0.4	-3.0	-2.9
08	115.7	-0.4	-4.2	-4.2
09	127.9	-0.5	-5.1	-5.2
10	140.1	-0.5	-5.9	-6.0
11	152.2	-0.5	-6.4	-6.6
12	164.5	-0.5	-6.7	-6.8
13	176.7	-0.6	-6.7	-6.7
14	188.9	-0.6	-6.3	-6.2
15	201.1	-0.6	-5.7	-5.3
16	213.3	-0.6	-4.6	-4.1
17	225.6	-0.7	-3.3	-2.6
18	237.8	-0.7	-1.8	-1.0
19	250.0	-0.7	-0.0	+0.8
20	262.3	-0.7	+1.7	+2.5
21	274.5	-0.8	+3.5	+4.0
22	286.8	-0.8	+5.0	+5.2
23	299.0	-0.8	+6.2	+6.1
24	311.3	-0.8	+6.9	+6.6
25	323.5	-0.8	+7.3	+6.7
26	335.7	-0.9	+7.2	+6.4
27	347.9	-0.9	+6.7	+5.9
28	000.1	-0.9	+5.8	+5.1
29	012.4	-0.9	+4.7	+4.1
30	024.6	-1.0	+3.4	+2.9
31	036.7	-1.0	+2.0	+1.6

2009 SEPTEMBER

01	048.9	-1.0	+0.6	+0.2
02	061.1	-1.0	-0.8	-1.2
03	073.3	-1.0	-2.0	-2.6
04	085.5	-1.1	-3.1	-3.8
05	097.7	-1.1	-4.0	-4.9
06	109.8	-1.1	-4.7	-5.8
07	122.0	-1.1	-5.2	-6.4
08	134.2	-1.1	-5.4	-6.7
09	146.4	-1.2	-5.5	-6.6
10	158.6	-1.2	-5.3	-6.2
11	170.8	-1.2	-4.9	-5.4
12	183.0	-1.2	-4.3	-4.3
13	195.2	-1.2	-3.5	-3.0
14	207.4	-1.2	-2.5	-1.4
15	219.6	-1.2	-1.3	+0.3
16	231.8	-1.3	+0.1	+2.0
17	244.1	-1.3	+1.6	+3.5
18	256.3	-1.3	+3.0	+4.8
19	268.5	-1.3	+4.3	+5.8
20	280.8	-1.3	+5.4	+6.4
21	293.0	-1.3	+6.1	+6.6
22	305.2	-1.4	+6.5	+6.4
23	317.4	-1.4	+6.4	+5.9
24	329.6	-1.4	+5.9	+5.1
25	341.9	-1.4	+5.0	+4.1
26	354.1	-1.4	+3.9	+3.0
27	006.3	-1.4	+2.6	+1.7
28	018.4	-1.4	+1.2	+0.4
29	030.6	-1.5	-0.2	-1.0
30	042.8	-1.5	-1.5	-2.3

2009 OCTOBER

01	055.0	-1.5	-2.7	-3.5
02	067.1	-1.5	-3.6	-4.6
03	079.3	-1.5	-4.2	-5.5
04	091.5	-1.5	-4.6	-6.2
05	103.6	-1.5	-4.7	-6.5
06	115.8	-1.5	-4.6	-6.5
07	128.0	-1.5	-4.3	-6.2
08	140.1	-1.5	-3.8	-5.5
09	152.3	-1.5	-3.2	-4.4
10	164.5	-1.5	-2.5	-3.1
11	176.7	-1.5	-1.7	-1.5
12	188.8	-1.5	-0.8	+0.1
13	201.0	-1.5	+0.1	+1.7
14	213.2	-1.5	+1.2	+3.2
15	225.5	-1.5	+2.2	+4.6
16	237.7	-1.5	+3.2	+5.6
17	249.9	-1.5	+4.2	+6.2
18	262.1	-1.5	+4.9	+6.5
19	274.3	-1.5	+5.4	+6.4
20	286.5	-1.5	+5.6	+6.0
21	298.7	-1.5	+5.4	+5.2
22	310.9	-1.5	+4.8	+4.2
23	323.2	-1.5	+3.9	+3.1
24	335.3	-1.5	+2.8	+1.8
25	347.5	-1.5	+1.5	+0.5
26	359.7	-1.5	+0.1	-0.9
27	011.9	-1.5	-1.3	-2.2
28	024.1	-1.5	-2.6	-3.4
29	036.2	-1.5	-3.7	-4.5
30	048.4	-1.5	-4.5	-5.4
31	060.6	-1.5	-5.0	-6.1

2009 NOVEMBER

01	072.7	-1.5	-5.1	-6.5
02	084.8	-1.5	-4.9	-6.5
03	097.0	-1.5	-4.4	-6.2
04	109.1	-1.5	-3.7	-5.6
05	121.3	-1.5	-2.7	-4.5
06	133.4	-1.4	-1.7	-3.2
07	145.6	-1.4	-0.7	-1.7
08	157.7	-1.4	+0.3	-0.0
09	169.9	-1.4	+1.2	+1.6
10	182.1	-1.4	+2.1	+3.1
11	194.2	-1.4	+2.9	+4.5
12	206.4	-1.4	+3.6	+5.5
13	218.6	-1.3	+4.3	+6.3
14	230.8	-1.3	+4.8	+6.6
15	243.0	-1.3	+5.1	+6.6
16	255.2	-1.3	+5.2	+6.2
17	267.4	-1.3	+5.0	+5.5
18	279.6	-1.3	+4.6	+4.5
19	291.8	-1.3	+3.9	+3.3
20	304.0	-1.3	+2.9	+2.0
21	316.2	-1.3	+1.7	+0.7
22	328.4	-1.3	+0.4	-0.7
23	340.6	-1.2	-1.1	-2.1
24	352.7	-1.2	-2.5	-3.3
25	004.9	-1.2	-3.8	-4.4
26	017.1	-1.2	-4.9	-5.3
27	029.2	-1.2	-5.7	-6.1
28	041.4	-1.2	-6.1	-6.5
29	053.5	-1.1	-6.2	-6.6
30	065.6	-1.1	-5.8	-6.4

2009 DECEMBER

01	077.8	-1.1	-5.1	-5.8
02	089.9	-1.1	-4.0	-4.9
03	102.0	-1.0	-2.7	-3.6
04	114.2	-1.0	-1.3	-2.0
05	126.3	-1.0	+0.2	-0.3
06	138.4	-1.0	+1.6	+1.4
07	150.6	-0.9	+2.9	+3.0
08	162.7	-0.9	+3.9	+4.4
09	174.9	-0.9	+4.8	+5.5
10	187.0	-0.8	+5.4	+6.3
11	199.2	-0.8	+5.8	+6.7
12	211.4	-0.8	+5.9	+6.7
13	223.6	-0.8	+5.9	+6.4
14	235.8	-0.7	+5.6	+5.8
15	247.9	-0.7	+5.0	+4.8
16	260.1	-0.7	+4.3	+3.7
17	272.3	-0.7	+3.3	+2.4
18	284.5	-0.7	+2.2	+1.0
19	296.7	-0.7	+0.9	-0.5
20	308.9	-0.6	-0.5	-1.8
21	321.1	-0.6	-1.9	-3.1
22	333.2	-0.6	-3.3	-4.3
23	345.4	-0.6	-4.6	-5.3
24	357.6	-0.6	-5.7	-6.1
25	009.7	-0.5	-6.6	-6.6
26	021.9	-0.5	-7.1	-6.8
27	034.0	-0.5	-7.3	-6.7
28	046.2	-0.4	-7.0	-6.2
29	058.3	-0.4	-6.2	-5.4
30	070.4	-0.4	-5.1	-4.2
31	082.5	-0.3	-3.6	-2.7

2010 JANUARY

01	094.7	-0.3	-1.9	-1.0

02	106.8	-0.3	-0.0	+0.8
03	118.9	-0.2	+1.8	+2.5
04	131.0	-0.2	+3.5	+4.1
05	143.2	-0.2	+4.9	+5.3
06	155.3	-0.1	+6.0	+6.2
07	167.4	-0.1	+6.8	+6.7
08	179.6	-0.1	+7.1	+6.8
09	191.8	-0.0	+7.1	+6.6
10	203.9	-0.0	+6.8	+6.0
11	216.1	+0.0	+6.2	+5.1
12	228.3	+0.0	+5.4	+4.0
13	240.5	+0.1	+4.3	+2.8
14	252.7	+0.1	+3.2	+1.4
15	264.8	+0.1	+1.9	-0.1
16	277.0	+0.1	+0.5	-1.5
17	289.2	+0.2	-0.9	-2.9
18	301.4	+0.2	-2.3	-4.1
19	313.6	+0.2	-3.7	-5.1
20	325.8	+0.2	-4.9	-5.9
21	337.9	+0.2	-6.0	-6.5
22	350.1	+0.3	-6.5	-6.8
23	002.3	+0.3	-7.5	-6.8
24	014.4	+0.3	-7.7	-6.4
25	026.6	+0.3	-7.6	-5.7
26	038.7	+0.4	-7.1	-4.7
27	050.9	+0.4	-6.1	-3.3
28	063.0	+0.4	-4.8	-1.7
29	075.1	+0.5	-3.0	+0.0
30	087.2	+0.5	-1.1	+1.8
31	099.4	+0.5	+1.0	+3.4

2010 FEBRUARY

01	111.5	+0.6	+3.0	+4.8
02	123.6	+0.6	+4.8	+5.9
03	135.8	+0.6	+6.2	+6.5
04	147.9	+0.7	+7.2	+6.7
05	160.0	+0.7	+7.7	+6.6
06	172.2	+0.7	+7.8	+6.1
07	184.4	+0.8	+7.4	+5.3
08	196.5	+0.8	+6.7	+4.2
09	208.7	+0.8	+5.7	+3.0
10	220.9	+0.8	+4.5	+1.7
11	233.1	+0.9	+3.1	+0.3
12	245.3	+0.9	+1.7	-1.1
13	257.5	+0.9	+0.3	-2.5
14	269.7	+0.9	-1.1	-3.8
15	281.9	+0.9	-2.4	-4.8
16	294.1	+0.9	-3.6	-5.7
17	306.2	+1.0	-4.7	-6.4
18	318.4	+1.0	-5.7	-6.7
19	330.6	+1.0	-6.5	-6.7
20	342.8	+1.0	-7.0	-6.4
21	355.0	+1.0	-7.3	-5.8
22	007.2	+1.0	-7.3	-4.9
23	019.3	+1.1	-7.0	-3.7
24	031.5	+1.1	-6.3	-2.2
25	043.6	+1.1	-5.2	-0.6
26	055.8	+1.1	-3.8	+1.1
27	067.9	+1.2	-2.0	+2.8
28	080.1	+1.2	+0.0	+4.3

2010 MARCH

01	092.2	+1.2	+2.0	+5.4
02	104.3	+1.2	+3.9	+6.2
03	116.5	+1.2	+5.5	+6.6
04	128.6	+1.3	+6.7	+6.5
05	140.8	+1.3	+7.3	+6.0
06	152.9	+1.3	+7.5	+5.3
07	165.1	+1.3	+7.2	+4.3
08	177.3	+1.3	+6.5	+3.1
09	189.5	+1.4	+5.5	+1.8
10	201.6	+1.4	+4.2	+0.5
11	213.8	+1.4	+2.9	-0.9
12	226.0	+1.4	+1.4	-2.2
13	238.2	+1.4	+0.0	-3.5
14	250.4	+1.4	-1.3	-4.6
15	262.7	+1.4	-2.5	-5.5
16	274.9	+1.4	-3.6	-6.2
17	287.1	+1.4	-4.5	-6.5
18	299.3	+1.4	-5.2	-6.6
19	311.5	+1.4	-5.7	-6.4
20	323.7	+1.4	-6.1	-5.8
21	335.9	+1.4	-6.3	-4.9
22	348.1	+1.5	-6.2	-3.8
23	000.3	+1.5	-5.9	-2.4
24	012.5	+1.5	-5.4	-0.8
25	024.6	+1.5	-4.6	+0.8
26	036.8	+1.5	-3.4	+2.4
27	049.0	+1.5	-2.0	+3.9
28	061.1	+1.5	-0.3	+5.1
29	073.3	+1.5	+1.4	+5.9
30	085.4	+1.5	+3.1	+6.4
31	097.6	+1.5	+4.6	+6.5

2010 APRIL

01	109.8	+1.5	+5.7	+6.1
02	121.9	+1.5	+6.4	+5.4
03	134.1	+1.5	+6.6	+4.4
04	146.3	+1.5	+6.4	+3.2
05	158.4	+1.5	+5.7	+1.9
06	170.6	+1.5	+4.8	+0.6
07	182.8	+1.5	+3.5	-0.8
08	195.0	+1.5	+2.2	-2.1
09	207.2	+1.5	+0.8	-3.3
10	219.4	+1.5	-0.6	-4.4
11	231.7	+1.5	-1.8	-5.3
12	243.9	+1.5	-2.9	-6.0
13	256.1	+1.5	-3.8	-6.5
14	268.3	+1.5	-4.4	-6.6
15	280.6	+1.5	-4.9	-6.4
16	292.8	+1.5	-5.1	-5.9
17	305.0	+1.5	-5.2	-5.0
18	317.2	+1.5	-5.0	-3.9
19	329.5	+1.5	-4.7	-2.5
20	341.7	+1.5	-4.3	-0.9
21	353.9	+1.5	-3.7	+0.7
22	006.1	+1.5	-2.9	+2.3
23	018.3	+1.5	-1.9	+3.7
24	030.5	+1.4	-0.8	+5.0
25	042.7	+1.4	+0.5	+5.9
26	054.8	+1.4	+1.8	+6.4
27	067.0	+1.4	+3.0	+6.5
28	079.2	+1.4	+4.2	+6.3
29	091.4	+1.4	+5.0	+5.6
30	103.5	+1.4	+5.5	+4.7

2010 MAY

01	115.7	+1.4	+5.6	+3.5
02	127.9	+1.4	+5.3	+2.1
03	140.1	+1.4	+4.6	+0.7
04	152.3	+1.3	+3.7	-0.7
05	164.5	+1.3	+2.5	-2.0
06	176.7	+1.3	+1.2	-3.3
07	188.9	+1.3	-0.2	-4.4
08	201.1	+1.3	-1.5	-5.3
09	213.3	+1.3	-2.7	-6.0
10	225.5	+1.3	-3.7	-6.5
11	237.8	+1.3	-4.4	-6.6
12	250.0	+1.3	-4.9	-6.5
13	262.3	+1.2	-5.1	-6.0
14	274.5	+1.2	-5.0	-5.2
15	286.7	+1.2	-4.7	-4.1
16	299.0	+1.2	-4.2	-2.7
17	311.2	+1.2	-3.5	-1.2
18	323.5	+1.1	-2.7	+0.5
19	335.7	+1.1	-1.8	+2.1
20	347.9	+1.1	-0.9	+3.6
21	000.1	+1.1	+0.1	+4.9
22	012.3	+1.0	+1.1	+5.9
23	024.5	+1.0	+2.1	+6.5
24	036.7	+1.0	+3.0	+6.7
25	048.9	+1.0	+3.9	+6.5
26	061.1	+1.0	+4.5	+5.9
27	073.3	+0.9	+4.9	+5.0
28	085.5	+0.9	+5.0	+3.9
29	097.7	+0.9	+4.9	+2.5
30	109.9	+0.9	+4.4	+1.1
31	122.0	+0.8	+3.6	-0.4

2010 JUNE

01	134.2	+0.8	+2.6	-1.8
02	146.4	+0.8	+1.4	-3.1
03	158.6	+0.8	+0.1	-4.3
04	170.9	+0.8	-1.2	-5.2
05	183.1	+0.7	-2.5	-6.0
06	195.3	+0.7	-3.6	-6.5
07	207.5	+0.7	-4.6	-6.7
08	219.8	+0.7	-5.3	-6.7
09	232.0	+0.7	-5.7	-6.3
10	244.2	+0.6	-5.7	-5.6
11	256.5	+0.6	-5.4	-4.5
12	268.7	+0.6	-4.8	-3.2
13	281.0	+0.6	-3.9	-1.6
14	293.3	+0.5	-2.9	+0.0
15	305.5	+0.5	-1.6	+1.8
16	317.7	+0.5	-0.4	+3.4
17	330.0	+0.5	+0.9	+4.7
18	342.2	+0.4	+2.1	+5.8
19	354.4	+0.4	+3.2	+6.5
20	006.7	+0.4	+4.1	+6.8
21	018.9	+0.3	+4.8	+6.7
22	031.1	+0.3	+5.2	+6.2
23	043.3	+0.3	+5.5	+5.4
24	055.5	+0.2	+5.4	+4.3
25	067.7	+0.2	+5.1	+3.0
26	079.9	+0.2	+4.6	+1.5
27	092.1	+0.2	+3.9	+0.0
28	104.2	+0.1	+2.9	-1.4
29	116.4	+0.1	+1.8	-2.8
30	128.6	+0.1	+0.6	-4.0

2010 JULY

01	140.8	+0.1	-0.7	-5.1
02	153.0	+0.0	-2.0	-5.9
03	165.3	+0.0	-3.3	-6.5
04	177.5	+0.0	-4.5	-6.8
05	189.7	-0.0	-5.4	-6.8
06	201.9	-0.0	-6.1	-6.5
07	214.2	-0.1	-6.6	-5.9
08	226.4	-0.1	-6.6	-4.9
09	238.6	-0.1	-6.2	-3.7
10	250.9	-0.1	-5.5	-2.2
11	263.1	-0.2	-4.4	-0.6
12	275.4	-0.2	-3.0	+1.1
13	287.7	-0.2	-1.3	+2.8
14	299.9	-0.2	+0.4	+4.3
15	312.2	-0.3	+2.0	+5.5
16	324.4	-0.3	+3.6	+6.3
17	336.6	-0.3	+4.8	+6.7
18	348.9	-0.4	+5.8	+6.7
19	001.1	-0.4	+6.4	+6.3
20	013.3	-0.4	+6.7	+5.6
21	025.5	-0.5	+6.6	+4.5
22	037.7	-0.5	+6.2	+3.3
23	049.9	-0.5	+5.6	+1.9
24	062.1	-0.6	+4.8	+0.4
25	074.3	-0.6	+3.8	-1.0
26	086.5	-0.6	+2.7	-2.5
27	098.7	-0.6	+1.4	-3.7
28	110.9	-0.7	+0.2	-4.8
29	123.1	-0.7	-1.2	-5.7
30	135.3	-0.7	-2.5	-6.3
31	147.5	-0.7	-3.7	-6.7

2010 AUGUST

01	159.7	-0.7	-4.9	-6.8
02	171.9	-0.8	-5.9	-6.5
03	184.1	-0.8	-6.6	-6.0
04	196.3	-0.8	-7.7	-5.2
05	208.5	-0.8	-7.2	-4.1
06	220.8	-0.8	-7.0	-2.7
07	233.0	-0.9	-6.3	-1.2
08	245.3	-0.9	-5.1	+0.5
09	257.5	-0.9	-3.6	+2.2
10	269.8	-0.9	-1.7	+3.7
11	282.0	-0.9	+0.3	+5.0
12	294.3	-0.9	+2.3	+6.0
13	306.5	-1.0	+4.1	+6.5
14	318.7	-1.0	+5.7	+6.6
15	331.0	-1.0	+6.8	+6.3
16	343.2	-1.0	+7.5	+5.6
17	355.4	-1.1	+7.7	+4.7
18	007.6	-1.1	+7.5	+3.5
19	019.8	-1.1	+7.0	+2.1
20	032.0	-1.2	+6.1	+0.7
21	044.2	-1.2	+5.1	-0.7
22	056.4	-1.2	+3.9	-2.1
23	068.6	-1.2	+2.7	-3.4
24	080.8	-1.2	+1.4	-4.6
25	093.0	-1.3	+0.0	-5.5
26	105.2	-1.3	-1.3	-6.2
27	117.3	-1.3	-2.5	-6.6
28	129.5	-1.3	-3.7	-6.7
29	141.7	-1.3	-4.8	-6.5
30	153.9	-1.3	-5.8	-6.0
31	166.1	-1.3	-6.5	-5.3

2010 SEPTEMBER

01	178.3	-1.3	-7.1	-4.3
02	190.5	-1.3	-7.3	-3.0
03	202.7	-1.3	-7.2	-1.6
04	215.0	-1.3	-6.6	+0.0
05	227.2	-1.3	-5.6	+1.6
06	239.4	-1.3	-4.2	+3.2
07	251.7	-1.4	-2.4	+4.6
08	263.9	-1.4	-0.3	+5.6
09	276.1	-1.4	+1.8	+6.3
10	288.4	-1.4	+3.9	+6.5
11	300.6	-1.4	+5.6	+6.3
12	312.8	-1.4	+6.9	+5.7
13	325.1	-1.4	+7.7	+4.7
14	337.3	-1.5	+8.0	+3.5
15	349.5	-1.5	+7.8	+2.2
16	001.7	-1.5	+7.2	+0.8
17	013.9	-1.5	+6.3	-0.6
18	026.1	-1.5	+5.2	-2.0
19	038.3	-1.5	+3.9	-3.2
20	050.4	-1.5	+2.6	-4.4
21	062.6	-1.6	+1.3	-5.3
22	074.8	-1.6	+0.0	-6.0
23	087.0	-1.6	-1.3	-6.5
24	099.1	-1.6	-2.4	-6.6
25	111.3	-1.6	-3.5	-6.5
26	123.5	-1.6	-4.4	-6.0
27	135.7	-1.6	-5.2	-5.3
28	147.8	-1.6	-5.9	-4.3
29	160.0	-1.6	-6.4	-3.1
30	172.2	-1.6	-6.6	-1.7

2010 OCTOBER

01	184.4	-1.5	-6.6	-0.2
02	196.6	-1.5	-6.2	+1.4
03	208.8	-1.5	-5.3	+2.9
04	221.0	-1.5	-4.1	+4.3
05	233.2	-1.5	-2.6	+5.4
06	245.4	-1.5	-0.7	+6.1
07	257.7	-1.5	+1.3	+6.5
08	269.9	-1.5	+3.2	+6.3
09	282.1	-1.5	+4.9	+5.8
10	294.3	-1.5	+6.2	+4.9
11	306.5	-1.5	+7.1	+3.7
12	318.8	-1.5	+7.5	+2.4
13	331.0	-1.5	+7.3	+0.9
14	343.2	-1.5	+6.8	-0.5
15	355.3	-1.5	+5.9	-1.9
16	007.5	-1.5	+4.8	-3.2
17	019.7	-1.5	+3.6	-4.3
18	031.9	-1.5	+2.2	-5.2
19	044.0	-1.5	+0.9	-6.0
20	056.2	-1.5	-0.3	-6.4
21	068.4	-1.5	-1.5	-6.6
22	080.5	-1.5	-2.5	-6.5
23	092.7	-1.5	-3.4	-6.1
24	104.8	-1.4	-4.1	-5.4
25	117.0	-1.4	-4.7	-4.4
26	129.1	-1.4	-5.1	-3.2
27	141.3	-1.4	-5.3	-1.8
28	153.5	-1.4	-5.4	-0.3
29	165.6	-1.3	-5.2	+1.3
30	177.8	-1.3	-4.8	+2.8
31	190.0	-1.3	-4.1	+4.2

2010 NOVEMBER

01	202.1	-1.3	-3.1	+5.3
02	214.3	-1.2	-1.8	+6.1
03	226.5	-1.2	-0.4	+6.5
04	238.7	-1.2	+1.2	+6.5
05	250.9	-1.2	+2.8	+6.1
06	263.2	-1.2	+4.2	+5.3
07	275.4	-1.1	+5.3	+4.1
08	287.6	-1.1	+6.1	+2.7
09	294.8	-1.1	+6.4	+1.3
10	312.0	-1.1	+6.3	-0.3
11	324.2	-1.1	+5.8	-1.7
12	336.5	-1.1	+5.0	-3.1
13	348.5	-1.1	+4.0	-4.2
14	000.7	-1.1	+2.7	-5.2
15	012.9	-1.1	+1.4	-6.0
16	025.0	-1.1	+0.2	-6.5
17	037.2	-1.0	-1.1	-6.7
18	049.3	-1.0	-2.2	-6.7
19	061.5	-1.0	-3.1	-6.3
20	073.6	-1.0	-3.8	-5.7
21	085.7	-1.0	-4.3	-4.7
22	097.9	-0.9	-4.5	-3.5
23	110.0	-0.9	-4.6	-2.1
24	122.1	-0.9	-4.5	-0.5
25	134.3	-0.8	-4.2	+1.1
26	146.4	-0.8	-3.7	+2.6
27	158.6	-0.8	-3.0	+4.1
28	170.7	-0.7	-2.2	+5.3
29	182.9	-0.7	-1.2	+6.2
30	195.1	-0.6	-0.2	+6.7

2010 DECEMBER

01	207.2	-0.6	+1.0	+6.8
02	219.4	-0.6	+2.1	+6.4
03	231.6	-0.6	+3.2	+5.7
04	243.8	-0.5	+4.1	+4.6
05	256.0	-0.5	+4.8	+3.3
06	268.2	-0.5	+5.2	+1.8
07	280.4	-0.5	+5.3	+0.2
08	292.6	-0.4	+5.1	-1.3
09	304.7	-0.4	+4.6	-2.8
10	316.9	-0.4	+3.8	-4.0
11	329.1	-0.4	+2.8	-5.1
12	341.3	-0.4	+1.6	-5.9
13	353.4	-0.3	+0.3	-6.5
14	005.6	-0.3	-0.9	-6.8
15	017.8	-0.3	-2.2	-6.8
16	029.9	-0.3	-3.2	-6.5
17	042.0	-0.3	-4.1	-5.9
18	054.2	-0.2	-4.8	-5.1
19	066.3	-0.2	-5.1	-4.0
20	078.4	-0.2	-5.2	-2.6
21	090.6	-0.1	-4.9	-1.1
22	102.7	-0.1	-4.3	+0.6
23	114.8	-0.1	-3.5	+2.2
24	126.9	-0.0	-2.6	+3.7
25	139.1	-0.0	-1.4	+5.0
26	151.2	+0.0	-0.3	+6.0
27	163.3	+0.1	+0.9	+6.6
28	175.5	+0.1	+2.0	+6.8
29	187.7	+0.2	+2.9	+6.6
30	199.8	+0.2	+3.7	+6.0
31	212.0	+0.2	+4.4	+5.0

Bibliography

Baldwin, Ralph B. (1949). *The Face of the Moon.* Chicago: University of Chicago Press.
_____. (1963). *The Measure of the Moon.* Chicago: University of Chicago Press.
_____. (1970). "A New Method of Determining the Depth of the Lava in Lunar Maria." *Pub. Astron. Soc. of the Pacific, 82,* 857-864.
Basaltic Volcanism Study Project (1981). *Basaltic Volcanism on the Terrestrial Planets.* New York: Pergamon Press.
Batson, Raymond M. and Russell, Joel F., Eds. (1995). *Gazetteer of Planetary Nomenclature 1994.* U.S. Geological Survey Bulletin 2129. Washington: U.S. Government Printing Office.
Berry, Richard (1992). *Choosing and Using a CCD Camera.* Richmond, VA: Willmann-Bell.
Bowker, David E. and Hughes, J. Kenrick (1971). *Lunar Orbiter Photographic Atlas of the Moon.* NASA SP-206. Washington: U.S. National Aeronautics and Space Administration.
Boyce, Joseph M.; Dial, A.J., Jr.; and Sonderblom, L.A. (1974). "Ages of the Nearside Light Plains and Maria." *Proc. Lunar Sci. Conf., 5,* 11-23.
Bryan, W.B. (1973). "Wrinkle-ridges as Deformed Surface Crust on Ponded Mare Lava." *Proc. Lunar Sci. Conf. 4,* 93-106.
Bryan, W.B. and Adams, Mary-Linda (1973). "Some Volcanic and Structural Features of Mare Serenitatis." *Apollo 17 Preliminary Science Report.* (NASA SP-330). 30-9 - 30-12.
Buil, Christian (1991). *CCD Astronomy: Construction and Use of an Astronomical CCD Camera.* Tr. by E. and B. Davoust. Richmond, VA: Willmann-Bell.
Calame, O. (1967). "Determination of the Moon's Shape by the Photometry of its Terminator." In: Kopal, 1967a, 451-454.
Cameron, Winifred Sawtell and Padgett, Joe L. (1974). "Possible Lunar Ring Dikes." *The Moon,* 9, 249-294.
Carr, Michael H., Ed. (1984). *The Geology of the Terrestrial Planets.* NASA SP-469. Washington: NASA.
Cattermole, Peter (1989). *Planetary Volcanism: A Study of Volcanic Activity in the Solar System.* Chichester: Ellis Horwood Ltd.
Christiansen, Eric H. and Hamblin, W. Kenneth (1995). *Exploring the Planets.* 2nd Ed. Englewood Cliffs, NJ: Prentice-Hall.
Colton, George W.; Howard, Keith A.; and Moore, Henry J. (1972). "Mare Ridges and Arches in Southern Oceanus Procellarum." *Apollo 16 Preliminary Science Report.* (NASA SP-315), 29-90 - 29-93.
Cruikshank, Dale P.; Hartmann, William K.; and Wood, Craig A. (1973). "Moon: 'Ghost' Craters Formed During Mare Filling." *The Moon, 7,* 440-452.
Dantowitz, Ron (1998). "Sharper Images Through Video." *Sky & Telescope, 96,* No. 2 (August), 48-54.
DeHon, R.A. (1974). "Thickness of Mare Material in the Tranquillitatis and Nectaris Basins." *Proc. Lunar Sci. Conf. 5,* 53-59.
_____. (1975). "Mare Spumans and Mare Undarum: Mare Thickness and Basin Floor." *Proc. Lunar Sci. Conf. 6,* 2553-2561.
_____. (1977). "Mare Humorum and Mare Nubium: Basalt Thickness and Basin-Forming History." *Proc. Lunar Sci. Conf. 8,* 633-641.
_____. (1979). "Thickness of the Western Mare Basalts." *Proc. Lunar Planet Sci. Conf. 10,* 2935-2955.
DeHon, R.A. and Waskom, J.D. (1976). "Geologic Structure of the Eastern Mare Basins." *Proc. Lunar Sci. Conf. 7,* 2729-2746.
Dobbins, Thomas A. (1996). "Recording the Moon and Planets with a Video Camera." *J.B.A.A., 106,* no. 6 (Dec.), 309-314.
Dobbins, Thomas A.; Parker, Donald C.; and Capen, Charles F. (1988). *Introduction to Observing and Photographing the Solar System.* Richmond, VA: Willmann-Bell.
Dragesco, Jean (1995). *High Resolution Astrophotography.* Cambridge, UK: Cambridge University Press.
Dyer, Alan (1998). "Revealed! The Secret Films of Astrophotographers." *Sky & Telescope, 95,* No. 4 (April), 101-107.
Eggleton, Richard E. and Schaber, G.G. (1972). "Cayley Formation Interpreted as Basin Ejecta." *Apollo 16 Preliminary Science Report.* (NASA SP-315). 29-7 - 29-16.
El-Baz, Farouk and Roosa, S.A. (1972). "Significant Results from Apollo 14 Lunar Orbital Photography." *Proc. Lunar Sci. Conf. 3,* 63-83.
Firsoff, Val A. (1957). "Paleography of the Nubian and Imbrian Plains." *J.B.A.A., 67,* 130-134.
Frankel, Charles (1996). *Volcanoes of the Solar System.* Cambridge, UK: Cambridge University Press.
Gifford, A.W. and El-Baz, Farouk (1981). "Thickness of Lunar Mare Flow Fronts." *Moon and Planets, 24,* 391-398.
Greeley, Ronald (1974). "Volcanism as a Planetary Process." In: Greeley and Schultz, 1974, 295-322.
Greeley, Ronald and Schultz, Peter, Eds. (1974). *A Primer in Lunar Geology.* ("Comment Edition".) Mountain View, CA: U.S. National Aeronautics and Space Administration, Ames Research Center..
Greeley, Ronald and Spudis, Paul D. (1978). "Mare Volcanism in the Herigonius Region of the Moon." *Proc. Lunar Planet. Sci. Conf. 9,* 3333-3349.
Guest, John E.; Butterworth, Paul; Murray, John; and O'Donnell, William (1979). *Planetary Geology.* New York: John Wiley.

Hackman, Robert J. and Mason, Arnold C. (1961). *Engineer Special Study of the Surface of the Moon.* U.S.Geological Survey. Map I-351 (4 sheets). Washington: U.S. Geological Survey.

Hapke, Bruce W. (1966). "An Improved Theoretical Lunar Photometric Function." *Astron. J., 71*, 333-339.

Hartmann, William K. (1963a). "Radial Structures Surrounding Lunar Basins, I: The Imbrian System." *Comm. Lunar Planetary Lab.*, No. 24, 1-16 & Pl. 24.1-24.40.

_____. (1963b). "Radial Structures Surrounding Lunar Basins, II: Orientale and Other Systems; Conclusions." *Comm. Lunar Planetary Lab.*, No. 36, 175-192 & Pl. 36.1-36.26.

_____. (1966). "Lunar Basins, Lunar Lineaments, and the Moon's Far Side." *Sky and Telescope, 32*, 128-131.

Hartmann, William K. and Kuiper, Gerard P. (1962). "Concentric Structures Surrounding Lunar Basins." *Comm. Lunar Planetary Lab.*, No. 12, 51-66 & Pl. 12.1-12.77.

Hartmann, William K. and Wood, Craig A. (1971). "Moon: Origin and Evolution of Multi-ring Basins." *The Moon, 3,* 3-78.

Head, James W. III (1976). "Lunar Volcanism in Space and Time." *Rev. Geophys. Space Phys., 14*, 265-300.

_____. (1979). "Serenitatis Multi-Ringed Basin: Regional Geology and Basin Ring Interpretation." *The Moon and the Planets, 21,* 439-462.

_____. (1981). "Lava Flooding of Ancient Planetary Crusts: Geometry, Thickness and Volumes of Lunar Impact Basins." *Moon and Planets, 26,* 61-88.

Head, James W. III and Gifford, A.W. (1980). "Lunar Mare Domes: Classification and Modes of Origin." *The Moon and the Planets, 22,* 235-258.

Head, James W. III and Lloyd, Douglas (1971). "Near-Terminator Photography." *Apollo 14 Preliminary Science Report* (NASA SP-272), 297-300.

Head, James W. III and Lloyd, Douglas (1972a). "Near-Terminator Photography." *Apollo 15 Preliminary Science Report* (NASA SP-289), 25-95 - 25-101.

Head, James W. III and Lloyd, Douglas (1972b). "Low-Relief Features in Terrain of the Descartes Region and Other Areas: Near-Terminator Photography." *Apollo 16 Preliminary Science Report.* (NASA SP-315). 29-97 - 29-103.

Head, James W. III and Lloyd, Douglas (1973). "Appendix. Near-Terminator and Earthshine Photography." *Apollo 17 Preliminary Science Report.* (NASA SP-330). 4-33 - 4-39.

Head, James W. III and McCord, Thomas B. (1978). "Imbrian-age Highland Volcanism on the Moon: The Gruithuisen and Mairan Domes." *Science, 199,* 1433-1436.

Heiken, Grant H.; Vaniman, David T.; and French, Bevan M., Eds. (1991). *Lunar Sourcebook.* Cambridge, UK: Cambridge University Press.

Herring, Alika K. (1961). "Saucers in Ptolemaeus." *Journal, A.L.P.O., 15,* 62-64.

Hill, Harold (1991). *A Portfolio of Lunar Drawings.* Cambridge, UK: Cambridge University Press.

Hodges, Carroll Ann (1973). "Mare Ridges and Lava Lakes." *Apollo 17 Preliminary Science Report.* (NASA SP-330). 31-12 - 31-21.

Hodges, Carroll Ann and Wilhelms, Don E. (1978). "Formation of Lunar Basin Rings." *Icarus, 34,* 294-323.

Hörz, Frederick (1978), "How Thick are Lunar Mare Basalts?" *Proc. Lunar Planet Sci. Conf. 9,* 3311-3331.

Howard, Keith A. (1974). "Multi-Ring Basins and Mascons." In: Greeley and Schultz, 1974, 253-274.

Howard, Keith A. and Larsen, Bradley R. (1972). "Lineaments that are Artifacts of Lighting." *Apollo 15 Preliminary Science Report* (NASA SP-289), 25-58 - 25-62.

Howell, Steve B., Ed. (1992). *Astronomical CCD Observing and Reduction Techniques.* Astronomical Society of the Pacific Conference Series Volume 23. San Francisco: Astronomical Society of the Pacific.

Jacoby, George H., Ed. (1990). *CCDs in Astronomy.* Astronomical Society of the Pacific Conference Series Volume 8. San Francisco: Astronomical Society of the Pacific.

Jamieson, Harry D. and Phillips, James H. (1992). "Lunar Dome Catalog (April 30, 1992 Edition)." *Journal, A.L.P.O., 36,* No. 3 (Sept.), 123-129.

Kopal, Zdeněk, Ed. (1962a). *Physics and Astronomy of the Moon.* New York: Academic Press.

_____. (1962b). "Topography of the Moon." In: Kopal, 1962a, 231-282.

_____, ed. (1967a). *Physics and Astronomy of the Moon.* Dordrecht: D. Reidel.

_____. (1967b). "Terminator Photography in Oblique Illumination for Lunar Topographic Work." In: Kopal, 1967a, 407-413.

_____, Ed. (1974). *Physics and Astronomy of the Moon.* New York: Academic Press.

Lenham, A.P. (1954). "Lunar Ridges and 'Ghost Rings'." *J.B.A.A., 64,* 124-126.

Lloyd, Douglas and Head, James W. III (1972). "Lunar Surface Properties as Determined from Earthshine and Near-Terminator Photography." *Proc. Lunar Sci. Conf. 3,* 3127-3142.

Lucchitta, Baerbel Koesters (1976). "Mare Ridges and Related Highland Scarps—Results of Vertical Tectonism?" *Proc. Lunar Planet. Sci. Conf. 7,* 2761-2782.

_____. (1977). "Topography, Structure, and Mare Ridges in Southern Mare Imbrium and Northern Oceanus Procellarum." *Proc. Lunar Planet Sci. Conf. 8,* 2691-2703.

_____. (1978). *Geologic Map of the North Side of the Moon.* U.S. Geological Survey Map I-1062. Washington: U.S. Geological Survey.

McCauley, J.F. (1969). "The Cones and Domes of the Marius Hills Region." *Trans. Ame. Geophys. Union, 50,* 229.

McGetchin, Thomas R. and Head, James W. III (1973). "Lunar Cinder Cones." *Science, 180,* 68-71.

Malin, Michael C. (1978). "Surfaces of Mercury and the Moon: Effects of Resolution and Lighting Conditions on the Discrimination of Volcanic Features." *Proc. Lunar Planet Sci. Conf. 9,* 3395-3409.

Masursky, Harold; Colton, G.W.; and El-Baz, Farouk, Eds. (1978). *Apollo Over the Moon: A View from Orbit.* NASA SP-362. Washington: National Aeronautics and Space Administration.

Maxwell, Ted A. and Andre, Constance G. (1981). "The Balmer Basin: Regional Geology and Geochemistry of an Ancient Lunar Impact Basin." *Proc. Lunar Planet Sci. Conf. 12*, 715-725.

Maxwell, Ted A.; El-Baz, Farouk; and Ward, S.H. (1975). "Distribution, Morphology, and Origin of Ridges and Arches in Mare Serenitatis." *Geological Society of America Bulletin*, 86, 1273-1278.

Melosh, H.J. (1989). *Impact Cratering : A Geologic Process.* New York: Oxford University Press.

Melosh, H.J. and Whitaker, Ewen A. (1994). "Lunar Crater Chains." *Nature, 365*, 713.

Moore, Henry J. (1972). "Crater-Shadowing Effects at Low Sun Angles." *Apollo 15 Preliminary Science Report* (NASA SP-289), 25-92 - 25-94.

Mutch, Thomas A. (1970). *Geology of the Moon: A Stratigraphic View.* Princeton: Princeton University Press.

Neison, Edmund (1876). *The Moon. The Condition and Configurations of Its Surface.* London: Longmans, Green, and Co.

Neukum, Gerhard (1977). "Different Ages of Lunar Light Plains." *The Moon, 17*, 383-393.

Oberbeck, V.R.; Hörz, Frederick; Morrison, R.H.; Quaide, W.L.; and Gault, D.E. (1975). "On the Origin of the Lunar Smooth-Plains." *The Moon, 12*, 19-54.

Oberbeck, V.R.; Morrison, R.H.; Hörz, Frederick; Quaide, W.L.; and Gault, D.E. (1974). "Smooth Plains and Continuous Deposits of Craters and Basins." *Proc. Lunar Sci. Conf. 5*, 111-136.

Olivarez, Jose (1966). "The Tobias Mayer - Milichius - Hortensius Domes." *Journal, A.L.P.O., 19*, 195-198.

Phillips, James H. (1990). "The Marius Hills Region–A Unique Lunar Dome Field."*Journal, A.L.P.O., 34*, 104-108.

Pike, Richard J. (1980). *Geometric Interpretation of Lunar Craters.* Apollo-15 Orbital Investigations. U.S. Geological Survey Professional Paper 1046-C. Washington: U.S. Government Printing Office.

Price, Fred W. (1988). *The Moon Observer's Handbook.* Cambridge, UK: Cambridge University Press.

Quaide, W.L. (1965). "Rilles, Ridges and Domes–Clues to Maria History." *Icarus, 4*, 374-389.

Quaide, W.L. and Oberbeck, V.R. (1968). "Thickness Determinations of the Lunar Surface Layer from Lunar Impact Craters." *J. Geophys. Res., 73*, 5247-5270.

Rükl, Antonín (1990). *Hamlyn Atlas of the Moon.* London: Paul Hamlyn Publishing.

Runcorn, Stanley Keith and Urey, Harold C., Eds. (1972). *The Moon.* Dordrecht: D. Reidel.

Schaber, Gerald G. (1973a). "Lava Flows in Mare Imbrium: Geologic Evaluation from Apollo Orbital Photography." *Proc. Lunar Sci. Conf. 4*, 73-92.

_____. (1973b). "Eratosthenian Volcanism in Mare Imbrium: Source of Youngest Lava Flows." *Apollo 17 Preliminary Science Report.* (NASA SP-330). 30-17 - 30-25.

Schaber, Gerald G.; Boyce, Joseph M.; and Moore, Henry J. (1976). "The Scarcity of Mappable Flow Lobes on the Lunar Maria: Unique Morphology of the Imbrium Flows." *Proc. Lunar Sci. Conf. 7th*, 2783-2800.

Schultz, Peter H. (1974). "A Review of Lunar Surface Features." In: Greeley and Schultz, 1974, 91-133.

_____. (1976a). *Moon Morphology.* Austin: University of Texas Press.

_____. (1976b). "Floor-Fractured Lunar Craters." *The Moon, 13*, Nos. 3/4 (June/July), 241-273.

Scott, David H. (1973). "Mare Serenitatis Cinder Cones and Terrestrial Analogies." *Apollo 17 Preliminary Science Report.* (NASA SP-330). 30-7 - 30-8.

Scott, David H.; McCauley, John F.; and West, Mareta N. (1977). *Geologic Map of the West Side of the Moon.* U.S. Geological Survey Map I-1034. Washington: U.S. Geological Survey.

Sharpton, Virgil L. and Head, James W. III (1988). "Lunar Mare Ridges: Analysis of Ridge-Crater Intersections and Implications of the Tectonic Origin of Mare Ridges." *Proc. Lunar Planet. Sci. Conf. 18*, 307-317.

Shoemaker, Eugene M. (1962). "Interpretation of Lunar Craters." In: Kopal, 1962a, 283-359.

Short, Nicolas M. (1973). *Planetary Geology*, Englewood Cliffs, NJ: Prentice-Hall.

Short, N.M. and Forman, M.L. (1972). "Thickness of Impact Crater Ejecta on the Lunar Surface." *Modern Geology, 3*, 69-91.

Smith, Eugene I. (1973). "Identification, Distribution and Significance of Lunar Volcanic Domes." *The Moon, 6*, 3-31.

_____. (1974). "Rümker Hills: A Lunar Volcanic Dome Complex." *The Moon, 10*, 175-181.

Spudis, Paul D. (1993). *The Geology of Multi-Ring Impact Basins: The Moon and Other Planets.* Cambridge, UK: Cambridge University Press.

_____. (1996). *The Once and Future Moon.* Washington: Smithsonian Institution Press.

Spudis, Paul D.; Hawke, B. Ray; and Lucey, Paul G. (1988). "Materials and Formation of the Imbrium Basin." *Proc. Lunar Planet Sci. Conf. 18*, 155-168.

Spudis, Paul D.; Hawke, B. Ray; and Lucey, Paul G. (1989). "Geology and Deposits of the Lunar Nectaris Basin." *Proc. Lunar Planet Sci. Conf. 19*, 51-59.

Spudis, Paul D.; Reisse, Robert A.; and Gillis, Jeffrey J. (1994). "Ancient Multiring Basins on the Moon Revealed by Clementine Laser Altimetry." *Science, 266*, 1848-1851.

Stewart, H.E.; Waskom, J.D.; and DeHon, R.A. (1975). "Photogeology and Basin Configuration of Mare Smythii." *Proc. Lunar Sci. Conf. 6*, 2541-2551.

Strain, P.L. and El-Baz, Farouk (1979). "Smythii Basin Topography and Comparisons with Orientale." *Proc. Lunar Planet. Sci. Conf. 10*, 2609-2621.

Strom, Robert G. (1971). "Lunar Mare Ridges, Rings, and Volcanic Ring Complexes." *Modern Geology, 3*, 133-157. [Reprinted in: Runcorn and Urey, 1972, 187-215.]

Stuart-Alexander, Desiree E. (1978). *Geologic Map of the Central Far Side of the Moon.* U.S. Geological Survey Map I-1047. Washington: U.S. Geological Survey.

Stuart-Alexander, Desiree E. and Howard, Keith A. (1970). "Lunar Maria and Circular Basins–A Review." *Icarus, 12*, 440-456. [Reprinted in: Greeley and Schultz (1974), 275-291.]

Taylor, Stuart Ross. (1982). *Planetary Science: A Lunar Perspective*. Houston: Lunar and Planetary Institute.
United States. National Aeronautics and Space Administration. Manned Spacecraft Center. (1969). *Analysis of Apollo 8 Photography and Visual Observations*. NASA SP-201. [Terminator observations on pp. 4, 15-16.]
United States. Naval Observatory. Nautical Almanac Office (1998). *The Astronomical Almanac for the Year 1999*. Washington: U.S. Government Printing Office. Annual publication.
Viscardy, Georges (1985). *Atlas-Guide Photographique de la Lune*. Monte-Carlo: Association Franco-Monegasque d'Astronomie.
Wallis, Brad D. and Provin, Robert W. (1988). *A Manual of Advanced Celestial Photography*. Cambridge, U.K.: Cambridge University Press.
Westfall, John E. (1964). "A Generic Classification of Lunar Domes." *Journal, A.L.P.O., 18,* Nos. 1-2 (Jan.-Feb.), 15-20.
_____. (1966). "Lunar Terminator Deformations." *Journal, A.L.P.O., 19,* Nos. 5-6 (Feb.), 75-79.
_____. (1971). "Statistical Analysis of Lunar Dome Distribution." *Journal, A.L.P.O., 23,* Nos. 5-6 (Nov.), 91-98.
Whitford-Stark, James L. (1979). "Charting the Southern Seas–The Evolution of the Lunar Mare Australe." *Proc. Lunar Planet Sci. Conf. 10,* 2975-2994.
_____. (1981). "The Evolution of the Lunar Nectaris Multiring Basin." *Icarus, 48,* 393-427.
_____. (1990). "The Volcanotectonic Evolution of Mare Frigoris." *Proc. Lunar Planet Sci. Conf. 20,* 175-185.
Whitford-Stark, James L. and Fryer, R.J. (1975). "Origin of Mare Frigoris." *Icarus, 26,* 231-242.
Whitford-Stark, James L. and Head, James W. III (1977). "The Procellarum Volcanic Complexes: Contrasting Styles of Volcanism." *Proc. Lunar Sci. Conf. 8,* 2705-2724.
Whitford-Stark, James L. and Head, James W. III (1980). "Stratigraphy of Oceanus Procellarum Basalts: Sources and Styles of Emplacement." *J. Geophys. Res., 85,* 6579-6609.
Wichman, R.W. and Schultz, P.H. (1996). "Crater-Centered Laccoliths on the Moon: Modeling Intrusion Depth and Magmatic Pressure at the Crater Taruntius." *Icarus, 122,* No. 1 (July), 193-199.
Wichman, R.W. and Wood, Craig A. (1995). "The Davy Crater Chain: Implications for Tidal Disruption in the Earth-Moon System and Elsewhere." *Geophys. Res. Lett., 22,* 583-586.
Wilhelms, Don E. (1970). *Summary of Lunar Stratigraphy–Telescopic Observations*. U.S.G.S. Professional Paper 599-F. Washington: U.S. Government Printing Office.
_____. (1987). *The Geologic History of the Moon*. U.S. Geological Survey Professional Paper 1348. Washington: U.S. Government Printing Office.
Wilhelms, Don E. and El-Baz, Farouk (1977). *Geologic Map of the East Side of the Moon*. U.S. Geological Survey Map I-948. Washington: U.S. Geological Survey.
Wilhelms, Don E.; Howard, K.A.; and Wilshire, H.G. (1979). *Geologic Map of the South Side of the Moon*. U.S. Geological Survey Map I-1162. Washington: U.S. Geological Survey.
Wilhelms, Don E. and McCauley, John F. (1971). *Geologic Map of the Near Side of the Moon*. U.S. Geological Survey Map I-703. Washington: U.S. Geological Survey.
Wilkins, H. Percy and Moore, Patrick (1955). *The Moon*. New York: The Macmillan Company.
Will, Matthew J. (1998). "Artistic Philosophies in Lunar Drawing." *Journal, A.L.P.O., 40,* No. 2 (April), 82-87.
Young, Richard A. (1972). "Lunar Volcanism: Mare Ridges and Sinuous Rilles." *Apollo 16 Preliminary Science Report*. (NASA SP-315). 29-79 - 29-80.
Young, Richard A.; Brennan, William J.; Wolfe, Robert W.; and Nichols, Douglas J. (1973a). "Analysis of Lunar Mare Geology from Apollo Photography." *Proc. Lunar Planet Sci. Conf. 4,* 52-71.
Young, Richard A.; Brennan, William J.; Wolfe, Robert W.; and Nichols, Douglas J. (1973b). "Volcanism in the Lunar Maria." *Apollo 17 Preliminary Science Report*. (NASA SP-330). 31-1 - 31-11.
Zisk, S.H.; Hodges, Carroll Ann; Moore, Henry J.; Shorthill, Richard W.; Thompson, Thomas W.; and Wilhelms, Don E. (1977). "The Aristarchus-Harbinger Region of the Moon: Surface Geology and History from Recent Remote-Sensing Observations." *The Moon, 17,* 59-99.
Zuber, Maria T.; Smith, David E.; Lemoine, Frank G.; and Neumann, Gregory A. (1994). "The Shape and Internal Structure of the Moon from the Clementine Mission." *Science, 266,* 1839-1843.

Organizations   These organizations either concentrate entirely upon the Moon, or have Sections that do so:

American Lunar Society. P.O. Box 209, East Pittsburgh, PA 15112, USA.
Association of Lunar and Planetary Observers (Membership Secretary: P.O. Box 171302, Memphis, TN 38187-130, USA)
  Eclipse Section. Michael D. Reynolds. Chabot Observatory and Science Center, 10902 Skyline Boulevard, Oakland, CA 94619, USA.
  Lunar Section: Lunar Topograhical Studies. William M. Dembowski. 219 Old Bedford Pike, Windber PA 15963, USA.
  Lunar Section: Lunar Transient Phenomena. David O. Darling. 416 W. Wilson St., Sun Prairie, WI 53590-2114, USA.
  Lunar Section: Selected Areas Program. Dr. Julius L. Benton, Jr. 305 Surrey Road, Savannah, GA 31410, USA.
The British Astronomical Association. Burlington House, Piccadilly, London W1V 9AG, UK.
  Lunar Section. Alan Wells, 135 Elmdon Lane, Marston Green, Birmingham B37 7DN, UK.

Mosaic Feature Index

Notes: Eastern longitudes are positive, western negative (where east and west follow the International Astronomical Union convention with Mare Crisium near the east limb). Northern latitudes are positive, southern negative. Terminator mosaics are indicated by colongitude in degrees, followed by section (C, center; N, north; S, south); Q indicates a quadrant mosaic in the "Near Full" mosaics (where E represents east and W is west). This index lists only the instances where features are *named* on the mosaic insets; in order to avoid congestion, there are often mosaics at similar colongitudes where a specific feature is shown but is not indicated by name.

1. Crater Index

Name	Long. °	Lat. °	Diam. km	Depth m	Terminator Mosaic(s)
Abbot	+54.8	+5.6	10.8	2070	315-C, 124-C
Abel	+84.6	−34.5	115.7	1190	291-S, Q4-SE
Abenezra	+12.0	−21.0	42.0	3730	351-S, 357-S, 154-S, 161-S, 167-S
Abetti	+22.7	+19.9	8.0	80	343-C, 143-C, 148-C
Abulfeda	+13.9	−13.8	62.0	1230	351-C/S, 154-C/S, 161-S. 167-C
Acosta	+60.1	−5.6	13.1	2900	105-C, 112-C, 119-C
Adams	+68.7	−31.8	66.2	4190	303-S, Q4-SE, 099-C, 105-C, 112-C
Agatharchides	−30.9	−19.8	48.6	1180	036-C/S, 040-C/S, 048-C/S, 201-C, 205-S
Agrippa	+10.5	+4.1	46.0	3000	351-C, 357-C, 161-C, 167-C
Airy	+5.7	−18.1	36.9	2410	357-S, 003-S, 161-S, 167-S, 174-S
Al-Bakri	+20.2	+14.3	12.4	1040	343-N/C, 148-C, 154-C, 161-C
Al-Marrakushi	+55.8	−10.4	8.3	1110	309-C/S, 119-C, 124-C
Albategnius	+4.1	−11.2	135.7	3200	357-C, 003-C/S, 161-C/S, 167-C, 174-C
Alexander	+13.4	+40.2	81.6	410	351-N, 154-N, 161-N
Alfraganus	+19.0	−5.4	20.4	3830	343-C, 148-C, 154-C, 161-C
Alhazen	+71.9	+15.9	32.8	2170	291-N, 303-C, Q1-E, Q4-E, 099-N, 105-N
Aliacensis	+5.2	−30.7	79.8	3680	357-S, 003-S, 161-S, 167-S, 174-S
Almanon	+15.2	−16.8	49.3	2480	351-S, 154-S, 161-S, 167-C
Alpetragius	−4.5	−16.0	39.9	3900	009-S, 018-C/S, 174-S, 17/9-C, 183-S
Alphonsus	−2.7	−13.5	118.5	2730	003-S, 009-C/S, 174-C/S, 179-C, 183-C
Ameghino	+57.0	+3.3	9.7	1870	309-C, 112-S, 119-C
Ammonius	−0.8	−8.5	8.5	1860	003-C, 174-C, 179-C
Amontons	+46.8	−5.4	3.3	200	181-C
Amundsen	+81.7	−85.1	95.2	5870	323-S, 343-S, Q3-S, Q4-S, 099-S, 105-S, 112-S
Anaxagoras	−10.1	+73.4	50.6	3060	009-N, 018-N, 024-N, 029-N, 036-N, 100-N, 106-N, 143-N, 161-N, 174-N, 179-N, 183-N, 191-N
Anaximander	−51.4	+66.9	67.7	1710	055-N, 061-N, 067-N, 075-N, 080-N, 201-N, 205-N, 213-N, 225-N
Anaximander B	−60.7	+67.8	77.7	-----	067-N
Anaximander D	−49.8	+65.3	89.1	-----	067-N
Anaximenes	−44.1	+72.4	79.9	880	055-N, 061-N, 075-N, 080-N, 191-N, 205-N, 213-N, 233-N
Andêl	+12.3	−10.4	33.7	1350	351-C, 357-C, 154-C, 161-C, 167-C
Ångström	−41.7	+29.9	9.8	2030	048-N, 055-N, 205-N, 213-N
Ansgarius	+79.6	−12.7	93.5	4190	291-S, Q4-E, 099-C
Anville	+49.5	+1.8	10.8	2220	315-C, 119-N/C, 124-C, 131-C
Apianus	+7.9	−26.9	66.3	2080	351-S, 357-S, 003-S, 161-S, 167-S
Apollonius	+61.0	+4.5	51.8	2750	303-C, 309-C, 099-C, 105-N, 112-C, 119-C
Arago	+21.4	+6.2	26.0	2680	343-C, 148-C, 154-C
Aratus	+4.5	+23.6	10.2	1880	357-C, 003-C, 161-C, 167-N
Archimedes	−4.0	+20.7	82.6	1600	003-N, 009-N, 174-N, 179-N, 183-N
Archytas	+5.0	+58.7	31.7	2940	357-N, 003-N, 009-N, 154-N, 161-N, 167-N, 174-N
Argelander	+5.8	−16.6	34.1	2980	357-S, 003-S, 161-S, 167-C, 174-S
Ariadaeus	+17.3	+4.6	11.2	1830	343-C, 154-C, 161-C

Name	Long. °	Lat. °	Diam. km	Depth m	Terminator Mosaic(s)
Aristarchus	−47.2	+23.6	45.3	3150	048-N, 055-N, 061-N, 067-N, 213-N, 225-N
Aristillus	+1.2	+33.9	55.3	3300	357-N, 003-N, 009-N, 167-N, 174-N, 179-N, 183-N
Aristoteles	+17.3	+50.1	87.3	3500	343-N, 351-N, 143-N, 148-N, 154-N, 161-N
Arnold	+36.0	+66.8	94.9	950	323-N, 328-N, 334-N, 343-N, 106-N, 119-N, 124-N, 131-N, 135-N, 143-N, 148-N
Arrhenius	−91.6	−56.5	41.2	2750	Q3-SW
Artsimovich	−36.7	+27.6	8.2	2010	040-N, 205-N, 213-N
Aryabhata	+35.2	+6.2	22.2	650	328-C, 135-C, 143-C
Arzachel	−1.9	−18.2	96.8	3610	003-S, 009-S, 174-S, 179-C, 183-S
Asada	+49.9	+7.2	12.8	2250	315-C, 124-N/C
Asclepi	+25.5	−55.1	42.5	1300	343-S, 143-S, 148-S, 154-S
Aston	−87.4	+32.9	41.8	2070	Q2-NW
Atlas	+44.4	+46.6	87.4	2050	315-N, 323-N, 328-N, 112-N, 119-N, 124-N, 131-N, 135-N
Atwood	+57.8	−5.9	29.2	2500	309-C/S, 315-C, 112-C, 119-C
Autolycus	+1.5	+30.7	39.2	3430	357-N, 003-N, 009-N, 167-N, 174-N, 179-N
Auwers	+17.2	+15.0	23.4	1070	343-N, 154-C, 161-C
Auzout	+64.0	+10.2	31.8	3850	303-C, 309-C, 099-N, 105-N, 112-N
Avery	+81.4	−1.3	10.8	2080	Q1-E, Q4-E
Azophi	+12.8	−22.1	47.7	3730	351-S, 357-S, 154-S, 161-S, 167-C/S
Baade	−82.3	−44.8	55.2	5240	080-S, Q3-SW
Babbage	−56.8	+59.5	143.8	1390	061-N, 067-N, 225-N, 233-N, 237-N
Back	+80.6	+1.2	35.8	2900	291-N/S, Q1-E, Q4-E
Bacon	+19.1	−51.0	69.6	2490	343-S, 135-S, 143-S, 148-S, 154-S, 161-S
Baillaud	+37.9	+74.7	89.5	1760	328-N, 334-N, 124-N, 131-N, 135-N, 143-N, 148-N
Bailly	−69.2	−67.0	283.8	4130	055-S, 061-S, 067-S, 075-S, 080-S, Q3-SW, 213-S, 225-S, 233-S, 237-S, 247-S
Baily	+30.3	+49.7	26.8	520	334-N, 343-N, 131-N, 135-N, 143-N, 148-N
Balboa	−83.0	+19.1	73.1	2140	Q2-NW
Ball	−8.4	−35.8	39.5	2810	009-S, 018-S, 179-S, 183-S
Balmer	+70.6	−20.1	111.8	1960	291-S, 303-C/S, 099-C, 105-C, 112-C
Banachiewicz	+80.1	+4.9	92.0	1680	291-N/S, Q1-E, Q4-E, 099-N
Bancroft	−6.4	+28.1	13.1	2490	009-N, 179-N, 183-N
Banting	+16.4	+26.6	5.4	1130	343-N, 154-N/C, 161-N/C
Barkla	+67.2	−10.7	43.2	2910	303-C/S, Q4-E, 099-C, 105-C, 112-C
Barnard	+85.8	−30.2	120.0	2660	Q4-SE
Barocius	+16.8	−44.9	82.3	3520	343-S, 351-S, 357-S, 148-S, 164-S, 161-S
Barrow	+7.6	+71.2	92.8	2380	357-N, 003-N, 009-N, 143-N, 148-N, 154-N, 161-N, 167-N
Bartels	−89.8	+24.5	45.	2570	Q2-NW
Bayer	−35.0	−51.5	47.4	2440	036-S, 040-S, 048-S, 055-S, 201-S, 205-S, 213-S
Beals	+88.0	+37.3	47.6	3150	Q1-NE
Beaumont	+28.8	−18.1	53.2	550	334-C/S, 343-C, 135-S, 146-S, 148-S
Beer	−9.1	+27.1	9.4	1670	018-N, 179-N, 183-N
Behaim	+79.7	−16.6	55.9	3330	291-S, Q4-SE/E, 099-C
Beketov	+29.2	+16.2	8.6	1070	334-C, 135-C, 143-C, 148-C
Bêla	+2.3	+24.7	4.7	690	174-N/C
Belkovich	+85.4	+61.2	197.6	2800	Q1-NE
Bellot	+48.2	−12.5	17.2	3120	315-C/S, 124-C, 131-C/S
Bernoulli	+60.6	+35.0	47.3	4080	303-N, 309-C, 099-N, 105-N, 112-N, 119-N
Berosus	+69.8	+33.5	74.2	4500	291-N, 303-N, Q1-E/NE, 099-N, 105-N
Berzelius	+50.9	+36.5	50.9	2990	315-N, 112-N, 119-N, 124-N
Bessarion	−37.3	+14.9	10.2	2000	040-N/C, 048-N/C, 055-N/C, 205-C, 213-N/C
Bessel	+17.9	+21.7	15.5	1770	343-N, 154-C, 161-C
Bettinus	−45.0	−63.5	71.4	3690	048-S, 055-S, 061-S, 067-S, 075-S, 080-S, 201-S, 205-S, 213-S
Bianchini	−34.4	+48.8	38.9	3100	036-N, 040-N, 048-N, 055-N, 067-N, 201-N, 205-N, 213-N

Name	Long.	Lat.	Diam.	Depth	Terminator Mosaic(s)
	°	°	km	m	
Biela	+51.3	−54.9	76.2	5550	309-S, 315-S, 323-S, 099-C/S, 105-C/S, 112S, 119-S, 124-S, 131-S
Bilharz	+56.4	−5.8	45.1	1750	309-C/S, 315-C, 119-C, 124-C
Billy	−50.0	−13.8	45.7	1290	055-C/S, 061-C/S, 067-C, 213-C, 225-C/S
Biot	+51.0	−22.7	12.9	900	315-C/S, 119-C/S, 124-C
Birmingham	−10.6	+64.8	97.2	830	018-N, 024-N, 029-N, 161-N, 174-N, 179-N, 183-N
Birt	−8.6	−22.4	16.8	3470	009-S, 018-S, 179-C/S, 183-S
Black	+80.4	−9.2	19.6	3170	291-S, Q4-E, 099-C
Blagg	+1.4	+1.2	5.4	920	003-C, 174-C
Blancanus	−21.5	−63.6	105.3	5970	024-S, 029-S, 036-S, 040-S, 048-S, 183-S, 191-S, 201-S
Blanchinus	+2.5	−25.4	62.8	1160	357-S, 003-S, 167-S, 174-S
Bobillier	+15.4	+19.6	6.2	1260	154-C, 161-C
Bode	−2.4	+6.7	18.6	3480	003-C, 009-C, 174-C, 179-C, 183-C
Boethius	+72.3	+5.6	10.0	2050	105-N
Boguslawsky	+43.4	−73.0	97.3	3990	323-S, 334-S, 343-S, Q3-S, Q4-S, 099-S, 105-S, 112-S, 119-S, 124-S, 131-S, 135-S
Bohnenberger	+40.1	−16.2	33.0	2400	323-C, 328-C, 131-C/S, 135-C/S
Bohr	−85.9	+13.0	73.1	4090	Q2-NW
Boltzmann	−90.4	−75.1	61.4	3580	Q3-S
Bombelli	+56.2	+5.3	10.3	1800	315-C, 119-C
Bond, G.	+36.4	+32.4	20.0	2780	323-N, 328-N/C, 131-N, 135-N/C, 143-N
Bond, W.	+3.5	+65.3	157.5	1890	357-N, 003-N, 143-N, 148-N, 154-N, 161-N, 167-N, 174-N
Bonpland	−17.3	−8.3	59.8	730	024-C, 029-C, 191-C
Boole	−79.8	+63.5	57.0	4170	080-N, Q2-N, 247-N
Borda	+46.6	−25.2	44.1	3640	315-S, 323-C/S, 328-S, 124-C/S, 131-C/S, 135-S
Borel	+26.4	+22.4	4.9	1090	334-N/C, 148-C
Born	+66.8	−6.1	14.8	2600	099-C, 105-C, 112-C
Boscovich	+11.1	+9.8	46.0	1780	351-C, 357-C, 154-C, 161-C, 167-C
Boss	+88.8	+45.8	47.0	3660	Q1-NE
Bouguer	−36.0	+52.4	23.1	3210	040-N, 048-N, 055-N, 201-N, 205-N, 213-N
Boussingault	+54.6	−70.4	131.1	5640	315-S, 323-S, 328-S, 334-S, Q4-S, 099-S, 105-S, 112-S, 119-S, 124-S, 131-S
Boussingault A	+54.0	−69.9	71.6	-----	119-S
Boussingault E	+46.7	−67.2	97.9	-----	105-S
Bowen	+9.1	+17.6	9.1	1100	357-C
Brackett	+23.5	+17.8	9.3	310	154-C
Brayley	−36.9	+20.9	14.5	2840	040-N, 048-N, 205-C, 213-N
Breislak	+18.3	−48.2	49.6	3520	343-S, 357-S, 143-S, 148-S, 154-S, 161-S
Brenner	+39.0	−39.2	87.4	1800	323-S, 328-S, 131-S, 135-S
Brewster	+34.7	+23.3	10.0	2340	328-N, 131-N/C, 135-N/C, 143-N/C
Brianchon	−81.5	+74.6	125.7	5360	Q2-N, 237-N
Briggs	−68.9	+26.4	38.6	1200	067-N, 075-N, 080-N, Q2-NW, 233-N, 237-C, 247-N/C
Brisbane	+68.5	−49.1	44.9	2900	303-S, Q4-SE, 099-C, 105-C/S, 112-C/S
Brown	−17.9	−46.4	32.0	1810	027-S, 029-S, 183-S, 191-S
Bruce	+0.4	+1.2	6.7	1270	003-C, 174-C, 179-C
Buch	+17.7	−38.9	53.8	1440	343-S, 143-S, 148-S, 154-S, 161-S
Bullialdus	−22.2	−20.7	58.9	3510	024-S, 029-S, 191-C/S, 201-C
Bunsen	−84.5	+41.5	61.1	1830	Q2-NW
Burckhardt	+56.5	+31.1	56.9	5920	303-N/C, 309-C, 315-C, 099-N, 105-N, 112-N, 119-N
Bürg	+28.2	+45.0	39.6	3070	334-N, 343-N, 131-N, 135-N, 143-N, 148-N, 154-N
Burnham	+7.3	−13.9	22.3	760	357-C/S, 161-S, 167-C
Büsching	+20.0	−38.0	52.3	1810	343-S, 143-S, 148-S, 154-S
Byrd	+10.4	+85.5	83.2	1380	357-N, 003-N, 029-N, Q1-N, 100-N, 106-N, 135-N, 143-N, 154-N, 161-N
Byrgius	−65.3	−24.7	89.8	2130	067-C, 075-C, 080-C/S, 233-C, 237-S

Name	Long. °	Lat. °	Diam. km	Depth m	Terminator Mosaic(s)
Cabeus	−35.6	−84.9	95.0	5710	048-S, 055-S, 067-S, 075-S, 080-S, Q3-S, Q4-S, 099-S, 105-S, 167-S, 179-S, 191-S, 201-S
Cajal	+31.1	+12.6	8.9	1810	327-C, 334-C, 135-C, 143-C, 148-C
Calippus	+10.7	+38.9	30.6	2690	357-N, 154-N, 161-N
Cameron	+46.0	+6.2	10.9	2190	315-C, 323-N/C, 124-N/C, 131-C
Campanus	−27.7	−28.0	48.0	2000	029-S, 036-S, 040-C/S, 201-S, 205-S
Cannon	+81.2	+19.9	56.6	2830	Q1-E, Q4-E, 099-N
Capella	+34.9	−7.7	44.7	3500	328-C/S, 131-C, 135-C/S, 143-C
Capuanus	−26.7	−34.1	59.7	1600	024-S, 029-S, 036-S, 040-S, 191-S, 201-S, 205-S
Cardanus	−72.4	+13.2	49.5	3440	075-N/C, 080-N/C, Q2-NW, 233-N/C, 247-C
Carlini	−24.1	+33.8	11.4	2200	024-N, 029-N, 036-N, 191-N, 201-N, 205-N
Carmichael	+40.4	+19.6	19.9	3800	323-N, 328-C, 131-N/C, 135-N/C
Carpenter	−50.8	+69.4	59.7	4170	061-N, 067-N, 075-N, 080-N, Q2-N, 205-N, 213-N, 225-N, 233-N
Carrel	+26.7	+10.7	15.9	2190	334-C, 143-C, 148-C, 154-C
Carrillo	+81.0	−2.1	17.9	2700	Q1-E, Q4-E
Carrington	+62.0	+44.0	29.6	2730	303-N, 309-N, 099-N, 112-N
Cartan	+59.3	+4.2	17.4	1850	309-C, 105-N, 119-C
Casatus	−30.5	−72.6	111.3	5120	024-S, 029-S, 036-S, 040-S, 0408-S, 055-S, 061-S, 080-S, Q3-S, 183-S, 191-S, 201-S, 205-S
Cassini	+4.5	+40.2	56.5	1240	357-N, 003-N, 161-N, 167-N, 174-N
Catalán	−89.0	−45.7	14.3	2110	Q3-SW
Catharina	+23.6	−18.1	97.5	3130	343-C, 143-S, 148-S, 154-S
Cauchy	+38.6	+9.6	12.8	2670	323-N/C, 328-C, 131-N/C, 135-C
Cavalerius	−66.9	+5.1	63.9	3600	067-N/C, 075-N/C, 080-N/C, 233-N/C, 237-C, 247-C
Cavendish	−53.8	−24.6	56.2	2550	055-C/S, 061-C/S, 067-C, 075-C/S, 225-C, 233-C
Cayley	+15.1	+3.9	14.3	3130	351-C, 154-C, 161-C
Celsius	+20.1	+34.1	36.4	2110	343-C, 143-S
Censorinus	+32.7	−0.4	3.8	1040	334-C, 135-C, 143-C
Cepheus	+45.8	+40.8	39.8	4590	315-N, 323-N, 112-N, 119-N, 124-N, 131-N, 135-N
Chacornac	+31.7	+29.8	51.0	1450	328-N/C, 334-N, 131-N, 135-N/C, 143-N, 148-N
Challis	+9.4	+79.6	55.6	1730	357-N, 003-N, 009-N, Q1-N, 135-N, 143-N, 148-N, 154-N, 161-N, 174-N
Chappe	−91.6	−61.5	59.2	2860	Q3-SW
Chevallier	+51.2	+44.9	52.3	360	309-N, 315-N, 106-N, 112-N, 119-N, 124-N
Ching-te	+30.0	+20.0	4.1	770	135-N/C
Chladni	+1.1	+4.0	13.6	2630	003-C, 174-C
Cichus	−21.2	−33.3	39.4	2760	024-S, 029-S, 040-S, 191-S, 201-S
Clairaut	+13.9	−47.7	75.2	2160	351-S, 357-S, 143-S, 148-S, 154-S, 161-S, 167-S
Clausius	−43.9	−36.9	22.3	2260	048-S, 055-S, 213-S
Clavius	−14.4	−58.4	225.3	4900	009-S, 018-S, 024-S, 029-S, 036-S, 174-S, 179-S, 183-S, 191-S
Cleomedes	+55.5	+27.7	126.1	3860	303-N/C, 309-C, 315-C, 105-N, 112-N, 119-N/C, 124-N
Cleostratus	−76.4	+60.3	62.9	4040	075-N, 080-N, Q2-N, 237-N
Clerke	+29.8	+21.7	6.8	1440	334-N, 135-N/C, 143-N/C, 148-C
Colombo	+46.0	−15.0	78.2	2530	315-C/S, 323-C, 328-C, 124-C/S, 131-C/S, 135-C/S
Condon	+60.4	+1.8	28.1	1850	309-C, 105-N, 112-C, 119-C
Condorcet	+69.7	+12.2	74.3	2650	303-C, Q1-E, Q4-E, 099-N, 105-N
Conon	+2.0	+21.7	20.8	2930	357-C, 003-C, 167-N, 174-C
Cook	+48.9	−17.5	46.6	1120	315-C/S, 323-C/S, 119-C/S, 124-C, 131-C/S
Copernicus	−20.0	+9.7	93.0	3800	018-C, 024-C, 029-C, 191-C, 201-C
Cremona	−88.2	+67.2	87.4	4080	Q2-N, 247-N
Crile	+46.0	+14.2	9.2	2020	315-C
Crozier	+50.8	−13.5	22.4	1300	315-C, 124-C
Crüger	−66.7	−16.7	44.8	530	067-C, 075-C, 080-C, 233-C, 237-C/S, 247-C/S
Curie	+91.8	−23.0	158.	3850	Q4-SE

Name	Long. °	Lat. °	Diam. km	Depth m	Terminator Mosaic(s)
Curtis	+56.8	+14.6	2.9	610	309-C, 315-C, 119-C
Curtius	+4.4	−67.2	94.7	3750	351-S, 357-S, 003-S, 009-S, 018-S, 148-S, 154-S, 161-S, 167-S, 174-S, 179-S
Cusanus	+70.1	+71.9	63.1	4610	291-N, 303-N, 309-N, 315-N, Q1-N, 100-N, 106-N
Cuvier	+9.9	−50.3	75.3	3020	351-S, 357-S, 003-S, 143-S, 148-S, 154-S, 161-S, 169-S
Cyrillus	+24.1	−13.4	93.4	3400	343-C, 143-C/S, 148-C, 154-C/S, 161-S
Cysatus	−6.2	−66.2	48.8	4250	009-S, 019-S, 167-S, 174-S, 179-S
d'Arrest	+14.6	+2.2	30.0	1900	351-C, 154-C, 161-C
da Vinci	+44.2	+9.6	31.3	1000	315-N/C, 323-N/C, 131-C
Daguerre	+33.6	−11.8	45.4	120	328-C/S, 135-C/S, 143-C/S
Dale	+83.0	−9.6	19.4	2250	Q4-E
Dalton	−84.3	+17.1	58.0	2150	Q2-NW
Daly	+59.5	+5.7	16.5	2860	309-C, 315-C, 105-N, 112-N/C, 119-C
Damoiseau	−61.1	−4.9	36.5	1250	061-C, 067-C, 075-C, 080-C, 233-C, 237-C/S
Damoiseau M	−61.3	−5.1	58.4	-----	067-C
Daniell	+31.2	+35.4	26.0	1900	328-N, 334-N, 343-N, 131-N, 135-N, 143-N, 148-N
Darney	−23.6	−14.6	15.2	2620	024-C, 029-S, 036-C/S, 040-C/S, 191-C, 201-C
Darwin	−69.2	−20.1	131.6	1360	075-C, 080-C, 233-C, 237-C/S, 247-C/S
Daubrée	+14.7	+15.7	14.0	1590	351-C, 154-C, 161-C
Davy	−8.1	−11.8	34.7	820	009-C, 018-C, 179-C, 183-C
Dawes	+26.3	+17.2	17.4	2310	334-C, 343-C, 143-C, 148-C, 154-C
de Gasparis	−50.7	−25.8	31.7	1080	055-C/S, 061-C/S, 067-C, 225-C/S
de la Rue	+52.9	+59.0	135.8	2610	309-N, 315-N, 323-N, 100-N, 105-N, 106-N, 112-N, 119-N, 124-N
de Morgan	+14.9	+3.3	10.2	1870	351-C, 154-C, 161-C
de Sitter	+38.5	+79.8	64.6	3370	334-N, 351-N, 357-N, Q1-N, 100-N, 106-N, 119-N, 124-N, 131-N, 135-N, 143-N
de Vico	−60.4	−19.7	19.2	1620	061-C, 067-C, 075-C, 080-S, 233-C
Debes	+51.7	+29.5	30.8	1050	309-N, 315-N/C, 112-N, 119-N, 124-N
Dechen	−68.3	+46.2	11.8	2290	075-N, 080-N, 233-N, 237-N
Delambre	+17.5	−1.9	53.0	3620	343-C, 154-C, 161-C
Delaunay	+2.5	−22.3	49.6	2520	003-S, 009-S, 167-C, 174-S
Delisle	−34.7	+30.0	25.4	2420	036-N, 040-N, 048-N, 201-N, 205-N, 213-N
Delmotte	+60.2	+27.1	32.8	4360	303-N/C, 105-N, 112-N
Deluc	−2.8	−55.0	46.9	5210	003S, 009-S, 018-S, 161-S, 167-S, 174-S, 179-S
Dembowski	+7.3	+2.8	29.3	280	357-C, 161-C, 167-C
Democritus	+35.1	+62.3	39.1	4300	328-N, 334-N, 343-N, 106-N, 119-N, 124-N, 135-N, 143-N, 148-N
Demonax	+56.8	−78.4	113.8	5000	323-S, 328-S, 334-S, 343-S, Q3-S, Q4-S, 099-S, 105-S, 112-S, 119-S, 124-S, 131-S
Desargues	−72.8	+70.2	82.6	2510	075-N, 080-N, Q2-N, 237-N, 247-N
Descartes	+15.7	−11.7	48.5	850	351-C, 154-C, 161-C/S
Deseilligny	+20.6	+21.1	6.4	1250	343-N, 154-C
Deslandres	−5.4	−32.3	234.4	1580	009-S, 018-S, 174-S, 179-S, 183-S
Dionysius	+17.3	+2.8	17.6	1200	343-C, 154-C, 161-C
Diophantus	−34.3	+27.6	18.1	3020	036-N, 040-N, 048-N, 201-N, 205-N/C, 213-N
Dollond	+14.4	−10.5	11.1	2040	351-C, 154-C, 161-C
Donati	+5.2	−20.7	36.0	2070	357-S, 003-S, 167-C, 174-S
Doppelmayer	−41.4	−28.5	63.8	1100	048-C/S, 055-C/S, 205-S, 213-C/S
Dove	+31.5	−46.7	30.4	1740	328-S, 334-S, 343-S, 131-S, 135-S, 143-S
Draper	−21.7	+17.6	8.8	1740	024-C, 029-C, 191-N/C, 201-N
Drebbel	−49.0	−40.9	30.2	3270	048-S, 055-S, 061-S, 067-C/S, 075-S, 213-S, 225-S
Drude	−91.8	−38.5	25.	4120	Q3-SW
Drygalski	−84.7	−79.7	162.5	5170	080-S, Q3-S, 099-S, 213-S, 225-S, 233-S, 247-S

Name	Long. °	Lat. °	Diam. km	Depth m	Terminator Mosaic(s)
Dubiago	+70.0	+4.4	45.7	2700	303-C, Q1-E, Q4-E, 099-N, 105-N
Dunthorne	−31.7	−30.1	15.7	2780	036-S, 201-S, 205-S
Eckert	+58.4	+17.3	2.8	430	116-C
Eddington	−71.7	+21.5	134.1	1300	075-N/C, 080-N/C, Q2-NW, 247-N/C
Egede	+10.6	+48.6	37.1	430	351-N, 357-N, 154-N, 161-N, 167-N
Eichstädt	−78.5	−22.6	49.2	4520	080-C, Q3-W
Eimmart	+64.7	+24.0	46.3	1560	303-C, Q4-E, 099-N, 105-N, 112-N
Einstein	−87.1	+16.6	189.5	1850	Q2-NW
Elger	−29.8	−35.3	20.5	1260	036-S, 040-S, 205-S
Elmer	+84.2	−10.3	17.9	3130	Q4-E
Encke	−36.6	+4.6	29.4	750	040-N/C, 048-N/C, 055-N/C, 205-C, 213-N/C
Encke T	−38.0	+3.3	95.5	1030	040-C
Endymion	+56.4	+53.6	125.1	4070	309-N, 315-N, 323-N, 328-N, 100-N, 105-N, 106-N, 112-N, 119-N, 124-N
Epigenes	−4.8	+67.4	55.2	3210	009-N, 018-N, 024-N, 029-N, 143-N, 154-N, 161-N, 167-N, 174-N, 179-N, 183-N
Epimenides	−30.2	−40.9	27.0	2370	035-S, 040-S, 048-S, 201-S, 205-S
Eppinger	−25.8	−9.4	6.1	1260	029-C, 191-C, 201-C, 205-C/S
Eratosthenes	−11.3	+14.5	58.3	3430	009-C, 018-C, 024-C, 179-C, 183-C, 191-C
Esclagon	+42.1	+21.4	15.7	590	323-C, 131-N/C
Euclides	−29.6	−7.4	13.1	2570	036-C, 040-C, 201-C, 205-C
Euctemon	+31.2	+76.4	62.1	1630	334-N, 343-N, 351-N, 357-N, Q1-N, 106-N, 119-N, 124-N, 131-N, 135-N, 143-N, 148-N
Eudoxus	+16.3	+44.2	67.1	4350	343-N, 351-N, 143-N, 148-N, 154-N, 161-N
Euler	−29.2	+23.3	26.6	2240	029-C, 036-N, 040-N, 048-N, 201-N, 205-N/C
Fabbroni	+29.2	+18.6	10.8	2150	343-N, 143-C, 148-C
Fabricius	+42.0	−42.8	78.1	5020	323-S, 328-S, 131-S, 135-S
Fahrenheit	+61.7	+13.1	6.6	1500	303-C, 105-N
Faraday	+8.7	−42.5	69.5	4090	351-S, 357-S, 003-S, 161-S, 167-S
Faustini	+77.0	−81.8	38.8	------	Q3-S, Q4-S
Fauth	−20.1	+6.3	12.1	1970	024-C, 029-C, 191-C, 201-C
Faye	+3.8	−21.4	36.7	2700	357-S, 003-S, 167-C, 174-S
Fedorov	−37.0	+28.2	4.8	170	040-N, 205-N
Fermat	+19.8	−22.6	38.5	2100	343-C, 148-S, 154-S
Fernelius	+5.0	−38.0	69.5	3280	003-S, 161-S, 167-S, 174-S
Feuillée	−9.4	+27.4	9.6	1890	018-N, 179-N, 183-N
Finsch	+21.3	+23.5	4.5	100	154-C
Firmicus	+63.3	+7.2	56.4	2140	303-C, 309-C, 315-C, Q4-E, 099-N, 105-N, 112-N
Flammarion	−3.7	−3.4	74.5	1510	009-C, 179-C, 183-C
Flamsteed	−44.3	−4.5	20.6	2160	048-C, 053-C, 061-C, 213-C, 225-C
Flamsteed P	−44.0	−3.2	112.2	990	048-C
Focas	−93.8	−33.7	22.	3420	Q3-W
Fontana	−56.6	−16.1	31.3	1130	061-C, 067-C, 225-C, 233-C
Fontenelle	−18.8	+63.3	38.0	1500	018-N, 024-N, 029-N, 036-N, 161-N, 179-N, 183-N, 191-N
Foucault	−39.9	+50.4	24.4	3400	040-N, 048-N, 055-N, 061-N, 205-N, 213-N
Fourier	−53.0	−30.3	51.3	2980	055-C/S, 061-C/S, 067-C, 075-C/S, 225-C/S
Fra Mauro	−17.0	−6.0	94.4	830	018-C, 024-C, 029-C, 191-C
Fracastorius	+33.1	−21.3	123.6	1480	328-S, 334-S, 343-C, 135-S, 143-S
Franck	+35.6	+22.6	11.8	2600	328-N/C, 131-N/C 135-N/C, 143-N/C
Franklin	+47.7	+38.8	56.3	3820	315-N, 323-N, 328-N, 106-N, 112-N, 119-N, 124-N, 131-N, 135-N
Franz	+40.3	+16.5	25.5	590	323-N/C, 131-N/C, 135-C
Fraunhofer	+59.1	−39.5	56.8	2160	303-S, 309-S, 105-C, 112-C, 119-S
Fredholm	+46.6	+18.4	14.7	2560	323-N, 124-N

Name	Long. °	Lat. °	Diam. km	Depth m	Terminator Mosaic(s)
Furnerius	+60.4	−36.3	125.2	3920	303-S, 309-S, 315-S, 099-C, 105-C, 112-C, 119-S, 124-C
Galen	+5.0	+21.9	10.0	1500	357-C, 003-C, 161-C, 167-N
Galilaei	−62.8	+10.5	15.5	2210	067-N/C, 075-N/C, 080-N/C, Q2-NW, 233-N/C, 237-C
Galle	+22.3	+55.8	21.0	2550	343-N, 351-N, 124-N, 131-N, 135-N, 143-N, 148-N, 154-N, 161-N
Galvani	−84.0	+49.6	74.7	2470	Q2-N/NW
Gambart	−15.2	+1.0	25.6	1100	018-C, 024-C, 191-C
Gardner	+33.8	+17.8	17.9	2930	328-C, 131-N/C, 135-N/C, 143-C
Gärtner	+34.6	+59.0	101.9	930	328-N, 334-N, 124-N, 131-N, 135-N, 143-N, 148-N
Gassendi	−39.9	−17.5	110.4	1420	036-C/S, 040-C/S, 048-C/S, 055-C/S, 205-S, 213-C/S
Gaudibert	+37.8	−11.0	28.6	620	328-C/S, 131-C/S, 135-C/S, 143-C
Gauricus	−12.5	−33.8	79.0	2370	018-S, 024-S, 183-S, 191-S
Gauss	+78.4	+36.3	176.9	2640	291-N, Q1-N/E, 099-N
Gay-Lussac	−20.8	+13.9	26.0	830	024-C, 029-C, 191-C, 201-C
Geber	+13.9	−19.5	45.7	3510	351-S, 357-S, 154-S, 161-S, 167-C
Geissler	+76.5	−2.6	17.8	2970	Q1-E
Geminus	+56.6	+34.4	85.5	4760	303-N, 309-N, 315-N, 099-N, 105-N, 112-N, 119-N
Gemma Frisius	+13.4	−34.3	87.7	5160	351-S, 357-S, 148-S, 154-S, 161-S, 167-S
Gerard	−79.5	+44.2	73.8	1040	080-N, Q2-N/W, 247-N
Gibbs	+84.3	−18.4	80.9	3960	291-S, Q4-SE
Gilbert	+76.2	−3.6	107.3	3700	291-S, Q1-E, Q4-E, 099-N/C
Gill	+74.6	−63.4	64.3	3930	Q4-S, 099-C/S, 105-C/S
Gioja	+1.9	+85.3	41.7	1630	352-N, 009-N, 018-N, 024-N, 029-N, Q1-N, 106-N, 135-N, 143-N, 148-N, 154-N, 161-N, 174-N, 183-N
Glaisher	+49.3	+13.2	16.4	3290	315-C, 119-N/C, 124-N
Glushko	−77.6	+8.1	42.5	------	080-N/C
Goclenius	+45.0	−10.0	60.0	2200	315-C/S, 323-C, 124-C, 131-C/S, 135-C/S
Goddard	+89.1	+14.9	84.9	1710	Q1-E, Q4-E
Godin	+10.2	+1.8	34.8	3200	351-C, 357-C, 161-C, 167-C
Goldschmidt	−3.0	+73.1	124.9	3420	357-N, 003-N, 009-N, 018-N, 024-N, 029-N, 143-N, 161-N, 167-N, 174-N, 179-N, 183-N
Golgi	−60.0	+27.8	5.4	1000	061-N
Goodacre	+14.1	−32.7	46.3	3190	351-S, 357-S, 148-S, 154-S, 161-S, 167-S
Gould	−17.2	−19.2	34.2	190	018-C/S, 024-C/S, 029-S, 191-C
Graff	−87.3	−42.5	40.2	4970	Q3-SW
Greaves	+52.8	+13.2	13.9	2410	309-C, 112-N, 119-N/C, 124-N
Grimaldi	−68.3	−5.2	221.9	1520	067-C, 075-C, 080-C, Q3-W, 233-C, 237-C, 247-C
Grove	+33.0	+40.4	26.8	2370	328-N, 334-N, 131-N, 135-N, 143-N, 148-N
Gruemberger	−10.0	−66.9	93.6	5140	018-S, 024-S, 161-S, 167-S, 174-S, 179-S, 183-S, 191-S
Gruithuisen	−39.8	+33.0	15.2	1870	040-N, 048-N, 055-N, 061-N, 205-N, 213-N
Guericke	−14.1	−11.5	58.4	740	018-C, 024-C, 029-C/S, 183-C, 191-C
Gum	+88.6	−40.4	51.	820	Q4-SE
Gutenberg	+41.3	−8.7	70.8	1750	323-C, 328-C, 131-C, 135-C/S
Gyldén	+0.3	−5.4	47.3	1130	003-C, 009-C, 174-C, 179-C
Hagecius	+46.6	−59.8	76.4	3970	315-S, 323-S, 328-S, 099-S, 105-C/S, 112-S, 119-S, 124-S, 131-S
Hahn	+73.6	+31.2	84.2	4390	291-N, 303-C, Q1-E/NE, 099-N, 105-N
Haidinger	−25.0	−39.2	21.0	2330	024-S, 029-S, 036-S, 040-S, 191-S, 201-S, 205-S
Hainzel	−33.5	−41.3	70.0	4430	036-S, 040-S, 048-S, 055-S, 201-S, 205-S, 213-S
Hainzel A	−33.9	−40.2	53.1	2600	036-S
Hainzel C	−32.8	−41.1	37.6	3590	036-S
Haldane	+84.1	−1.7	40.1	320	Q4-E
Hale	+88.9	−74.2	80.3	4200	Q4-S
Hall	+36.9	+33.7	39.3	1140	323-N, 328-N, 131-N, 135-S, 143-N
Halley	+5.8	−8.1	36.0	2470	357-C, 003-C, 161-C, 167-C, 174-C

Name	Long. °	Lat. °	Diam. km	Depth m	Terminator Mosaic(s)
Hamilton	+84.2	−42.7	57.2	2810	Q4-SE
Hanno	+71.1	−56.3	56.2	3320	291-S, 303-S, Q4-SE, 099-C, 105-C/S
Hansen	+72.6	+14.0	39.6	2770	291-N, 303-C, Q1-E, Q4-E, 099-N, 105-N
Hansteen	−51.9	−11.5	44.8	1390	055-C, 061-C/S, 067-C, 225-C, 233-C
Harding	−71.3	+43.4	22.6	2100	075-N, 080-N, Q2-NW, 233-N, 237-N, 247-N
Hargreaves	+64.1	−2.2	16.2	1550	105-C
Harpalus	−43.4	+52.6	40.5	3600	040-N, 048-N, 055-N, 061-N, 067-N, 075-N, 205-N, 213-N, 225-N
Hartwig	−80.4	−7.3	80.5	2210	Q3-W
Hase	+62.5	−29.4	83.2	3520	303-C/S, 309-S, 105-C, 112-C
Hausen	−88.2	−65.4	166.7	6020	Q3-SW, 233-S, 247-S
Hayn	+82.5	+64.7	87.1	4480	291-N, Q1-NE, 100-N
Hecataeus	+79.1	−21.3	127.2	5280	291-S, Q4-SE, 099-C
Hédervári	+85.5	−82.5	69.7	3910	Q3-S, Q4-S
Hedin	−76.8	+3.1	143.4	1250	075-C, 080-N/C, Q3-W, 247-C
Heinrich	−15.4	+24.8	7.4	1460	018-N, 191-N
Heinsius	−17.7	−39.5	66.4	2650	024-S, 029-S, 191-S
Heis	−31.9	+32.4	14.0	1910	036-N, 040-N, 201-N, 205-N, 213-N
Helicon	−23.0	+40.4	23.6	1910	024-N, 029-N, 036-N, 040-N, 191-N, 201-N
Hell	−7.8	−32.4	33.3	2200	009-S, 018-S, 179-S, 183-S
Helmholtz	+64.0	−68.1	94.9	4410	303-S, 309-S, 315-S, 328-S, Q4-S, 099-S, 105-S, 112-S
Henry	−56.9	−24.0	41.4	2630	061-C/S, 067-C, 075-C/S, 080-S, 233-C
Henry Frères	−58.9	−23.6	41.7	3750	061-C, 067-C, 075-C/S, 080-S, 233-C
Heraclitus	+6.2	−49.2	90.4	4260	351-S, 357-S, 003-S, 161-S, 167-S, 174-S
Hercules	+39.2	+46.8	67.0	2320	323-N, 328-N, 119-N, 124-N, 131-N, 135-N, 143-N
Herigonius	−34.0	−13.4	15.4	2100	036-C, 040-C/S, 048-C, 205-C/S, 213-C
Hermann	−57.3	−0.8	15.5	1710	061-N, 067-N/C, 075-N/C, 080-C, 225-C, 233-C, 237-C
Hermite	−60.1	+85.6	108.8	5140	Q1-N, 100-N, 106-N, 135-N, 143-N, 154-N, 161-N, 174-N, 183-N, 205-N
Herodotus	−49.7	+23.2	34.7	1300	055-N, 061-N, 067-N, 075-N, 225-N/C
Herschel, C.	−31.3	+34.5	13.4	1850	036-N, 040-N, 201-N, 205-N, 213-N
Herschel, J.	−41.2	+62.1	156.0	900	040-N, 048-N, 055-N, 061-N, 205-N, 213-N
Herschel, W.	−2.1	−5.7	40.6	3330	003-C, 009-C, 174-C, 179-C, 183-C
Hesiodus	−16.3	−29.4	42.6	450	024-S, 029-S, 191-S
Hevelius	−67.5	+2.2	117.9	1930	067-N/C, 075-N/C, 080-N/C, 233-C, 237-C, 247-C
Hill	+40.8	+20.9	16.2	3830	323-N, 131-N/C, 135-N/C
Hind	+7.3	−7.9	29.1	2980	351-C, 357-C, 161-C, 167-C
Hippalus	−30.2	−24.8	57.8	900	029-S, 036-S, 040-C/S, 048-C/S, 201-S, 205-S
Hipparchus	+4.8	−5.5	150.5	1100	357-C, 003-C, 161-C, 167-C, 174-C
Hirayama	+93.5	−5.9	139.	5100	Q1-E
Holden	+62.5	−19.1	47.1	1770	303-C/S, 309-S, 105-C, 112-C
Hommel	+32.9	−54.6	125.3	5050	328-S, 334-S, 343-S, 131-S, 135-S, 143-S, 148-S
Hooke	+54.9	+41.1	36.8	1470	309-N, 105-N, 112-N, 119-N, 124-N
Horrebow	−40.9	+58.6	25.4	3010	040-N, 048-N, 055-N, 061-N, 080-N, 201-N, 205-N, 213-N
Horrocks	+5.9	−4.0	30.5	2980	357-C, 003-C, 161-C, 167-C, 174-C
Hortensius	−27.9	+6.5	14.6	2860	029-C, 036-C, 040-N/C, 201-C, 205-C
Hubble	+86.5	+22.2	82.0	3160	Q1-E, Q4-E
Huggins	−1.4	−41.0	63.3	1850	003-S, 009-S, 167-S, 174-S, 179-S
Humboldt	+80.8	−27.2	201.3	5160	291-S, Q4-SE, 099-C
Huxley	−4.5	+20.2	4.0	770	179-N, 183-C
Hyginus	+6.3	+7.8	9.7	780	357-C, 161-C, 167-C
Hypatia	+22.6	−4.2	33.6	1500	343-C, 143-C, 148-C, 154-C
Ibn Battuta	+50.3	−7.0	11.5	1510	315-C, 124-C
Ibn Rushd	+21.7	−11.7	30.0	1160	343-C, 148-C, 154-C

Name	Long. °	Lat. °	Diam. km	Depth m	Terminator Mosaic(s)
Ideler	+22.3	−49.2	38.9	2330	343-S, 135-S, 143-S, 148-S, 154-S
Inghirami	−68.7	−47.5	91.1	2810	067-C/S, 075-S, 080-S, 233-S, 237-S, 247-S
Isidorus	+33.4	−8.0	39.1	2500	328-C/S, 135-C/S, 143-C, 148-C
Ivan	−43.3	+26.8	5.6	590	055-N
Jacobi	+11.4	−56.7	68.0	3490	351-S, 357-S, 003-S, 018-S, 143-S, 148-S, 154-S, 161-S, 167-S
Jansen	+28.6	+13.6	22.9	340	334-C, 135-C, 143-C, 148-C
Jansky	+89.8	+8.6	71.8	6180	Q1-E
Janssen	+41.5	−44.9	189.5	2740	32-S, 328-S, 124-S, 131-S, 135-S, 143-S
Jeans	+91.4	−55.8	79.	3800	Q4-SE
Jehan	−31.9	+20.7	4.7	730	036-N/C
Jenkins	+78.0	+0.4	37.9	3040	291-N/S, Q1-E, Q4-E, 099-N
Joliot	+92.7	+25.6	143.	2860	Q1-E, Q4-E
Joy	+6.6	+25.0	5.7	1200	357-C
Julius Caesar	+15.3	+9.0	90.8	1270	351-C, 154-C, 161-C, 167-C
Kaiser	+6.6	−36.4	53.1	1060	357-S, 003-S, 161-S, 167-S, 174-S
Kane	+26.0	+63.0	54.7	570	334-N, 343-N, 351-N, 131-N, 135-N, 143-N, 148-N, 154-N
Kant	+20.2	−10.7	31.6	3720	343-C, 148-C, 154-C
Kapteyn	+70.6	−10.8	49.3	3710	291-S, 303-C/S, Q4-E, 099-C, 105-C
Kästner	+79.2	−6.9	118.9	3200	291-S, Q1-E, Q4-E, 099-C
Keldysh	+43.6	+51.2	32.7	2720	323-N, 105-N, 106-N, 112-N, 119-N, 124-N, 131-N, 135-N
Kepler	−38.0	+8.1	31.6	2700	040-N/C, 048-N/C, 055-N/C, 205-C, 213-N/C
Kies	−22.6	−26.3	44.1	390	024-S, 029-S, 036-S, 191-S, 201-S
Kiess	+84.1	−6.4	62.7	800	Q1-E, Q4-E
Kinau	+15.1	−60.8	41.6	2120	351-S, 357-S, 143-S, 148-S, 154-S, 161-S, 167-S
Kirch	−5.6	+39.3	11.7	1830	009-N, 018-N, 179-N, 183-N
Kircher	−45.3	−67.1	72.5	4400	048-S, 055-S, 061-S, 067-S, 075-S, 080-S, 201-S, 205-S, 213-S
Kirchoff	+38.9	+30.3	24.6	2590	323-N, 131-N, 135-N/C
Klaproth	−26.0	−69.7	118.8	2770	024-S, 029-S, 036-S, 040-S, 048-S, 183-S, 191-S, 201-S, 205-S
Klein	+2.6	−11.9	44.3	1460	357-C, 003-C/S, 167-C, 174-C, 179-C
Knox-Shaw	+80.2	+5.4	12.4	2530	Q1-E
König	−24.6	−24.2	22.2	2440	024-S, 029-S, 040-C/S, 191-S, 201-C/S, 205-S
Kopff	−89.6	−17.4	43.0	2000	Q3-W
Krafft	−72.6	+16.6	51.3	3520	075-N/C, 080-N/C, Q2-NW, 233-N, 237-C, 247-N/C
Krasnov	−80.2	−29.9	43.9	4560	080-C, Q3-W, 247-S
Kreiken	+84.6	−9.0	19.4	360	Q4-E
Krieger	−45.6	+29.0	22.9	950	048-N, 055-N, 061-N, 213-N, 225-N
Krishna	+11.4	+24.5	4.4	470	351-N, 161-C, 167-N
Krogh	+65.7	+9.4	19.4	1960	303-C, Q4-E, 099-N, 105-N, 112-N
Krusenstern	+5.9	−26.2	47.4	2140	357-S, 161-S, 167-S, 174-S
Kuiper	−22.7	−9.8	6.8	1550	024-C, 029-C, 036-C, 191-C, 201-C
Kundt	−11.6	−11.6	10.6	2410	018-C, 024-C, 179-C, 183-C, 191-C
Kunowsky	−32.5	+3.2	18.4	860	036-C, 040-N/C, 048-N/C, 201-C, 205-C, 213-C
L'allemand	−84.5	−14.4	12.4	2260	Q3-W
la Caille	+1.1	−23.8	67.7	1710	003-S, 009-S, 167-C/S, 174-S, 179-S
la Condamine	−28.1	+53.4	37.3	1510	029-N, 036-N, 040-N, 191-N, 201-N, 205-N
la Pérouse	+76.3	−10.7	77.7	4180	291-S, 303-C/S, Q4-E, 099-C, 105-C
Lacroix	−58.9	−37.9	36.2	2360	061-S, 067-C, 075-C/S, 233-C/S, 237-S
Lade	+10.1	−1.4	55.7	820	351-C, 357-C, 161-C, 167-C
Lagalla	−22.5	−44.6	85.1	1390	024-S, 029-S, 040-S, 191-S, 201-S, 205-S
Lagrange	−71.5	−33.0	159.7	1150	075-C/S, 080-C/S, 237-S, 247-S
Lalande	−8.6	−4.4	24.1	2590	009-C, 018-C, 179-C, 183-C
Lamarck	−69.7	−22.9	109.0	1110	075-C, 080-C, 247-C/S
Lambert	−21.0	+25.8	30.3	2600	024-N, 029-N/C, 191-N, 201-N

Name	Long. °	Lat. °	Diam. km	Depth m	Terminator Mosaic(s)
Lamé	+64.5	−14.7	84.4	4180	303-C/S, 099-C, 105-C, 112-C
Lamèch	+13.1	+42.7	13.1	1460	351-N, 357-N, 154-N, 161-N
Lamont	+23.3	+5.1	71.8	90	343-C, 148-C, 154-C
Landsteiner	−14.8	+31.3	6.6	1360	018-N, 191-N
Langley	−86.4	+51.1	39.5	1350	Q2-N/NW
Langrenus	+61.2	−8.9	132.0	4500	303-C/S, 309-C, 315-C, 099-C, 105-C, 112-C, 119-C
Lansberg	−26.6	−0.3	40.0	2750	024-C, 029-C, 036-C, 040-C, 201-C, 205-C
Lassell	−7.8	−15.5	23.2	900	009-S, 018-C/S, 179-C, 183-C/S
Lavoisier	−80.6	+38.1	68.0	770	080-N, Q2-NW, 247-N
Lawrence	+43.3	+7.4	23.5	850	323-N/C, 131-C, 135-C
le Gentil	−76.4	−74.4	121.1	6140	067-S, 075-S, 080-S, Q3-SW/S, 213-S, 225-S, 233-S, 247-S
le Monnier	+30.5	+26.5	60.8	2400	334-N, 131-N, 135-N/C, 143-N/C, 148-N
le Verrier	−20.6	+40.3	21.1	2100	018-N, 024-N, 029-N, 036-N, 191-N, 201-N
Leakey	+37.4	−3.2	12.8	1800	328-C/S, 131-C, 135-C/S
Lee	−40.6	−30.7	41.7	1340	048-C/S, 055-C/S, 205-S, 213-C/S
Legendre	+70.1	−28.9	78.9	3680	291-S, 303-C/S, 099-C, 105-C, 112-C
Lehmann	−56.0	−40.0	53.2	1290	055-S, 061-S, 067-C, 225-S, 233-C/S
Lepaute	−33.6	−33.3	16.4	2070	048-S, 201-S, 205-S
Letronne	−42.4	−10.4	119.3	1190	048-C, 055-C, 213-C
Lexell	−4.1	−35.5	62.8	1590	009-S, 018-S, 174-S, 179-S, 183-S
Liapunov	+90.0	+26.5	65.6	2950	Q1-E, Q4-E
Licetus	+6.7	−47.1	74.7	3530	351-S, 357-S, 003-S, 161-S, 167-S, 174-S
Lichtenberg	−67.5	+31.8	20.7	2770	067-N, 075-N, 080-N, Q2-NW, 233-N, 237-C, 247-N
Lick	+52.9	+12.4	31.4	300	309-C, 315-C, 119-N/C
Liebig	−48.2	−24.3	38.2	2360	048-C/S, 055-C/S, 061-C/S, 067-C, 213-C, 225-C/S
Lilius	+6.1	−54.4	61.1	3680	351-S, 357-S, 003-S, 018-S, 143-S, 148-S, 154-S, 161-S, 167-S, 174-S
Lindbergh	+52.9	−5.4	13.8	1890	315-C, 119-C, 124-C
Lindenau	+24.9	−32.3	53.4	2930	343-S, 135-S, 143-S, 148-S
Lindsay	+13.0	−7.0	32.0	620	351-C, 154-C, 161-C, 167-C
Linné	+11.8	+27.7	2.4	600	351-N, 154-N, 161-N/C, 167-N
Liouville	+73.6	+2.7	15.9	2910	303-C/S
Lippershey	−10.3	−25.9	6.8	1360	018-S, 179-S, 183-S, 191-S
Littrow	+31.4	+21.6	30.7	1230	334-N/C, 131-N/C, 135-N/C, 148-C
Lockyer	+36.7	−46.2	34.4	1530	323-S, 328-S, 135-S, 143-S
Loewy	−32.7	−22.6	24.0	1090	036-C/S, 040-C/S, 048-C/S, 201-C, 213-C/S
Lohrmann	−67.3	−0.5	33.9	1310	067-C, 075-N/C, 233-C, 237-C, 247-C
Lohse	+60.2	−13.7	41.8	2120	303-C/S, 309-C/S, 105-C, 112-C, 119-C
Longomontanus	−21.7	−49.6	145.3	4810	018-S, 024-S, 029-S, 036-S, 040-S, 183-S, 191-S, 201-S, 205-S
Louise	−34.2	+28.5	1.5	260	036-N
Louville	−45.9	+43.9	36.2	1870	048-N, 055-N, 061-N, 205-N, 213-N, 225-N
Lubbock	+41.7	−4.0	14.5	300	323-C, 328-C, 131-C, 135-C/S
Lubiniezky	−23.9	−17.8	43.9	780	024-C/S, 029-S, 040-C/S, 191-C, 201-C
Lucian	+36.8	+14.3	7.0	1540	328-C, 131-N/C, 135-C, 143-C
Luther	+24.2	+33.2	9.5	1900	334-N, 343-N, 143-N, 148-N, 154-N
Lyell	+40.6	+13.6	32.2	2560	323-N/C, 328-C, 131-N/C, 135-C
Lyot	+84.1	−50.2	141.0	2940	Q4-SE
Maclaurin	+68.0	−2.0	50.2	3380	303-C/S, 099-N, 105-C, 112-C
Maclear	+20.1	+10.5	20.3	610	343-C, 148-C, 154-C, 161-C
Macmillan	−7.8	+24.2	7.5	360	009-C, 018-N, 179-N, 183-N/C
Macrobius	+46.0	+21.2	64.0	4050	315-N/C, 323-N, 328-C, 124-N, 131-N/C, 135-C
Mädler	+29.7	−11.1	27.1	2800	328-C/S, 334-C, 343-C, 135-C/S, 143-C/S, 148-C
Maestlin	−40.7	+4.9	7.1	1650	048-N/C, 055-N/C
Maestlin R	−41.5	+3.5	60.8	-----	048-N/C

Name	Long. °	Lat. °	Diam. km	Depth m	Terminator Mosaic(s)
Magelhaens	+44.1	−11.9	38.5	1700	323-C, 131-C/S, 135-C/S
Maginus	−6.2	−50.0	162.6	5050	003-S, 009-S, 018-S, 161-S, 167-S, 174-S, 179-S, 183-S, 191-S
Main	+10.4	+80.9	51.3	1930	357-N, 003-N, 009-N, Q1-N, 135-N, 143-N, 148-N, 154-N, 161-N
Mairan	−43.4	+41.6	41.0	2670	048-N, 055-N, 061-N, 205-N, 213-N, 225-N
Malapert	+12.9	−84.9	53.8	3290	067-S, 075-S, Q3-S, Q4-S, 099-S, 105-S, 112-S, 119-S, 124-S, 167-S
Mallet	+54.1	−45.4	58.3	4920	315-S, 323-S, 112-C/S, 119-S, 124-S
Manilius	+9.1	+14.4	38.6	3060	351-C, 357-C, 161-C, 167-C
Manners	+20.0	+4.6	15.1	1720	343-C, 148-C, 154-C, 161-C
Manzinus	+26.7	−67.6	98.1	5000	334-S, 343-S, 357-S, Q4-S, 105-S, 112-S, 119-S, 124-S, 131-S, 135-S, 143-S, 148-S, 154-S
Maraldi	+34.8	+19.3	39.9	1500	328-N/C, 131-N/C, 135-N/C, 143-C
Marco Polo	−2.0	+15.4	24.4	360	003-C, 009-C, 174-C, 179-C
Marinus	+76.4	−39.4	58.1	2960	291-S, Q4-SE, 099-C
Marius	−50.8	+11.9	41.1	1500	055-N/C, 061-N/C, 067-N/C, 075-N/C, 225-N/C
Markov	−62.5	+53.3	41.3	3380	061-N, 067-N, 075-N, 080-N, 225-N, 233-N, 237-N
Marth	−29.3	−31.1	6.1	1070	036-S, 205-S
Maskelyne	+30.0	+2.2	23.8	2500	334-C, 135-C, 143-C, 148-C
Maskelyne F	+35.3	+4.1	19.8	580	143-C
Mason	+30.4	+42.6	37.4	1390	334-N, 131-N, 135-N, 143-N, 148-N
Maunder	−93.8	−14.6	53.5	3200	Q3-W
Maupertuis	−27.2	+49.6	45.7	1350	029-N, 036-N, 040-N, 191-N, 201-N, 205-N
Maurolycus	+14.0	−41.8	114.2	4730	353-S, 351-S, 357-S, 148-S, 154-S, 161-S, 167-S
Maury	+39.7	+37.1	17.5	3280	323-N, 328-N, 131-N, 135-N
Mayer, C.	+17.4	+63.3	38.0	2360	343-N, 351-N, 357-N, 003-N, 106-N, 135-N, 143-N, 148-N, 154-N, 161-N
Mayer, T.	−29.2	+15.5	33.0	2920	036-N/C, 040-N/C, 201-N/C, 205-C
Mayer, T. C	−26.0	+12.2	15.6	2510	205-C
McClure	+50.3	−15.3	23.7	1180	315-C, 119-C/S, 124-C
McDonald	−20.9	+30.4	7.1	1600	024-N, 029-N, 191-N, 201-N
Mee	−35.0	−43.7	132.3	2680	036-S, 040-S, 048-S, 055-S, 201-S, 205-S, 213-S
Menelaus	+15.9	+16.3	27.4	2600	343-N, 351-C, 154-C, 161-C
Menzel	+36.9	+3.4	3.6	500	328-C
Mercator	−26.1	−29.2	46.7	1100	024-S, 029-S, 036-S, 040-C/S, 201-S, 205-S
Mercurius	+66.2	+46.6	67.6	3600	303-N, 309-N, 099-N, 105-N, 112-N
Mersenius	−49.2	−21.5	81.5	2680	048-C, 055-C/S, 061-C/S, 067-C, 213-C, 225-C/S
Mersenius P	−47.8	−20.2	41.9	1200	055-C/S
Messala	+59.8	+39.1	124.1	2070	303-N, 309-N, 315-N, 105-N, 112-N, 119-N
Messier	+47.6	−1.9	11.6	1900	315-C/S, 323-C, 328-C, 124-C, 131-C
Metius	+43.3	−40.3	87.6	4120	323-S, 328-S, 124-C/S, 131-S, 135-S
Meton	+19.1	+73.7	121.5	1830	334-N, 343-N, 351-N, 357-N, 003-N, 106-N, 135-N, 143-N, 148-N, 154-N, 161-N
Milichius	−30.2	+10.0	12.9	2520	036-C, 040-N/C, 048-N/C, 201-C, 205-C
Miller	+0.7	−39.4	60.7	3550	003-S, 009-S, 167-S, 174-S, 179-S
Mitchell	+20.1	+49.7	30.1	1040	343-N, 351-N, 143-N, 148-N, 154-N
Moigno	+28.9	+66.3	36.5	760	334-N, 135-N, 143-N, 148-N
Moltke	+24.2	−0.6	6.5	1300	343-C, 143-C, 148-C, 154-C
Monge	+47.6	−19.2	36.9	1540	315-C/S, 323-C/S, 124-C/S, 131-C/S
Montanarí	−20.6	−45.8	77.0	2010	024-S, 029-S, 040-S, 191-S, 201-S
Moretus	−6.6	−70.3	114.4	5240	003-S, 009-S, 018-S, 024-S, 040-S, 067-S, 161-S, 167-S, 174-S, 179-S, 183-S, 191-S
Morley	+64.6	−2.8	14.4	2250	309-C, 105-C, 112-C
Moseley	−90.1	+20.9	90.	3110	Q2-NW
Mösting	−5.8	−0.7	26.5	2800	009-C, 018-C, 179-C, 183-C

Name	Long. °	Lat. °	Diam. km	Depth m	Terminator Mosaic(s)
Mouchez	−26.8	+78.2	81.6	1380	018-N, 024-N, 029-N, 036-N, 161-N, 174-N, 183-N, 191-N, 201-N, 205-N
Müller	+2.1	−7.6	21.8	2030	003-C, 174-C
Murchison	−0.1	+5.1	57.9	870	003-C, 174-C, 179-C
Mutus	+30.1	−63.6	77.6	3580	334-S, 343-S, 105-S, 112-S, 119-S, 124-S, 131-S, 135-S, 143-S, 148-S
Nansen	+83.1	+81.0	110.0	4270	309-N, 328-N, Q1-N, 100-N
Naonobu	+57.9	−4.7	34.8	1870	309-C/S, 315-C, 112-C, 119-C
Nasireddin	+0.2	−41.1	51.4	3350	003-S, 009-S, 018-S, 161-S, 167-S, 174-S, 179-S
Natasha	−31.1	+20.0	11.4	290	036-N/C, 040-N/C, 201-N, 205-C
Nasmyth	−56.2	−50.5	76.8	1060	055-S, 061-S, 067-C/S, 075-S, 225-S, 233-S
Naumann	−62.0	+35.4	9.6	1840	061-N, 067-N, 075-N, 080-N, 233-N, 237-N/C
Neander	+40.0	−31.3	52.2	3400	323-C/S, 328-S, 131-S, 135-S
Nearch	+39.1	−58.4	75.5	4200	323-S, 328-S, 334-S, 343-S, 119-S, 124-S, 131-S, 135-S, 143-S
Neison	+25.0	+68.2	53.0	1600	334-N, 106-N, 124-N, 131-N, 135-N, 143-N, 148-N, 154-N
Neper	+84.5	+8.8	141.9	3970	291-N, Q1-E, Q4-E
Neumayer	+71.2	−71.2	76.2	3770	303-S, Q4-S, 099-S, 105-S, 112-S
Newcomb	+43.7	+29.8	39.4	2180	323-N, 124-N, 131-N
Newton	−17.5	−76.4	75.2	5540	024-S, 029-S, 036-S, 040-S, 048-S, 067-S, Q3-S, 167-S, 174-S, 179-S, 183-S, 191-S
Nicholson	−84.9	−26.3	35.8	3570	Q3-W
Nicolai	+25.9	−42.4	42.0	2420	334-S, 343-S, 135-S, 143-S, 148-S, 154-S
Nicollet	−12.5	−22.0	15.2	2030	018-S, 024-S, 179-C/S, 183-S, 191-C/S
Nielsen	−51.8	+31.8	9.7	1960	055-N, 061-N, 067-N, 075-N, 225-N, 233-S
Nobile	+51.3	−85.2	69.0	3740	Q3-S, Q4-S
Nobili	+75.0	+0.2	38.4	3807	291-N/S, Q1-E, Q4-E, 099-N, 105-C
Nöggerath	−45.7	−48.8	30.8	1940	048-S, 055-S, 061-S, 067-C/S, 213-S
Nonius	+3.8	−34.8	69.7	2990	357-C, 003-S, 161-S, 167-S, 174-S
Norman	−30.3	−11.8	10.3	2150	036-C, 040-C, 201-C, 205-C/S
Oenopides	−63.7	+56.9	68.9	1390	067-N, 075-N, 080-N, 233-N, 237-N, 247-N
Oersted	+47.2	+43.0	42.2	940	315-N, 323-N, 112-N, 119-N, 124-N, 131-N
Oken	+75.8	−43.7	71.9	2550	291-S, Q4-SE, 099-C, 105-C
Olbers	−75.7	+7.1	71.0	1940	080-N/C, Q2-NW, 247-C
Opelt	−17.5	−16.3	51.0	280	018-C/S, 024-C, 029-S, 191-C
Oppolzer	−0.5	−1.5	42.9	1500	003-C, 174-C, 179-C
Orontius	−3.8	−40.3	108.7	3370	003-S, 009-S, 018-S, 174-S, 179-S, 183-S
Palisa	−7.1	−9.5	33.4	540	009-C, 018-C, 179-C, 183-C
Palitzsch	+64.5	−28.0	41.0	2870	303-C/S, 105-C, 112-C
Pallas	−1.7	+5.5	49.6	1500	003-C, 009-C, 174-C, 179-C, 183-C
Palmieri	−47.7	−28.7	40.5	970	048-C/S, 055-C/S, 061-C/S, 067-C/S, 225-C/S
Parrot	+3.4	−14.6	67.8	2080	357-S, 003-S, 167-C, 174-C/S
Parry	−15.8	−7.8	46.1	870	018-C, 024-C, 029-C, 183-C, 191-C
Pascal	−68.0	+74.3	102.2	4630	075-N, 080-N, Q2-N, 237-N
Peary	+14.2	+87.4	84.3	3500	357-N, 003-N, 009-N, 029-N, Q1-N, 100-N, 106-N, 135-N, 143-N, 154-N, 161-N, 183-N
Peirce	+53.4	+18.3	18.5	2160	309-C, 315-C, 112-N, 119-N/C, 124-N
Peirescius	+67.6	−46.5	61.7	3680	303-S, Q4-SE, 099-C, 105-C, 112-C/S
Pentland	+11.5	−64.6	56.2	4010	351-S, 357-S, 018-S, 143-S, 154-S, 167-S
Petavius	+60.4	−25.3	176.6	3330	303-C/S, 309-S, 315-C, 099-C, 105-C, 112-C, 119-C/S, 124-C
Petermann	+66.6	+74.2	73.2	3330	309-N, 315-N, 328-N, Q1-N, 100-N, 106-N
Peters	+29.5	+68.1	15.2	500	334-N, 131-N, 135-N, 143-N, 148-N
Petit	+63.5	+2.3	5.7	1070	105-N
Petrov	+88.0	−61.4	50.	2140	Q4-S/SE

Name	Long. °	Lat. °	Diam. km	Depth m	Terminator Mosaic(s)
Pettit	−85.8	−27.6	36.0	3910	Q3-W
Phillips	+76.0	−26.6	101.0	3390	291-S, 303-C/S, Q4-SE, 099-C, 105-C
Philolaus	−32.3	+72.0	70.9	3680	029-N, 036-N, 040-N, 048-N, 055-N, 067-N, 080-N, Q2-N, 179-N, 183-N, 191-N, 201-N, 205-N, 213-N
Philolaus C	−32.6	+70.9	92.6	-----	205-N
Phocylides	−57.3	−52.9	113.9	3000	055-S, 061-S, 067-C/S, 075-S, 080-S, 225-S, 233-S
Piazzi	−67.9	−36.2	101.1	1620	067-C, 075-C/S, 080-C/S, 233-C, 237-S, 247-S
Piazzi Smyth	−3.2	+41.9	12.8	2530	003-N, 009-N, 167-N, 174-N, 179-N, 183-N
Picard	+54.8	+14.6	23.0	2320	309-C, 315-C, 112-N, 119-N/C, 124-N
Piccolomini	+32.3	−29.8	89.3	4200	328-S, 334-S, 343-S, 135-S, 143-S, 148-S
Pickering	+7.0	−2.8	16.0	2740	357-C, 167-C
Pictet	−7.4	−43.6	62.4	1780	009-S, 018-S, 179-S, 183-S
Pilâtre	−86.8	−60.7	68.0	4680	080-S, Q3-SW, 247-S
Pingré	−73.9	−58.7	90.4	4050	075-S, 080-S, Q3-SW, 233-S, 237-S, 247-S
Pitatus	−13.5	−29.8	104.8	680	018-S, 024-S, 179-S, 183-S, 191-S
Pitiscus	+30.8	−50.3	82.3	4630	334-S, 343-S, 131-S, 135-S, 143-S, 148-S
Plana	+28.2	+42.2	44.2	1800	334-N, 343-N, 131-N, 135-N, 143-N, 148-N
Plato	−9.2	+51.4	100.0	2000	009-N, 018-N, 024-N, 040-N, 048-N, 067-N, 174-N, 179-N, 183-N, 191-N
Playfair	+8.5	−23.5	47.9	3370	351-S, 357-S, 161-S, 167-C/S
Plinius	+23.6	+15.4	43.2	3070	334-C, 343-N/C, 143-C, 148-C, 154-C
Plutarch	+79.2	+24.1	68.1	4760	291-N, Q1-E, Q4-E, 099-N
Poczobutt	−99.3	+57.5	209.	3120	Q2-N
Poisson	+10.6	−30.4	44.4	1730	351-S, 357-S, 161-S, 167-S
Polybius	+25.6	−22.5	42.6	2050	334-S, 343-C, 135-S, 143-S, 148-S
Pomortsev	+66.9	+0.7	23.4	1460	105-N/C, 112-C
Poncelet	−53.7	+75.7	65.0	1180	061-N, 205-N
Pons	+21.6	−25.3	35.2	1960	343-C/S, 148-S, 154-S
Pontanus	+14.5	−28.4	53.8	1800	351-S, 357-S, 148-S, 154-S, 161-S, 167-S
Pontécoulant	+65.9	−58.6	91.0	5600	303-S, 309-S, Q4-S/SE, 099-C, 105-C/S, 112-S
Porter	−10.1	−56.1	51.6	4940	018-S, 024-S, 029-S, 174-S, 179-S, 183-S, 191-S
Posidonius	+29.9	+31.8	100.5	1370	328-N/C, 334-N, 343-N, 131-N, 135-N/C, 143-N, 148-N
Prinz	−44.1	+25.5	51.8	1020	048-N, 055-N, 061-N, 225-N
Proclus	+46.8	+16.1	28.0	4040	315-N/C, 323-N/C, 124-N
Proctor	−5.1	−46.4	52.0	1980	003-S, 009-S, 018-S, 179-S, 183-S
Protagoras	+7.3	+55.8	21.7	2600	357-N, 003-N, 009-N, 143-N, 154-N, 161-N, 167-N
Ptolemaeus	−1.6	−9.0	153.2	2400	003-C, 009-C, 174-C, 179-C, 183-C
Puiseux	−39.0	−27.8	24.7	340	040-C/S, 048-C/S, 213-C/S
Purbach	−1.9	−25.5	117.8	2980	003-S, 009-S, 018-S, 174-S, 179-S, 183-S
Pythagoras	−62.4	+63.4	128.1	5250	067-N, 075-N, 080-N, Q2-N, 213-N, 225-N, 233-N, 237-N, 247-N
Pytheas	−20.6	+20.6	19.6	2540	024-N/C, 029-C, 191-N, 201-N
Rabbi Levi	+23.6	−34.8	81.0	2480	343-S, 135-S, 143-S, 148-S, 154-S
Ramsden	−31.8	−32.9	24.5	1900	036-S, 040-S, 048-S, 201-S, 205-S
Rankine	+71.5	−3.9	8.7	2700	291-S, 099-N/C, 105-C
Rayleigh	+89.3	+29.3	101.6	1070	Q1-E/NE
Réaumur	+0.7	−2.4	52.8	820	003-S, 009-C, 174-C, 179-C
Regiomontanus	−1.0	−28.5	117.8	3610	003-S, 009-S, 174-S, 179-S
Regnault	−87.0	+54.0	49.8	2520	Q2-N
Reichenbach	+48.0	−30.3	71.4	3200	315-S, 323-C/S, 124-C/S, 131-S
Reimarus	+60.3	−47.6	48.4	3050	309-S, 105-C/S, 112-C/S, 119-S
Reiner	−55.1	+7.0	29.9	2960	055-C, 061-C, 067-N/C, 075-N/C, 080-N/C, 225-C, 233-N/C
Reinhold	−22.8	+3.3	47.5	2700	024-C, 029-C, 036-C, 191-C, 201-C, 205-C
Repsold	−77.4	+51.2	106.7	1930	075-N, 080-N, Q2-N/NW, 237-N, 247-N

Name	Long. °	Lat. °	Diam. km	Depth m	Terminator Mosaic(s)
Respighi	+71.9	+2.8	20.1	3090	105-N
Rhaeticus	+4.9	+0.0	45.6	1200	357-C, 003-C, 161-C, 167-C, 174-C
Rheita	+47.2	−37.1	70.0	2730	315-S, 323-S, 328-S, 124-C/S, 131-S, 135-S
Riccioli	−74.2	−3.1	151.6	2850	075-C, 080-C, Q3-W, 247-C
Riccius	+26.5	−36.9	70.6	1720	334-S, 343-S, 135-S, 143-S, 148-S, 154-S
Riemann	+87.9	+39.4	133.2	490	Q1-NE
Ritchey	+8.5	−11.1	29.0	1400	351-C, 357-C, 161-C, 167-C
Ritter	+19.2	+2.0	30.9	1300	343-C, 148-C, 154-C, 161-C
Robinson	−46.0	+59.0	24.1	3120	055-N, 061-N, 080-N, 205-N, 213-N, 225-N
Rocca	−72.8	−13.0	79.9	3520	075-C, 080-C, 247-C
Römer	+36.4	+25.4	39.6	3500	328-N/C, 131-N/C, 135-N/C, 143-N/C
Röntgen	−91.4	+33.0	126.	3250	Q2-NW
Rosenberger	+43.1	−55.4	95.6	3240	323-S, 328-S, 119-S, 124-S, 131-S, 135-S
Ross	+21.7	+11.7	26.3	1840	343-C, 143-C, 148-C, 154-C
Rosse	+35.0	−18.0	12.0	2420	328-S, 131-C/S, 135-S, 143-S
Rost	−33.6	−56.4	48.7	2460	036-S, 040-S, 048-S, 055-S, 061-S, 201-S, 205-S, 213-S
Rothmann	+27.8	−30.9	43.0	4220	334-S, 343-S, 135-S, 143-S, 148-S
Rozhdestvenskiy	−159.1	+85.8	179.	5020	161-N, 179-N
Russell	−75.2	+26.7	99.1	850	075-N, 080-N, Q2-NW, 247-N/C
Rutherfurd	−12.1	−61.0	50.8	3120	018-S, 024-S, 029-S, 174-S, 179-S, 183-S, 191-S
Sabatier	+79.0	+13.2	9.3	1630	291-N, Q1-E, 099-N
Sabine	+20.1	+1.4	30.3	1400	343-C, 148-C, 154-C
Sacrobosco	+16.7	−23.6	95.9	3800	343-C, 148-S, 154-S, 161-S
Santbech	+44.1	−20.9	64.4	3680	315-C/S, 323-C/S, 328-S, 124-C/S, 131-C/S, 135-S
Santos-Dumont	+47.6	+27.8	8.7	2000	003-N, 167-N
Sarabhai	+21.0	+24.8	7.6	1730	343-C, 154-C
Sasserides	−9.3	−39.1	90.4	1760	009-S, 018-S, 179-S, 183-S, 191-S
Saunder	+8.8	−4.3	44.6	640	351-C, 357-C, 161-C, 167-C
Saussure	−3.8	−43.3	55.6	1880	003-S, 009-S, 018-S, 174-S, 179-S, 183-S
Scheele	−37.8	−9.4	4.8	760	040-C, 213-C
Scheiner	−27.8	−60.5	110.4	5070	024-S, 029-S, 036-S, 040-S, 048-S, 183-S, 191-S, 201-S, 205-S
Schiaparelli	−58.6	+23.4	24.3	2190	061-N, 067-N, 075-N/C, 080-N/C, 233-N, 237-C
Schickard	−54.6	−44.4	226.7	3080	048-S, 055-S, 061-S, 067-C/S, 075-S, 080-S, 213-S, 225-S, 233-C/S, 237-S
Schiller	−39.5	−51.5	112.8	2500	040-S, 048-S, 055-S, 061-S, 080-S, 201-S, 205-S, 213-S
Schlüter	−83.5	−6.0	90.8	2980	Q3-W
Schmidt	+18.7	+1.0	11.4	2310	343-C, 148-C, 154-C, 161-C
Schomberger	+25.2	−76.8	85.0	5760	343-S, 351-S, Q3-S, Q4-S, 099-S, 105-S, 112-S, 119-S, 124-S, 135-S, 143-S, 148-S
Schorr	+88.6	−19.2	54.0	3700	Q4-SE
Schröter	−6.9	+2.6	34.5	830	009-C, 018-C, 179-C, 183-C
Schubert	+84.0	+2.9	54.0	3350	291-N/S, Q1-E, Q4-E, 099-N
Schumacher	+60.6	+42.3	60.6	1810	303-N, 309-N, 105-N, 112-N
Schwabe	+45.6	+65.0	25.4	1320	323-N, 328-N, 105-N, 119-N, 124-N, 131-N
Scoresby	+14.1	+77.7	55.8	4100	334-N, 351-N, 357-N, 003-N, 009-N, Q1-N, 100-N, 106-N, 119-N, 135-N, 143-N, 148-N, 154-N, 161-N
Scott	+45.1	−81.9	97.8	5570	323-S, 343-S, Q3-S, Q4-S, 099-S, 105-S, 112-S, 119-S, 124-S, 131-S
Secchi	+43.6	+2.4	24.5	1350	315-C, 323-N/C, 328-C, 131-C, 135-C
Seeliger	+3.0	−2.2	8.5	1810	003-C, 167-C, 174-C
Segner	−48.5	−58.9	67.1	1500	048-S, 055-S, 061-S, 067-S, 213-S, 225-S
Seleucus	−66.5	+21.1	43.3	2870	067-N, 075-N/C, 080-N/C, Q2-NW, 233-N, 237-C, 247-N/C
Seneca	+80.2	+27.0	63.3	2980	291-N, Q1-E/NE, Q4-E, 099-N
Shackleton	+0.0	−89.9	19.4	------	Q3-S, Q4-S, 099-S

Name	Long. °	Lat. °	Diam. km	Depth m	Terminator Mosaic(s)
Shaler	−85.1	−33.0	44.0	3310	Q3-W
Shapley	+56.9	+9.4	23.0	4040	309-C, 112-N, 119-C
Sharp	−40.2	+45.8	39.6	3850	048-N, 055-N, 061-N, 067-N, 205-N, 213-N
Sheepshanks	+17.0	+59.2	24.6	3190	343-N, 351-N, 357-N, 143-N, 148-N, 154-N, 161-N
Shi Shen	+104.1	+76.0	43.	2160	Q1-N
Short	−7.4	−74.5	70.8	3370	009-S, 018-S, 024-S, Q4-S, 161-S, 167-S, 174-S, 179-S, 191-S
Shuckburgh	+52.8	+42.6	38.5	1310	309-N, 315-N, 105-N, 106-N, 112-N, 119-N, 124-N
Silberschlag	+12.5	+6.2	13.4	2540	351-C, 154-C, 161-C, 167-C
Simpelius	+15.1	−72.9	70.4	5780	343-S, 351-S, 357-S, 018-S, Q4-S, 105-S, 112-S, 119-S, 143-S, 148-S, 154-S, 161-S, 167-S
Sinas	+31.6	+8.9	12.4	2260	328-C, 334-C, 135-C, 143-C, 148-C
Sirsalis	−60.4	−12.5	44.0	4730	061-C, 067-C, 075-C, 080-C, 233-C, 237-C/S
Sirsalis A	−61.3	−12.7	49.0	2100	067-C
Smithson	+53.6	+2.4	7.9	950	124-C
Snellius	+55.7	−29.3	82.7	3830	309-S, 119-C/S, 124-C
Somerville	+65.0	−8.3	15.4	3080	303-C/S, 105-C, 112-C
Sömmering	−7.5	+0.1	29.7	1320	009-C, 018-C, 179-C, 183-C
Sosigenes	+17.6	+8.7	17.8	1500	343-C, 154-C, 161-C
South	−50.5	+57.4	97.6	970	055-N, 061-N, 225-N
Spallanzani	+24.7	−46.3	32.3	2270	334-S, 343-S, 135-S, 143-S, 148-S, 154-S
Spörer	−1.8	−4.3	27.5	320	003-C, 174-C, 179-C
Spurr	−1.2	+27.9	11.2	100	009-N, 174-N, 179-N
Stadius	−13.7	+10.4	64.4	650	018-C, 024-C, 191-C
Steinheil	+46.5	−48.6	66.6	5570	315-S, 323-S, 328-S, 124-S, 131-S
Stevinus	+54.2	−32.5	74.5	3820	309-S, 315-S, 112-C, 119-S, 124-C/S
Stewart	+66.8	+2.4	13.4	1090	105-N, 112-C
Stiborius	+32.1	−34.4	43.1	3750	334-S, 343-S, 135-S, 143-S
Stöfler	+6.0	−41.1	136.6	2760	351-S, 357-S, 003-S, 161-S, 167-S, 174-S
Stokes	−88.5	+52.5	50.2	1830	Q2-N/NW
Strabo	+54.5	+62.0	55.1	4420	309-N, 315-N, 323-N, 328-N, 100-N, 105-N, 106-N, 112-N, 119-N
Street	−10.5	−46.5	57.8	2670	018-S, 029-S, 179-S, 183-S, 191-S
Struve	−76.4	+22.8	183.4	2290	075-N/C, 080-N/C, Q2-NW, 247-C
Suess	−47.7	+4.4	9.2	1920	048-C, 055-N/C, 061-C, 067-N/C, 225-N/C
Sulpicius Gallus	+11.6	+19.6	11.9	2170	351-C, 154-C, 161-C, 167-N
Swasey	+89.7	−5.5	25.0	450	Q4-E
Swift	+53.4	+19.4	10.5	2040	309-C, 315-C, 112-N, 119-N/C, 124-N
Sylvester	−82.0	+82.7	58.4	2980	Q1-N, Q2-N, 106-N, 161-N, 174-N, 183-N, 205-N
Tacchini	+85.7	+5.1	39.1	3550	Q1-E, Q4-E
Tacitus	+19.0	−16.2	39.9	2840	343-C, 148-S, 154-S
Tacquet	+19.2	+16.7	7.0	1100	343-N, 148-C, 154-C
Tannerus	+22.0	−56.4	28.6	2280	343-S, 143-S, 148-S, 154-S
Taruntius	+46.6	+5.5	56.0	1100	315-C, 323-N/C, 328-C, 124-C, 131-C
Taylor	+16.7	−5.4	36.2	3100	343-C, 154-C, 161-C
Tebbutt	+53.2	+9.4	31.9	600	309-C, 315-C, 119-N/C, 124-N/C
Tempel	+11.9	+3.8	48.2	1250	351-C, 154-C, 161-C, 167-C
Thales	+50.3	+61.7	31.6	4540	315-N, 323-N, 328-N, 100-N, 105-N, 106-N, 112-N, 119-N, 124-N
Theaetetus	+6.1	+37.0	24.8	2830	357-N, 003-N, 161-N, 167-N, 174-N
Thebit	−4.0	−22.0	54.7	3270	009-S, 018-S, 174-S, 179-C/S, 183-S
Theiler	+82.9	+13.3	8.6	690	Q4-E
Theon Junior	+15.8	−2.4	18.6	3580	343-C, 154-C, 161-C
Theon Senior	+15.4	−0.8	18.2	3470	154-C, 161-C
Theophilus	+26.4	−11.4	100.0	4100	334-C, 343-C, 135-C/S, 143-C/S, 148-C, 154-C

Name	Long. °	Lat. °	Diam. km	Depth m	Terminator Mosaic(s)
Theophrastos	+39.1	+17.5	8.2	1690	323-N, 131-N/C, 135-C
Timaeus	−0.6	+62.9	32.5	3230	003-N, 009-N, 018-N, 143-N, 154-N, 161-N, 167-N, 174-N, 179-N, 183-N
Timocharis	−13.1	+26.7	33.5	3000	018-N, 024-N, 183-N, 191-N
Tisserand	+48.1	+21.4	36.5	2930	315-N/C, 119-N/C, 124-N
Tolansky	−16.0	−9.5	13.4	980	018-C, 024-C, 029-C, 183-C, 191-C
Torricelli	+28.4	−4.8	20.4	2500	334-C, 343-C, 135-C/S, 143-C, 148-C
Toscanelli	−47.6	+27.9	7.6	1300	055-N
Townley	+63.2	+3.4	16.9	1870	309-C, 105-N, 112-C
Tralles	+52.7	+28.4	43.2	3040	309-C, 315-N/C, 105-N, 112-N, 119-N, 124-N
Triesnecker	+3.6	+4.2	26.0	2760	357-C, 003-C, 167-C, 174-C
Trouvelot	+5.8	+49.4	9.0	930	357-N, 003-N, 161-N, 167-N
Turner	−13.2	−1.4	11.8	2630	018-C, 024-C, 179-C, 183-C, 191-C
Tycho	−11.2	−43.2	84.7	4600	009-S, 018-S, 029-S, 179-S, 183-S, 191-S
Ukert	+1.4	+7.7	24.2	2800	357-C, 003-C, 167-C, 174-C, 179-C
Ulugh Beigh	−81.3	+32.7	56.5	580	080-N, Q2-NW
Urey	+87.6	+27.9	40.8	3340	Q1-E
Väisälä	−47.9	+25.9	8.3	1530	055-N
van Albada	+64.5	+9.3	25.7	3940	303-C, Q4-E, 105-N, 112-N
Van Biesbroeck	−45.3	+28.4	9.5	1610	055-N
Van Vleck	+78.2	−1.8	31.3	3030	291-S, Q1-E, Q4-E
Vasco da Gama	−83.7	+14.2	89.9	2780	Q2-NW
Vega	+63.4	−45.4	75.6	3940	303-S, 309-S, Q4-SE, 099-C, 105-C/S, 112-C/S, 119-S
Vendelinus	+61.8	−16.3	147.1	2200	303-C/S, 309-C/S, 315-C, 105-C, 112-C, 119-C
Vera	−43.8	+26.3	4.9	560	048-N, 055-N
Very	+25.4	+25.6	5.2	930	334-N, 343-N, 143-N/C, 148-N, 154-N/C
Vieta	−56.3	−29.2	87.0	4830	055-C/S, 061-C/S, 067-C, 075-C/S, 225-C, 233-C
Virchow	+83.8	+9.9	16.8	1500	Q4-E
Vitello	−37.5	−30.4	44.6	1700	036-S, 040-C/S, 048-C/S, 055-C/S, 205-S, 213-C/S
Vitruvius	+31.3	+17.7	30.1	1880	334-C, 135-N/C, 143-C, 148-C
Vlacq	+38.8	−53.3	89.2	3830	323-S, 328-S, 334-S, 131-S, 135-S
Vogel	+5.9	−15.1	34.0	2400	357-S, 161-S, 167-C
Volta	−84.4	+54.0	109.4	3580	Q2-N/NW
von Behring	+71.7	−7.8	38.8	4200	Q4-E, 099-C, 105-C, 112-N/C
von Braun	−77.6	+41.0	61.8	------	080-N, Q2-NW
Voskresenskiy	−86.4	+27.9	46.6	2510	Q2-NW
Wallace	−8.7	+20.2	28.3	160	009-C, 018-C, 179-C, 183-C
Wallach	+32.3	+4.8	6.5	1140	334-C, 135-C, 143-C
Walter	+0.7	−33.0	136.0	4130	003-S, 009-S, 167-S, 174-S, 179-S
Wargentin	−60.4	−49.5	84.4	300	061-S, 067-C/S, 075-S, 225-S, 233-S, 237-S
Warner	+87.3	−4.1	34.8	250	Q4-E
Watt	+48.6	−49.6	66.3	3450	315-S, 323-S, 328-S, 119-S, 124-S, 131-S
Watts	+46.3	+8.8	15.2	970	315-N/C, 323-C, 124-N/C, 131-C
Webb	+60.0	−0.9	22.6	1850	303-C/S, 309-C, 315-C, 105-C, 112-C, 119-C
Weierstrass	+77.2	−1.3	32.0	2580	291-S, Q1-E, Q4-E, 099-N
Weigel	−38.8	−58.2	35.6	2360	048-S, 055-S, 061-S, 067-S, 201-S, 205-S, 213-S
Weinek	+37.1	−27.5	32.4	3370	328-S, 131-S, 135-S, 143-S
Weiss	−19.5	−31.9	63.0	1000	024-S, 029-S, 191-S
Werner	+3.0	−27.8	70.2	4200	357-S, 003-S, 167-S, 174-S, 179-S
Wexler	+90.2	−69.1	50.	4630	Q4-S
Whewell	+13.7	+4.2	13.6	2270	351-C, 154-C, 161-C
Wichmann	−38.1	−7.5	10.6	1970	040-C, 048-C, 205-C, 213-C
Widmännstätten	+85.6	−6.1	46.7	720	Q1-E

Name	Long. °	Lat. °	Diam. km	Depth m	Terminator Mosaic(s)
Wildt	+75.9	+9.0	11.3	2870	099-N
Wilhelm	−20.8	−43.1	107.3	3240	024-S, 029-S, 040-S, 191-S, 201-S
Wilkins	+19.7	−29.5	59.3	2770	343-S, 143-S, 148-S, 154-S
Williams	+37.2	+41.9	36.1	830	323-N, 328-N, 131-N, 135-N, 143-N
Wilson	−42.4	−69.2	69.5	4430	048-S, 055-S, 061-S, 067-S, 191-S, 201-S, 205-S, 213-S
Winthrop	−44.5	−10.7	17.3	380	048-C, 055-C
Wöhler	+31.5	−38.3	26.2	2060	334-S, 343-S, 131-S, 143-S, 148-S
Wolf	−16.6	−22.8	26.4	790	018-S, 024-S, 029-S, 191-S
Wollaston	−47.0	+30.6	9.8	2220	048-N, 055-N, 061-N, 213-N, 225-N
Wright	−86.4	−31.6	35.9	4030	Q3-W
Wrottesley	+56.8	−23.9	57.1	4230	309-S, 315-C, 105-C, 112-C, 119-C/S, 124-C
Wurzelbauer	−15.9	−33.9	82.0	2770	018-S, 024-S, 183-S, 191-S
Xenophanes	−80.4	+57.4	110.9	3140	Q2-N, 247-N
Yakovkin	−78.7	−54.4	34.8	1640	080-S, Q3-SW
Yangel'	+4.7	+17.0	8.2	600	357-C, 174-C
Yerkes	+51.7	+14.6	35.2	620	309-C, 315-C, 119-N/C, 124-N
Young	+50.9	−41.5	71.7	2000	315-S, 119-S, 124-C/S, 131-S
Zach	+5.3	−60.9	70.8	4460	351-S, 357-S, 003-S, 009-S, 018-S, 143-S, 148-S, 154-S, 161-S, 167-S, 174-S, 179-S
Zagut	+22.1	−31.9	84.2	3440	343-S, 143-S, 148-S, 154-S
Zähringer	+40.2	+5.5	11.5	2200	323-N, 328-C, 131-C, 135-C
Zeno	+72.8	+45.1	65.2	4420	291-N, Q1-NE, 099-N, 105-N
Zinner	−58.9	+26.6	3.6	720	061-N
Zöllner	+18.8	−8.0	42.8	2500	343-C, 148-C, 154-C, 161-C
Zucchius	−50.9	−61.5	64.2	3820	048-S, 055-S, 061-S, 067-S, 075-S, 080-S, 201-S, 205-S, 213-S, 225-S
Zupus	−52.2	−17.2	35.2	1010	055-C, 061-C/S, 067-C, 225-C, 233-C

2. Non-Crater Index

Name* (feature type)	Long. °	Lat. °	Terminator Mosaic(s)
Aestatis, Lacus	−69	−15	075-C, 080-C, 237-C, 248-C/S
Aestuum, Sinus	−9	+11	003-C, 009-C, 018-C, 179-C, 183-C
Agarum, Promontorium	+66	+14	303-C, Q4-E, 099-N, 105-N, 112-N
Agassiz, Promontorium	+2	+42	357-N, 003-N, 009-N, 161-N, 167-N, 174-N, 179-N, 183-N
Agricola, Montes	−54	+29	055-N, 061-N
Aldrovandi, Dorsa	+28	+24	148-N/C
Alpes, Montes	−1	+46	357-N, 003-N, 009-N, 161-N, 167-N, 174-N, 179-N, 183-N
Alpes, Vallis	+3	+48	357-N, 003-N, 009-N, 161-N, 167-N, 174-N
Altai, Rupes	+23	−24	334-S, 343-C, 143-S, 148-S, 154-S
Ampère, Mons	−4	+19	009-C, 179-N, 183-C
Anguis, Mare	+68	+23	303-C, 099-N, 105-N
Apennine Bench*	0	+25	179-N
Apenninus, Montes	−4	+19	357-N/C, 003-N/C, 009-C, 167-N, 174-N/C, 179-N/C, 183-C
Arago α* (dome)	+21	+7	154-C
Arago β* (dome)	+20	+7	154-C
Archimedes, Montes	−5	+25	003-N/C, 009-N/C, 018-N, 174-N, 179-N, 183-N/C
Archerusia, Promontorium	+22	+17	343-N, 143-C, 148-C, 154-C
Argaeus, Mons	+29	+19	334-C, 143-C, 148-C
Ariadaeus, Rima	+14	+6	351-C, 154-C, 161-C, 167-N
Aristarchus Plateau*	−51	+26	225-N, 233-N
Asperitatis, Sinus	+27	−4	334-C, 343-C, 135-C/S, 143-C, 148-C, 154-C

Name* (feature type)	Long.	Lat.	Terminator Mosaic(s)
Australe, Mare	+93°	−39°	291-S, Q4-SE, 099-C, 105-C
Autumni, Lacus	−84	−10	Q3-W
Azara, Dorsum	+19	+27	164-N, 161-N/C
Baade, Vallis	−76	−46	075-C, 080-S, Q3-SW, 247-S
Barlow, Dorsa	+31	+15	143-C, 148-C
Blanc, Mont	+1	+45	357-N, 003-N, 009-N, 167-N, 174-N, 179-N, 183-N
Bohr, Vallis	−87	+12	Q2-NW
Bonitas, Lacus	+44	+23	124-N, 131-N/C
Bouvard, Vallis	−83	−38	080-C/S, Q3-W, Q3-SW, 247-S
Bradley, Mons	+1	+22	357-C, 003-C, 009-C, 167-N, 174-C, 179-N, 183-C
Buckland, Dorsum	+13	+20	167-N
Bullialdus W (valley)	−26	−19	205-S
Burnet, Dorsa	−57	+20	237-C
Carpatus, Montes	−24	+14	018-C, 024-C, 029-C, 036-N/C, 191-C, 201-N/C, 205-C, 213-N/C
Cato, Dorsa	+47	+1	131-C
Caucasus, Montes	+10	+38	351-N, 357-N, 161-N, 167-N, 174-N
Cauchy, Rupes	+37	+9	328-C, 135-C, 143-C
Cauchy τ* (dome)	+37	+7	143-C
Cayeux, Dorsum	+51	+2	124-C
Cognitum, Mare	−23	−10	024-C, 029-C, 036-C, 040-C, 201-C, 205-C/S
Concordiae, Sinus	+43	+11	131-N/C, 135-C
Cordillera, Montes	−82	−18	080-C, Q3-W, 237-C/S, 247-C
Crisium, Mare	+59	+17	303-C, 309-C, 313-C, Q4-E, 099-N, 105-N, 112-N, 119-N, 124-N
Cushman, Dorsum	+49	+1	131-C
Delisle, Mons	−36	+30	036-N, 040-N, 048-N, 201-N, 205-N/C, 213-N
Deville, Promontorium	+1	+43	357-N, 003-N, 009-N, 167-N, 174-N, 179-N, 183-N
Doloris, Lacus	+9	+17	351-C, 357-C, 161-C, 167-N/C
Epidemiarum, Palus	−28	−32	024-S, 029-S, 036-S, 040-S, 201-S, 205-S
Excellentiae, Lacus	−44	−35	048-S, 055-S, 213-S
Ewing, Dorsa	−39	−10	036-C
Fecunditatis, Mare	+51	−8	309-C/S, 315-C, 112-C, 119-N/C, 124-C, 131-C, 135-C/S
Felicitatis, Lacus	+5	+19	357-C, 167-N, 174-C
Fidei, Sinus	+2	+18	357-C, 003-C, 167-N, 174-C
Fresnel, Promontorium	+5	+29	357-N, 167-N, 174-N
Frigoris, Mare	+1	+56	323-N, 328-N, 334-N, 343-N, 351-N, 357-N, 003-N, 009-N, 018-N, 024-N, 029-N, 036-N, 040-N, 048-N, 106-N, 112-N, 124-N, 131-N, 135-N, 143-N, 148-N, 154-N, 161-N, 167-N, 174-N, 179-N, 183-N, 191-N, 201-N, 205-N, 213-N
Gast, Dorsum	+9	+24	167-N
Gaudii, Lacus	+13	+16	154-C, 161-C
Geikie, Dorsa	+52	−5	124-N
Grabau, Dorsum	−16	+29	191-N
Gruithuisen Delta, Mons	−40	+36	040-N, 048-N, 205-N, 213-N
Gruithuisen Gamma, Mons	−40	+37	040-N, 048-N, 205-N, 213-N
Hadley, Mons	+5	+26	357-N, 003-N, 167-N, 174-N, 179-N
Hadley Delta, Mons	+4	+26	357-N, 167-N, 174-N/C, 179-N
Haemus, Montes	+9	+20	357-C, 154-C, 161-C, 167-N
Hansteen, Mons	−50	−12	055-C, 061-C, 067-C, 225-C
Harbinger, Montes	−41	+27	040-N, 048-N, 205-N/C, 213-N
Harker, Dorsa	+64	+14	112-N
Heim, Dorsum	−30	+32	201-N, 205-N
Heraclides, Promontorium	−33	+40	036-N, 040-N, 048-N, 055-N, 201-N, 205-N, 213-N

Name* (feature type)	Long.	Lat.	Terminator Mosaic(s)
	°	°	
Herodotus, Mons	−53	+28	061-N
Hesiodus, Rima	−20	−30	201-S
Hiemalis, Lacus	+14	+15	154-C, 161-C
Higazy, Dorsum	−17	+28	191-N
Honoris, Sinus	+18	+12	154-C, 161-C
Humboldtianum, Mare	+82	+57	291-N, 303-N, Q1-NE, 100-N, 105-N
Humorum, Mare	−39	−24	036-C/S, 040-C/S, 048-C/S, 055-C/S, 201-S, 205-S, 213-C/S, 225-C/S
Huygens, Mons	−3	+20	003-C, 009-C, 174-C, 179-N, 183-C
Hyginus, Rima	+8	+7	357-C, 161-C, 167-C
Hypatia, Rimae	+22	0	154-C
Imbrium, Mare	−16	+33	357-N, 003-N, 009-N/C, 018-N/C, 024-N/C, 029-N/C, 036-N/C, 040-N/C, 048-N, 161-N, 167-N, 174-N, 179-N, 183-N/C, 191-N/C, 201-N, 205-N/C, 213-N
Inghirami, Vallis	−72	−44	075-C, 080-S, 233-S, 237-S, 247-S
Insularum, Mare	−31	+8	018-C, 024-C, 029-C, 036-C, 040-N/C, 048-N/C, 179-C, 183-C, 191-C, 201-C, 205-C, 213-N/C
Iridum, Sinus	−32	+44	029-N, 036-N, 040-N, 048-N, 191-N, 201-N, 205-N, 213-N
Jura, Montes	−34	+47	029-N, 036-N, 040-N, 201-N, 205-N, 213-N
Kelvin, Promontorium	−33	−27	036-S, 040-C/S, 048-C/S, 201-S, 205-C, 213-C/S
Kies π* (dome)	−24	−26	201-S
Krafft, Catena	−72	+15	075-N/C, 080-N/C, 247-C
La Hire, Mons	−26	+28	024-N, 029-N, 036-N, 191-N, 201-N, 205-N/C
Laplace, Promontorium	−26	+46	024-N, 029-N, 036-N, 040-N, 048-N, 191-N, 201-N, 205-N
Lavinium, Promontorium*	+49	+15	119-N/C, 124-N
Lenitatis, Lacus	+12	+14	351-C, 154-C, 161-C, 167-C
Liebig, Rupes	−48	−24	048-C/S
Lister, Dorsa	+24	+20	148-C, 154-C
Lunicus, Sinus	−1	+32	003-N, 009-N, 167-N, 174-N, 179-N, 183-N
Maraldi, Mons	+35	+20	131-N/C, 135-N/C, 143-C
Marginis, Mare	+86	+13	291-N, Q1-E, Q4-E
Marius Hills*	−55	+12	061-N/C, 225-N/C, 233-N/C
Mawson, Dorsa	+56	−5	124-N
Medii, Sinus	+2	+2	003-C, 009-C, 167-C, 174-C, 179-C
Mendel-Rydberg Basin*	−94	−50	237-S, 247-C
Mercator, Rupes	−22	−31	024-S, 029-S
Moro, Mons	−20	−12	024-C, 029-C/S, 191-C
Mortis, Lacus	+27	+45	328-N, 334-N, 343-N, 131-N, 135-N, 143-N, 148-N, 154-N
Nectaris, Mare	+35	−15	323-C, 328-C/S, 334-C/S, 343-C, 131-C/S, 135-C/S, 143-C/S, 148-C/S
Nicol, Dorsum	+23	+18	154-C
Nubium, Mare	−17	−21	009-S, 018-C/S, 024-C/S, 029-S, 036-C/S, 040-C, 179-C/S, 183-C/S, 191-C/S, 201-C/S
Odii, Lacus	+7	+19	357-C, 161-C, 167-N
Olivium, Promontorium*	+49	+16	119-N/C, 124-N
Oppel, Dorsum	+53	+19	124-N
Orientale, Mare	−93	−19	Q3-W
Orientale Basin*	−95	−20	237-C, 247-C
Owen, Dorsum	+11	+25	167-N
Palitzsch, Vallis	+64	−26	099-C, 105-C, 124-C
Penck, Mons	+22	−10	148-C, 154-C
Perseverantiae, Lacus	+62	+8	099-N, 112-N
Petavius, Rimae	+59	−26	105-C, 112-C

Name* (feature type)	Long.	Lat.	Terminator Mosaic(s)
Pico, Mons	−9	+46	009-N, 018-N, 024-N, 174-N, 179-N, 183-N, 191-N
Piton, Mons	−1	+41	003-N, 009-N, 167-N, 174-N, 179-N, 183-N
Procellarum, Oceanus	−57	+18	036-C, 040-C, 048-N/C, 055-N/C, 061-N/C, 067-N/C, 075-N/C, 080-N/C, Q2-NW, 201-C, 213-N/C, 225-N/C, 233-N/C, 237-N/C, 247 N/C
Putredinis, Palus	0	+26	003-N/C, 009-N, 174-N/C, 179-N
Pyrenaeus, Montes	+41	−16	323-C, 131-C/S, 135-C/S
Recta, Rupes	−8	−22	009-S, 018-S, 179-C/S, 183-S
Recti, Montes	−20	+48	018-N, 024-N, 029-N, 191-N, 201-N
Reiner Gamma	−59	+8	061-C, 067-N/C, 075-N/C, 080-N/C, 233-N/C, 237-C
Rheita, Vallis	+52	−42	315-S, 323-S, 328-S, 119-S, 124-C/S, 131-S
Riphaeus, Montes	−28	−8	029-C, 036-C, 040-C, 201-C, 205-C/S
Rook, Montes	−82	−21	Q3-W
Roris, Sinus	−57	+54	055-N, 061-N, 067-N, 075-N, 080-N, Q2-N/NW, 225-N, 233-N, 237-N, 247-N
Rümker, Mons	−58	+41	061-N, 067-N, 225-N, 233-N, 237-N
Schneckenberg* (hills)	+6	+10	174-C
Schröteri, Vallis	−51	+26	055-N, 061-N
Scilla, Dorsum	−60	+33	237-C
Secchi, Montes	+43	+3	131-C, 135-C
Serenitatis, Mare	+18	+28	334-N/C, 343-N, 351-N/C, 357-N, 135-N/C, 143-N/C, 148-N/C, 154-N/C, 161-N/C, 167-N
Sirsalis, Rimae	−62	−16	067-C, 233-C
Smirnov, Dorsa	+25	+27	334-N, 148-N, 154-N/C
Smythii, Mare	+88	+1	291-N/S, Q1-E, Q4-E
Snellius, Vallis	+56	−31	315-S, 323-C/S
Somnii, Palus	+45	+14	323-N/C, 328-C, 131-N/C, 135-C
Somniorum, Lacus	+29	+38	328-N, 334-N, 343-N, 131-N, 135-N, 143-N, 148-N, 154-N
Spei, Lacus	+65	+43	099-N, 105-N
Spitzbergen, Montes	−5	+35	003-N, 009-N, 018-N, 179-N, 183-N
Spumans, Mare	+65	+1	303-C/S, Q1-E, Q4-E, 099-N, 105-C, 112-C
Successus, Sinus	+59	+1	112-C, 119-C
Sylvester, Catena	−86	+81	205-N, Q1-N, Q2-N
Taenarium, Promontorium	−8	−19	018-S, 179-C
Taurus, Montes	+41	+28	323-N, 328-N/C, 131-N/C, 135-N/C
Temporis, Lacus	+58	+46	105-N, 106-N, 112-N, 119-N, 124-N
Teneriffe, Montes	−12	+47	009-N, 018-N, 024-N, 029-N, 179-N, 183-N, 191-N
Termier, Dorsum	+58	+11	119-C
Tetyaev, Dorsa	+64	+20	112-N
Timoris, Lacus	−27	−39	029-S, 036-S, 040-S, 201-S, 205-S
Tranquillitatis, Mare	+31	+8	323-N/C, 328-C, 334-C, 343-C, 131-N/C, 135-C, 143-C, 154-C
Undarum, Mare	+68	+7	303-C, Q1-E, Q4-E, 099-N, 105-N, 112-N/C
Usov, Mons	+63	+12	099-N, 105-N, 112-N
Valentine Dome*	+10	+31	351-N, 167-N
Vaporum, Mare	+4	+13	351-C, 357-C, 003-C, 167-N/C, 174-C
Veris, Lacus	−86	−16	Q3-W
Vinogradov, Mons	−32	+22	036-N/C, 040-N, 201-N, 205-N/C, 213-N
Vitruvius, Mons	+31	+19	135-N/C, 143-C, 148-C
von Cotta, Dorsum	+12	+23	167-N
Wolf, Mons	−7	+17	009-C, 179-N/C
Zirkel, Dorsum	−24	+28	024-N, 029-N, 201-N, 205-N

* Name not listed in *Gazeteer of Planetary Nomenclature, 1994.*

Text Index

Subject Index

Albedo features, lunar 2–1
Altitude, lunar
  orbit, effect by 5–6
  season, effect by 5–7
  seeing, effect on 5
CCD (charge-coupled device) imaging, lunar
  advantages 20
  bit depth 21–22
  calibration
    dark frame 22
    flat-field frame 22
  CCD chip
    photosite 21
    pixel (picture element) 21
  disadvantages 20
  enhancement
    contrast stretching 23, 24
    geometric correction 23, 25
    mosaicing 23
    unsharp masking 23, 24
  exposure 23
  DN (brightness level) 21
  field of view 22, 23
  noise
    readout 22
    thermal 22
  Nyquist sampling theorem 23
  quantum efficiency 22
  spectral sensitivity 21
Chronology, lunar
  cratering rate 28
  geologic time scale 29
  heavy bombardment period 27
  period 28
    Copernican Period 28, 29
    Eratosthenian Period 28, 29
    Imbrian Period 28, 29
    Nectarian Period 28, 29
  system 28
  time-stratigraphic unit 28
Comets, possible impact by 27, 45
  Comet Shoemaker-Levy 9 27, 45
Coordinates, selenocentric
  latitude 1
  longitude 1, 3

Endogenic (processes, features) 42
Exogenic (processes, features) 42

Highlands regions, map 35

International Astronomical Union (IAU) 35, 48

Landforms, lunar (*see also* Individual Feature Index; Processes, lunar)
  calderas 43
  catenae 46
    map 41
  Cayley plains (Cayley Formation) 39, 51
  cinder cones 45, 53
  collapse features 42–43
  craters 30–43
    central peaks (central elevations) 30, 31
    central peak basins 33
    complex craters 30, 31–33
    crater-pits 42–43
    ejecta blanket 30, 31
    categories 30
    floors 30, 31, 33
    fractured-floor 39–42
      map 40
    ghost craters 39
    glacis (outer wall) 30, 31
    light-plains floor 39
    mare-flooded 38–39
    names 30
    peak ring basins 33
    rims 30, 31, 33
    secondary 42
      from Copernius 42, 43
    terraces 30, 31, 33
    simple craters 30–31
  depressions 43
  domes, lunar 42, 53–55
    catalog, A.L.P.O. 53
    highlands 52
    map 54
  Fra Mauro Formation 39
  highlands plateaus 51
  graben 44
  laccoliths 54
  lava tubes 42, 44
  light-plains material 39
  mare ridges (dorsa, ridges, wrinkle ridges) 52
    map 54
  maria 52–53
    lava flows 52
      lava-flow fronts 53
      thickness 53
  mons 48–50
  montes 48, 49, 50–51
  mountains 47–52
    IAU designations 48
  multiring basins 34–37
    disputed (11 basins) 34
    formation 9
    list (17 basins) 34
    map 35
    names 34, 35
    rings 34
  promontoria 48
  rilles, sinuous 42, 44–45
    map 41
  rilles, straight 43
    map 41
  rima, rimae (*see* rilles, straight; rilles, sinuous)
  ring dikes 39
  rupes 48, 53
  vallis, valles (basin-secondary chains, chains, secondary chains, valleys) 47–48
    map 41
Librations, lunar
  full phase, at; during 1999-2004, 9
  libration in latitude 4, 7
  libration in longitude 3, 7
  libratory zone (marginal zone) 4
  physical libration 4
  sample lunation, 2001 JAN 24-FEB 23, for 8
  topocentric libration 4
Lighting conditions, lunar
  age 1
  angle of emission 10
  angle of incidence 10
  brightness longitude 10
  colongitude 1, 3, 7
  phase 1
  phase angle 1, 10
  phase coefficient 1
  photometric function
    observed 10–12
    theoretical 10, 11
  relationship among parameters 3
  selenocentric solar latitude 3, 7
  solar altitude 3
  solar azimuth 3
  terminator 1

Orbit, lunar
  eccentricity (lunar distancce) 7
  nodes, movement of 5, 7

Perspective, effects on lunar observing
  foreshortening 4
  loss of detail, near limb 4–5
  relief displacement 4
Photography, lunar
  advantages 17
  camera 17
  film choice
    black-and-white 18
    color 18
  image amplification
    eyepiece projection 17–18
    Barlow lens 18
  scanning (digitizing) 18
    enhancement 18, 19
    color-differencing 18, 19
  T-adapter 17
Processes, lunar (*see also* Landforms, lunar)
  erosion 28
  impact cratering 26–27
    ejecta 26–27
      ejecta blanket 26
    impact melt 26
    primary cratering 27
    secondary cratering 27
    simultaneous impacts 43

tectonism 27–28
  compression 28
  extension 28
  faulting 42
  graben 28
  scarp (rupes) 28
volcanism 27
  cinder cone 27
  dome, lunar 27
  extrusive 27
  highlands 29, 51–52
  laccolith 27
  mare ridge (wrinkle ridge, dorsum) 27
  maria units 27
    lacus 27
    maria 27
    palus 27
    sinus 27
  sinuous rille 27
  summit crater 27

Telescopes
  apodizing ring 13
  central obstruction, effect on 13
  clock drive 14
  desirable properties 12
  mounting 13–14
  optical designs
    refractor 12, 13
    Cassegrain 12
    Maksutov 12
    Newtonian 12, 13
    Schiefspiegler 12
    Schmidt-Cassegrain 12

Video observing, lunar
  advantages 18
  camera 18, 20
  digitizing 20
    enhancement 20, 21
  format 20
  T-adapter 20
Visibility cycles, lunar
  comparison of 7–10
  Saros Cycle 7
Visual observation, lunar
  Barlow lens 14
  binocular viewer 14
  drawing 14–15
    schematic method 15–16
    shading method 17
    stipple method 15–16
    supporting data 15
    visual photometry method 15–17
  eyepiece 14
  filter 14
  magnification 14
  tonal scale 15–16

Individual Feature Index

Abulfeda, Catena 46
Agassiz, Promontorium 49
Agatharchides, Rima 37
Albategnius, crater 39
Alpes, Montes 49, 50
Alpes, Vallis 44, 45
Alphonsus, crater 39
Altai, Rupes 51
Ampère, Mons 49
Amundsen-Ganswindt Basin 34, 37
Anaximenes, crater 39, 40
Apennine Bench (Apennine Bench Formation) 49, 51
Apennine Scarp 45, 51
Apenninus, Montes 27, 49, 50
Archimedes, crater 28, 38
  age 29
Archimedes, Montes 49, 50
Ariadaeus, Rima 44
Aristarchus, crater 45
  age 29
Aristarchus, Rimae 45
Aristarchus Plateau 27
  age 29
Arnold, crater 39, 40
Arzachel, crater, age 29
Atlas, crater, age 29
Australe, Mare 37
Australe Basin, 37
  age 29, 34
  map 35
  ring diameters 34
Autumni, Lacus 37
Auwers, crater 42

Baade, Vallis 48
Baillaud, crater 39, 40
Bailly, crater
  age 29
  Clementine image 33
  description 33–34
Bailly Basin 37
  map 35
Balmer Basin 34, 37
Barrow, crater 39, 40
Beer, crater, appearance under varying lighting 31–32
Birmingham, crater 39, 40
Birt, crater 45
Birt E, crater 45
Birt F, crater 45
Birt, Rima 45
Blanc, Mont 50
Bohr, Vallis 48
Bond, W., crater 39, 40
Bonpland, crater 39
Boscovich P, valley 48
Bouvard, Vallis 48
Bradley, Mons 49
Brayley, Rima 45
Brigette, Catena 46

Bullialdus, crater 30, 31
Bullialdus A, crater 30, 31
Bullialdus B, crater 30, 31
Bullialdus W, valley 48
Byrd, crater 39, 40

Calippus, crater 50
Capella, Vallis 47
Carlini, crater 52
Carpatus, Montes 50
Cassini, crater, age 29
Caucasus, Montes 49, 50
Cauchy, Rupes 51
Clavius, crater 33, 39
  age 29
  sunrise image sequence 2
"Cobrahead" 42, 45
Cognitum, Mare, map 35
Cognitum Basin 51
Copernicus, crater 28
  age 29
Cordillera, Montes 50
Crisium, Mare
  age 29
  concentric ridges 53
Crisium Basin 47
  age 29, 34
  elongation 43
  inner ring 37, 48
  main ring 50
  map 35
  ring diameters 32

d'Arrest, crater 42
Daniell, crater 43
Darney χ, possible highlands dome 52
Darwin, crater 41
Davy, Catena 46, 47
Davy C, crater 46
Davy Y, crater 46
de Gasparis, crater 37
Delisle, Mons 50
Delisle, Rima 45
Deslandres, crater, age 29
Deville, Promontorium 49
Diophantus, Rima 45
Doppelmayer, crater 37, 38
Doppelmayer, Rimae 37
Dziewulski, Catena 46

Eratosthenes, crater 27
Euler, crater, low-sun view 31
Endymion, crater 38

Fecunditatis, Mare
  age 29
  highlands units within 48
Fecunditatis Basin
  age 29, 34
  map 35
  ring diameters 34
Feuillée, crater, appearance under varying lighting 31–32
Flammarion, crater 39
Flamsteed P, crater 39

Flamsteed-Billy Basin 34
Fra Mauro, crater 39
Fracastorius, crater 38
Fresnel, Promontorium 49
Frigoris, Mare
   age of flooding 52
   highlands units within 48
   map 35

Galilaei E, crater 45
Galvani, crater 42
Gassendi, crater 37, 39
Gauss, crater 33
   age 29
Gerard Q, crater 42
Grimaldi Basin
   age 29, 34
   map 35
   ring diameters 34
Gruithuisen Delta, highlands dome 52
Gruithuisen Gamma, highlands dome 52
Guericke, crater 39

Hadley, Mons 49
Hadley Delta, Mons 49
Haemus, Montes 44, 50
Hansteen, Mons, highlands dome 52
Harbinger, Montes 50
Hausen, crater 33
   age 29
Heim, Dorsum 52
Helicon, crater 52
Herigonius, Rimae 37
Herodotus, crater 45
Herschel, J., crater 39, 40
Hesiodus, crater 45
Hippalus, crater 37, 38
Hippalus, Rimae 37, 44
Hipparchus, crater 39
   age 29
Hommel, crater, age 29
Hortensius, crater 55
Hortensius σ, dome 55
Hortensius-Milichius-Tobias Mayer dome cluster 54, 55
Humboldt, crater 33, 38, 39, 45
   age 29
Humboldt, Catena 46
Humboldtianum Basin 37
   age 29, 34
   main ring 50
   map 35
   ring diameters 34
Humorum, Mare
   age 29
   concentric ridges 53
   Orbiter 4 photography 36
Humorum Basin 47, 51
   age 29, 34
   description 35, 37
   inner ring 37, 48
   main ring 51
   map 35
   Orbiter 4 photograph 36
   ring diameters 34
Huygens, Mons 49
Hyginus, crater 42–43
Hyginus, low dome 55
Hyginus, Rima 42–43, 44
Hypatia, crater 42

Imbrian Basin 27, 37, 46, 47
   age 29, 34
   intermediate ring 50
   inner ring 37, 48, 50
   main ring 50
   map 35
   mountains associated with 50
   ring diameters 34
Imbrian Flows, age 29
Imbrium, Mare
   age of flooding 52
   concentric ridges 53
   eastern, Orbiter photograph 49
   interior highlands units 50
   photograph, Apollo oblique 52
Ina, depression 43
Inghirami, Vallis 48
Insularum, Mare, map 35
Insularum Basin 34
   age 29
Iridum, Sinus 38, 50
   age 29

Janssen, crater, age 29
Jura, Montes 50

Kane, crater 39, 40
"Kant Plateau" 51
Kelvin, Rupes 37, 51
König, crater 30, 31
Kopff, crater 43
Krafft, Catena 46
Krieger, crater 45
Krishna, depression 43

La Hire, Mons 50, 52
Lambert R, crater 28, 39
Langley, crater 42
Laplace, Promontorium 52
Lavoisier, crater 42
Le Monnier, crater 38
Le Verrier, crater 52
Lee, crater 38
Letronne, crater 38
Lichtenberg flow, age 29, 52
Lick, crater 38
Liebig, Rupes 37, 51
Lister, Dorsa 53
Littrow, Catena 46
Longomontanus, crater, age 29
Lorentz Basin 34, 37
Lubiniezky F, crater 30, 31

Maginus, crater 39
   age 29
Mairan T, highlands dome 52
Marginis, Mare, map 35
Marginis Basin 34, 37
   age 29
Marius, Rima 45
Marius Hills 27, 45, 55
   age 29
Mayer, Tobias, crater 55
Maupertuis, crater, age 29
Mendel-Rydberg Basin 37
   age 34
   map 35
   ring diameters 34
Mercator, Rupes 51
Mersenius, crater 37, 38
Mersenius, Rimae 37
Messier, crater 43
Messier A, crater 43
Meton, crater 39, 40
Milichius, crater 55
Moretus, crater, image 33
Müller, crater 42
Müller, "Catena" 46, 47
Mutus-Vlacq Basin 34, 37

Natasha, crater 49, 50
Nectaris, Mare
   age 29
   concentric ridges 53
   highlands units within 48
Nectaris Basin 47
   age 29, 34
   main ring 49–50, 51
   map 35
   ring diameters 34
Neison, crater 39
"Newton", ghost ring 39
Nubium, Mare
   age 29
   highlands units within 48
   rim 51
Nubium Basin,
   age 29, 34
   map 35
   ring diameter 34

Orientale, Mare 37
   highlands units within 48
Orientale Basin 27, 37, 47, 48
   age 29, 34
   inner ring 50
   main ring 50
   map 35
   ring diameters 34

Palitzsch, Vallis 47
Palmieri, crater 37
Parry, crater 39
Petavius, crater 38
   age 29
   sample drawings 16
Piccolomini, crater, age 29
Pico, Mons 28, 48, 49, 50

Pico β, mountain 48, 49
Pierre, Catena 46
Pingré-Hausen Basin 34, 37
   age 29
Pitatus, crater 38, 45
Piton, Mons 48, 49
Piton γ, mountain 48, 49, 50
Plaskett, crater 46
Plato, crater 38
   age 29
Plato λ, mountain 49, 50
Posidonius, crater 39, 45
Procellarum, Oceanus
   age of flooding 52
   map 35
Procellarum ("Gargantuan") Basin 34
Ptolemaeus, crater 39
   age 29
Pyrenaeus, Montes 50–51

Ramsden, Rimae 43
Recta, Montes 50
Recti, Rupes 51
Repsold, crater 42
Rheita, Vallis 47
Riccioli, crater 38
Riphaeus, Montes 50
Ritchey, crater 42
Ritter, crater 43
Röntgen, crater 42
Rook, Montes 50
Rümker, Mons (dome complex) 53

Sabine, crater 43
Schickard, crater 38, 39
Schiller, crater 39, 43
Schiller-Zucchius Basin 37
   age 29, 34
   map 35
   ring diameters 34
Schrödinger Basin, age 29
Schröter, crater 42
Schröteri, Vallis 42, 45
Secchi, crater 42
Serenitatis, Mare
   age 29
   concentric ridges 53
   enhanced photograph 19
   highlands units within 48
Serenitatis Basin 47
   age 29, 34
   inner ring 37
   main ring 50
   map 35
   ring diameters 34
"Serpentine Ridge" (Dorsa Smirnov/Lister) 53
Sharp, crater, age 29
Sikorsky-Rittenhouse Basin 34, 37
Smirnov, Dorsa 53
Smythii, Mare 52
Smythii Basin
   age 29, 34
   map 35
   ring diameters 34
Snellus, Vallis 47
South, crater 39, 40
South Pole-Aitken Basin 37
   age 29, 33, 34
   map 35
   ring diameters 33, 34
Spitzbergen, Montes 49, 50
Stadius R, crater 43
Sulpicius Gallus, crater 44
Sulpicius Gallus, Rimae 44
Sylvester, crater 46
Sylvester, Catena 46

Taruntius, Rima 46
Taurus, Montes 48, 50
Teneriffe, Montes 49, 50
Timocharis, crater, appearance under varying lighting 31–32
Timocharis, Rima 46
Toscanelli, Rupes 45, 51
Tranquillitatis, Mare
   age 29
   highlands units within 48
   video frame 21
Tranquillitatis Basin
   age 29, 34
   map 35
   ring diameters 34
Triesnecker, Rimae 44
Tycho, crater, age 29

Ulugh Beigh, crater 42

Vaporum, Mare 52
Veris, Lacus 37
Vinogradov, Mons 49–50
Vitello, crater 37, 40
Vogel, crater 42
Volta, crater 42

Wallace, crater 28
Wargentin, crater 38
Werner-Airy Basin 34, 37
Wolf, crater 27, 43
Wolff, Mons 49

Xenophanes, crater 41, 42

Yuri, Rima 46

Zirkel, Dorsum 52
Zöllner, crater 42